Biodiversity Conservation and Poverty Alleviation

Conservation Science and Practice Series

Published in association with the Zoological Society of London

Wiley-Blackwell and the Zoological Society of London are proud to present our *Conservation Science and Practice* series. Each book in the series reviews a key issue in conservation today from a multidisciplinary viewpoint.

Books in the series can be single or multi-authored and proposals should be sent to:
   Ward Cooper, Senior Commissioning Editor. Email: ward.cooper@wiley.com

Each book proposal will be assessed by independent academic referees, as well as our Series Editorial Panel. Members of the Panel include:
   Richard Cowling, Nelson Mandela Metropolitan University, Port Elizabeth, South Africa
   John Gittleman, Institute of Ecology, University of Georgia, USA
   Andrew Knight, University of Stellenbosch, South Africa
   Nigel Leader-Williams, University of Cambridge, UK
   Georgina Mace, Imperial College London, Silwood Park, UK
   Daniel Pauly, University of British Columbia, Canada
   Stuart Pimm, Duke University, USA
   Hugh Possingham, University of Queensland, Australia
   Peter Raven, Missouri Botanical Gardens, USA
   Helen Regan, University of California, Riverside, USA
   Alex Rogers, Institute of Zoology, London, UK
   Michael Samways, University of Stellenbosch, South Africa
   Nigel Stork, Griffith University, Australia.

Previously published

Conservation Science and Practice Series

Biodiversity Conservation and Poverty Alleviation: Exploring the Evidence for a Link

Edited by

Dilys Roe
International Institute for Environment and Development, London, UK

Joanna Elliott
African Wildlife Foundation, Oxford, UK

Chris Sandbrook and Matt Walpole
United Nations Environment Programme World Conservation Monitoring Centre, Cambridge, UK

LIVING CONSERVATION

A John Wiley & Sons, Inc., Publication

Library of Congress Cataloging-in-Publication Data
Biodiversity conservation and poverty alleviation : exploring the evidence for a link / edited by Dilys Roe . . . [et al.].
   p. cm.
Includes bibliographical references and index.
ISBN 978-0-470-67479-6 (cloth) – ISBN 978-0-470-67478-9 (pbk.) 1. Biodiversity conservation – Economic aspects. 2. Poverty – Prevention – Environmental aspects. I. Roe, Dilys.
   QH75.B53225 2013
333.95′16 – dc23

                                                                    2012017200

Set in 9.5/11.5 pt Minion by Laserwords Private Limited, Chennai, India
Printed and bound in Malaysia by Vivar Printing Sdn Bhd

1   2013

Contents

Contributors

William M. (Bill) Adams Department of Geography, University of Cambridge, Downing Place, Cambridge CB2 3EN, UK; wa12@cam.ac.uk

Brian Belcher Centre for Livelihoods and Ecology, Royal Roads University, 2005 Sooke Rd., Victoria, BC V9B 5Y2, Canada; Brian.belcher@royalroads.ca
Centre for International Forestry Research, Bogor, Indonesia

Augustin Berghöfer Helmholtz Centre for Environmental Research-UFZ, Permoser-straße 15, 04318 Leipzig, Germany; augustin.berghoefer@ufz.de

Fikret Berkes Natural Resources Institute, University of Manitoba, Winnipeg, MB R3T 2N2, Canada; berkes@cc.umanitoba.ca

Jill Blockhus The Nature Conservancy, PO Box 303, Monson, ME 04464, USA; jblockhus@tnc.org

Jan Börner Center for Development Research, University of Bonn, Walter-Flex-Str .3, 53113 Bonn, Germany; jborner@uni-bonn.de

Katrina Brandon Independent Consultant, Beijing, China; katrinabrandon @gmail.com

Dan Brockington School of Environment and Development, University of Manchester, Arthur Lewis Building, Oxford Road, Manchester M13 9PL, UK; Dan.Brockington@manchester.ac.uk

Thomas M. Brooks NatureServe, 4600 N. Fairfax Dr., 7th Floor, Arlington, VA 22203, USA; tbrooks@natureserve.org
World Agroforestry Centre (ICRAF), University of the Philippines, Los Baños, the Philippines
School of Geography and Environmental Studies, University of Tasmania, Tasmania, Australia

Jock Campbell Integrated Marine Management Ltd, The Innovation Centre, University of Exeter, Exeter EX4 4RN, UK; j.campbell-imm@ex.ac.uk

Pippa Chenevix Trench Independent Consultant, Washington D.C., USA; Ptrench@me.com

Anna Davis Independent Consultant, Windhoek, Namibia; ad@iway.na

Lara Diez Nyae Nyae Development Foundation of Namibia (NNDFN), 9 Delius Street, Windhoek, Namibia; nndfn@iafrica.com.na

Richard W. Diggle World Wildlife Fund, Namibia, PO Box 9681, 19 Lossen Street, Windhoek, Namibia; rwdiggle@wwf.na

Willy Douma Hivos, Raamweg 16, 2596 HL The Hague, The Netherlands; w.douma@hivos.nl

Joanna Elliott African Wildlife Foundation, Oxford, UK; jelliott@awf.org

Claude Gascon National Fish and Wildlife Foundation, 1133 Fifteenth St. NW Suite 1100, Washington, DC 20005, USA; Claude.Gascon@NFWF.ORG

Holly K. Gibbs Nelson Institute for Environmental Studies, Department of Geography, Center for Sustainability and the Global Environment (SAGE), University of Wisconsin-Madison, Madison, WI 53706, USA; hkgibbs@wisc.edu

George Holmes School of Earth and Environment, University of Leeds, Woodhouse Lane, Leeds LS2 9JT, UK; G.holmes@leeds.ac.uk

Katherine Homewood Department of Anthropology, University College London, Gower Street, London WC1E 6BT, UK; k.homewood@ucl.ac.uk

Brian T. B. Jones Environment and Development Consultant, Windhoek, Namibia; bjones@mweb.com.na

Andreas Kontoleon University of Cambridge, Department of Land Economy, 19 Silver Street, Cambridge CB3 9EP, UK; ak219@cam.ac.uk

S. Neil Larsen Independent Consultant, The Nature Conservancy, Seattle, United States; sneillarsen@gmail.com

Keith Lawrence Conservation International, 2011 Crystal Drive, Suite 500, Arlington, VA 22202, USA; k.lawrence@conservation.org

Craig Leisher The Nature Conservancy, PO Box 303, Monson, ME 04464, USA; cleisher@tnc.org

Kathy MacKinnon IUCN/World Commission on Protected Areas Conservation Science Group, University of Cambridge, UK
Independent Consultant, Cambridgeshire, UK; kathy.s.mackinnon@gmail.com

Russell A. Mittermeier Conservation International, 2011 Crystal Drive, Suite 500, Arlington, VA 22202, USA; rmittermeier@conservation.org

Michael Mortimore Independent Consultant, Drylands Research, Somerset, UK; mike@mikemortimore.co.uk

Dilys Roe International Institute for Environment and Development, 80–86 Gray's Inn Road, London WC1X 8 NH, UK; dilys.roe@iied.org

Chris Sandbrook United Nations Environment Programme World Conservation Monitoring Centre, 219 Huntingdon Road, Cambridge CB3 0DL, UK; Chris.Sandbrook@unep-wcmc.org

M. Sanjayan The Nature Conservancy, PO Box 303, Monson, ME 04464, USA; msanjayan@tnc.org

Elizabeth R. Selig Conservation International, 2011 Crystal Drive, Suite 500, Arlington, VA 22202, USA; e.selig@conservation.org.

Pavan Sukhdev Green Indian States Pvt Ltd, G-175, Palam Vihar, Gurgaon, Haryana 122017, India; pavan@gistadvisory.com

Daudi Sumba African Wildlife Foundation, Ngong Road, Karen, P.O. Box 310, 00502, Nairobi, Kenya; dsumba@awfke.org

David H.L. Thomas BirdLife International, Wellbrook Court, Girton Road, Cambridge CB3 0NA, UK; david.thomas@birdlife.org

Philip Townsley Integrated Marine Management Ltd, The Innovation Centre, University of Exeter, Rennes Drive, Exeter EX4 4RN, UK; ptownsley@fastwebnet.it

Will Turner Conservation International, 2011 Crystal Drive Suite 500, Arlington, VA 22202, USA; w.turner@conservation.org

Bhaskar Vira Department of Geography, University of Cambridge, Downing Place, Cambridge CB2 3EN, UK; bv101@cam.ac.uk

Matt Walpole United Nations Environment Programme World Conservation Monitoring Centre, 219 Huntingdon Road, Cambridge CB3 0DL, UK; Matt.Walpole@unep-wcmc.org

Heidi Wittmer Helmholtz Centre for Environmental Research-UFZ, Permoserstr, 15, 04318 Leipzig, Germany; heidi.wittmer@ufz.de

Sven Wunder Center for International Forestry Research (CIFOR), Rua do Russel, 450/sala 601, Bairro: Gloria, CEP: 22.210-010, Rio de Janeiro, Brazil; swunder@cgiar.org

Preface and Acknowledgements

The links between environment and development have long been discussed, but only recently has this discussion focussed specifically on the possible links between biodiversity conservation and poverty alleviation. In 2002, the Convention on Biological Diversity (CBD) adopted a target "to achieve by 2010 a significant reduction of the current rate of biodiversity loss at the global, regional and national level *as a contribution to poverty alleviation* and to the benefit of all life on earth". The "2010 Target" was not met and it is clear that biodiversity loss is continuing apace–but even if conservation efforts were successful, would this really contribute to poverty alleviation? There is a diversity of opinion as to the nature and scale of biodiversity conservation, poverty alleviation links and the most appropriate mechanisms that can help to maximise them. Also, many generalisations and assumptions are made about these links. To explore the current state of knowledge and to challenge some of the prevailing myths and assumptions, a 2-day symposium titled 'Linking biodiversity conservation and poverty alleviation: what, why and how?' was organised at the Zoological Society of London (ZSL) in April 2010 by the editors of this book. This book is based on the presentations made during that symposium.

First and foremost, the editors would like to thank all the presenters at the meeting and the authors of the chapters contained in this book, for the care, rigour and patience with which they carried out their work. We are grateful to ZSL for hosting this meeting, and for the support offered throughout the organisation of the meeting and the development of this book. Joy Hayward and Linda Da Volls deserve a special mention for their dedication and the countless hours they worked to ensure the successful delivery of this event. We would like to thank Jon Hutton (UNEP-WCMC), Nigel Leader-Williams (University of Cambridge) and Matthew Hatchwell (Wildlife Conservation Society) for expertly chairing sessions during the symposium. A thank you also goes to Bill Adams (University of Cambridge), Willy Douma (Hivos), Katrina Brandon (Conservation International), Steve Bass (International Institute for Environment and Development [IIED]) and Jayant Sarnaik (Applied Environmental Research Foundation) who were engaging panellists in the closing session of the symposium. Finally, a mention goes to award winners from the Equator Initiative and everyone else who prepared and presented a poster during the symposium. Even though the posters have not been included in the present book, the poster sessions were a valuable and highly appreciated feature of the symposium.

Funding for the symposium and the outputs associated with it was provided through the generous support of a range of donors that support IIED including the Danish International Development Agency, the Department for International Development (United Kingdom), the Directorate-General for International Cooperation (the Netherlands), Irish Aid, the Norwegian Agency for Development Cooperation, the Swedish International Development Cooperation Agency and the Swiss Agency for Development and Cooperation; and through the African Wildlife Foundation and United Nations Environment Programme World Conservation Monitoring Centre (UNEP-WCMC). The symposium was organised under the umbrella of the Poverty and Conservation Learning Group (www.povertyandconservation.info) which is supported by the Arcus Foundation.

We are very grateful to Ward Cooper, Carys Williams and Kelvin Matthews at Wiley-Blackwell for their assistance during the production of this book and to our institutions – the International Institute for Environment and Development, African Wildlife Foundation and United Nations Environment Programme World Conservation Monitoring Centre – for providing the time and space for our editorial contributions. Before submitting this book to Wiley-Blackwell, a considerable effort was put into carefully reviewing each chapter to ensure accuracy and consistency of style. For this, we would like to thank Jessica Smith at UNEP-WCMC and also Abisha Mapendembe, who diligently checked every reference and numbered every box! Last, but not least, we would like to thank Alessandra Giuliani from IIED without whom production of this book would not have been possible. Alessandra coordinated the entire editorial process, acted as a focal point for communications with the authors and gently kept us all on track to completion.

Dilys Roe, Joanna Elliott, Chris Sandbrook and Matt Walpole
February 2012

Introduction

$$1$$

Linking Biodiversity Conservation and Poverty Alleviation: What, Why and Where?

Dilys Roe[1], Joanna Elliott[2], Chris Sandbrook[3] and Matt Walpole[3]

[1]International Institute for Environment and Development, London, UK
[2]African Wildlife Foundation, Oxford, UK
[3]United Nations Environment Programme World Conservation Monitoring Centre, Cambridge, UK

Biodiversity conservation and poverty alleviation: separate or linked challenges?

Biodiversity conservation and poverty alleviation are both important societal goals demanding increasing international attention. At first glance they may appear to be separate policy realms with little connection. The former is largely the concern of ministries of environment, conservation organisations and ecologists; the latter falls within the remit of ministries of finance and planning, development organisations and economists. The Convention on Biological Diversity (CBD), agreed in 1992, was drafted in response to escalating biodiversity loss and provides an international policy framework for biodiversity conservation activities worldwide. Similarly, the Organisation for Economic Co-operation and Development (OECD) International Development Targets of 1996 – reiterated as the Millennium Development Goals (MDGs) in 2000 – focus international development efforts on global poverty alleviation.

Biodiversity Conservation and Poverty Alleviation: Exploring the Evidence for a Link, First Edition.
Edited by Dilys Roe, Joanna Elliott, Chris Sandbrook and Matt Walpole.
© 2013 John Wiley & Sons, Ltd. Published 2013 by John Wiley & Sons, Ltd.

On closer inspection, although ostensibly aimed at very different communities of interest, both of these policy frameworks recognise links between their objectives:

- The preamble of the CBD acknowledges that "economic and social development and poverty eradication are the first and overriding priorities of developing countries". In 2002 the Conference of Parties (CoP) to the CBD agreed a Strategic Plan which included a target to "achieve by 2010 a significant reduction of the current rate of biodiversity loss ... as a contribution to poverty alleviation and to the benefit of all life on Earth" (CBD, 2002). The new *Strategic Plan for Biodiversity 2011–2020* has a mission to halt the loss of biodiversity thereby contributing to human well-being and poverty eradication (CBD, 2010). The 2010 CoP also adopted a decision on the "integration of biodiversity into poverty eradication and development" (Decision X/VI).
- Beyond the CBD, at the UN World Summit in 2005, the secretariats of the five major biodiversity conventions – CBD, the Convention on International Trade in Endangered Species of Wild Fauna and Flora (CITES), the Convention on Migratory Species of Wild Animals (CMS), the Ramsar Convention on Wetlands and the World Heritage Convention – issued a joint statement emphasizing the important role that biodiversity plays in the achievement of the MDGs: "Biodiversity can indeed help alleviate hunger and poverty, can promote good human health, and be the basis for ensuring freedom and equity for all" (Biodiversity Liaison Group, 2005).
- The seventh of the eight MDGs (MDG7) is to "ensure environmental sustainability" which originally included a sub-target to "reverse the loss of environmental resources" with biodiversity-related indicators (e.g. protected area coverage and forest land). The CBD "2010 Biodiversity Target" was included as a new target within MDG7 following the 2006 UN General Assembly (United Nations, 2006) with additional biodiversity indicators (United Nations, 2008).

The international policy statements described here contain an explicit assumption that conserving biodiversity (or reducing the rate of biodiversity loss) will help to tackle global poverty. There is, however, considerable variation in the potential nature and scale of biodiversity–poverty links.

Nadkarni (2000) describes six different relationships between poverty and environment: from a vicious cycle of poverty leading to environmental degradation and thence to more poverty; to a win–win scenario where environmental conservation contributes to poverty alleviation. The same is true of the underlying relationship between poverty and biodiversity. For example, the Millennium Ecosystem Assessment (MA) – an authoritative review of the state of the world's ecosystems which concluded in 2005 – highlights the different winners and losers from biodiversity use. It notes that the use (and loss) of biodiversity has actually benefitted many social groups (largely in allowing for current levels of food production). Meanwhile, as a result of that biodiversity loss, "people with low resilience to ecosystem changes – mainly the disadvantaged – have been the biggest losers and witnessed the biggest increase in not only monetary poverty but also relative, temporary poverty and the depth of poverty" (MA, 2005: 40). It suggests giving priority to protecting those elements of biodiversity, and the services it provides (Box 1.1), that are of particular importance to the well-being of poor and vulnerable people.

Box 1.1 **Biodiversity, ecosystem services and human well-being**

The conceptual framework of the Millennium Ecosystem Assessment (Figure 1.1) describes biodiversity as underpinning the delivery of a range of ecosystem services which in turn contribute to human well-being. The ecosystem services include provisioning services (e.g. food and fuel wood),

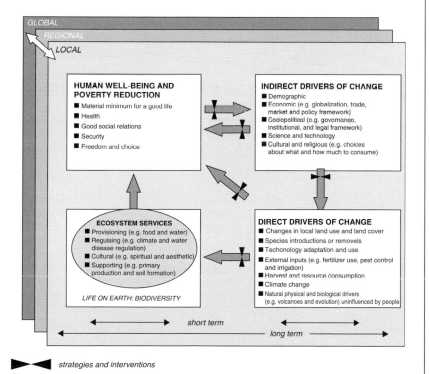

Figure 1.1 **Biodiversity, ecosystem services and human well-being: the conceptual framework of the Millennium Ecosystem Assessment. Source: Millennium Ecosystem Assessment (2003).**

regulating services (e.g. local climate control), cultural services (e.g. spirituality) and supporting services (e.g. soil formation). These services contribute to different dimensions of human well-being: basic material for a good life, health, good social relations and freedom of choice and actions. These dimensions of well-being are very closely aligned with dimensions of poverty – poverty being "a pronounced deprivation of wellbeing" (MA, 2003: 12). The MA framework also shows how ecosystem services are affected by external drivers – including economic, governance and demographic factors. Thus different development pathways have different implications for biodiversity and ecosystem services and different outcomes for human well-being.

In practice, different organisations – and individuals – have very different perspectives on the links between biodiversity conservation and poverty alleviation and their roles and responsibilities in addressing these links. A useful framework for thinking about this logic is provided by Adams *et al.* (2004), who propose a typology of four positions which can be adopted on the relationship between biodiversity conservation and poverty alleviation. In the first, poverty and conservation are seen as separate realms. In the second, poverty is seen as a critical constraint on conservation, meaning it must be tackled to achieve conservation goals. In the third, conservation activities must not compromise poverty. In the fourth, poverty alleviation is the goal, and is seen as being dependent on resource conservation. Recognising these fundamentally different value positions can help to explain the behaviour of different actors when they are faced with difficult trade-off decisions between conservation and development goals (Leader-Williams *et al.*, 2010).

Where is the evidence for biodiversity–poverty linkages? The objectives and structure of this book

Recognising the diversity of opinion as to the nature and scale of biodiversity–poverty links and the most appropriate mechanisms that can help to maximise them, this book is intended to explore the current state of knowledge on different aspects of this relationship. The book is based on a symposium held in April 2010 at the Zoological Society of London that brought together leading experts to discuss the nature of the links between biodiversity and poverty in different ecological contexts and for different groups of poor people. The symposium was built on the foundations of three fundamental questions:

- Is there a geographical overlap between biodiversity and poverty?
- Are poor people dependent on biodiversity?
- Is biodiversity conservation an effective mechanism for poverty alleviation?

Through the symposium and in the chapters of this book, a richer and more nuanced set of issues were explored:

- Which aspects of biodiversity are particularly useful or not useful to the poor?
- Does the relationship between biodiversity and poverty differ according to particular ecological conditions?
- How do particular conservation interventions differ in their poverty impacts?
- How do distributional and institutional issues affect the poverty impacts of interventions?
- How do broader issues such as climate change and the global economic system affect the biodiversity–poverty relationship at different scales?

This chapter sets the scene for the rest of the book by highlighting the broad policy statements that are made about the links between biodiversity and poverty, and the assumptions these statements imply. Lack of clarity over terminology is a major obstacle to better understanding the true nature of the relationship between biodiversity and poverty, and we highlight areas where confusion may lie. After this chapter the book is divided into five parts, informed by the set of questions posed here. Part I provides an overview of the relationship between biodiversity and poverty – particularly focusing on the delivery of ecosystem services and their role in supporting the poorest. In Part II we examine the different nature of the biodiversity–poverty relationship in different ecological contexts – from forests to farmland, and from drylands to coastal zones. In Part III we compare the poverty impacts of a wide range of common conservation interventions and approaches – including protected areas, enterprises and community-based conservation. In Part IV we assess distributional and institutional issues associated with biodiversity–poverty linkages, notably the use of payments for ecosystem services (PES) mechanisms, the distribution of benefits in pastoralist and other community areas and the use of local organisations as an entry point. Finally in Part V we look at the relationship between biodiversity and poverty in the context of larger scale drivers of environmental degradation – climate change, consumption and the nature of development itself. In the final chapter of Part V (Chapter 20), we present our conclusions with respect to the fundamental questions addressed by chapter authors, the emerging themes and their implications for policy makers and practitioners going forwards.

A note on terminology

The issue of definitions and terminology is critical. There exists a tendency to talk in generalisations – for example, that biodiversity conservation can contribute to poverty alleviation – without clearly defining either what we mean by these terms or how we are measuring impacts and outcomes. *Biodiversity* is defined by the CBD as "the *variability* among living organisms", but it is often used to refer to *amount* in terms of abundance of species and populations, or to specific elements of biodiversity rather than variety per se.

Biodiversity *conservation* is variously defined depending on different values, objectives and world views. These vary from place to place, culture to culture and even individual to individual. In general terms, however, it can be taken to mean the protection, maintenance and/or restoration of living natural resources to ensure their survival over the long term. The way in which biodiversity is conserved and managed also varies hugely from place to place, from strict preservation to commercial consumptive use – with much debate about the relative merits and effectiveness of these different approaches.

Poverty is another term with many different definitions. The simplest usually relate to some level of material wealth – for example, the Millennium Development Goal to "eradicate extreme poverty" refers to the billion-plus people whose income is less than US$1 a day. However, poor people often do not define themselves in cash income terms – indeed, the concept of cash is completely meaningless for some indigenous communities who live outside of the cash economy. In many cases, issues such as power and voice, opportunity and a healthy environment are valued more highly than money. It has therefore become increasingly recognised that poverty is multidimensional. The World Bank, for example, describes poverty as

> a pronounced deprivation in well being. . . . To be poor is to be hungry, to lack shelter and clothing, to be sick and not cared for, to be illiterate and not schooled. But for poor people, living in poverty is more than this. Poor people are particularly vulnerable to adverse events outside their control. They are often treated badly by the institutions of state and society and excluded from voice and power in these institutions. (2001: 15)

Similarly *poverty reduction* implies lifting people beyond a defined poverty line – transforming them from poor to non-poor. But often poverty is alleviated (i.e. some of the symptoms of poverty are addressed but people are not actually transformed from 'poor' to 'non-poor') or prevented (i.e. people are prevented from falling into – or further into – poverty) rather than actually being reduced (Figure 1.2).

Finally, there is a need to be clear that the relationship between biodiversity and poverty/poor people is different from the one between biodiversity *conservation interventions* and poverty/poor people. The difference may seem semantic but is crucial.

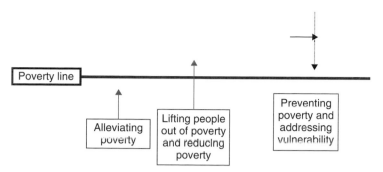

Figure 1.2   **Poverty reduction, alleviation and prevention. Source: Adapted from King and Palmer (2007).**

For example, because the rural poor depend on biodiversity for their day-to-day livelihoods, it is logical that if it is conserved (maintained, enhanced or restored) it can continue to provide livelihood support functions. However, the conservation *intervention* that is employed to reach that outcome may have a negative poverty impact.

Biodiversity, ecosystem services and poverty – the potential for synergies

In the first section of the book, authors explore the links between biodiversity, ecosystem services and poverty globally through spatial analysis, case studies and literature review. Spatial overlap is often presented as a rationale for why biodiversity conservation and poverty alleviation should be pursued together, although patterns of co-location are complex and vary depending on the elements of biodiversity and poverty being compared (Fisher & Christopher, 2007; Redford *et al.*, 2008; Sachs *et al.*, 2009; Roe *et al.*, 2010). However, spatial overlap, as we shall see, is not enough to guide policy makers as to the synergies. The nature of the specific linkages, and in particular their relation to property rights, has to be understood at the local level if policies are to reflect and address these linkages. Emerging efforts to map the distribution and flows of ecosystem services could be a valuable way of identifying where the connection between biodiversity (that in part underpins the supply of ecosystem services) and the poor (who in part depend on such services) is most acute. This, in turn, could help to identify where conservation action could have the greatest impact for the well-being of the poor.

In Chapter 2, Turner *et al.* use global data sets to investigate the overlaps between areas of high biodiversity value, areas supplying ecosystem service value and areas occupied by poor people. They find a strong relationship between biodiversity conservation priorities and ecosystem service provision – particularly those services used by the poor. The study confirms the difficulty of geographically separating efforts to manage biodiversity from those to reduce poverty, and finds that the most important places for effective PES mechanisms are also the same places where the poor depend on essential ecosystem services. If effective and equitable PES mechanisms were implemented comprehensively, the value of ecosystem services to the poor would exceed US$1 per person per day for 30% of an estimated 1.1 billion people living in poverty. The chapter concludes that, at the national level, ecosystem services may provide a substantial subsidy for the world's poorest countries, representing a significant contribution towards meeting the Millennium Development Goals.

Whether or not there is a geographical overlap between poverty and biodiversity says little about the nature and consequences of this link. What is arguably more important is to understand the multidimensional interactions and dependencies between biodiversity and poverty which cannot be easily captured in a two-dimensional map.

In Chapter 3, Wittmer *et al.* take the analysis of 'realised' ecosystem services one layer deeper, to examine the property rights of the human population using these services. They analyse the 'GDP of the poor' to estimate overall dependence of livelihoods on non-marketed natural resources and other ecosystem services, and demonstrate that standard measures of GDP fail to capture the costs to the rural poor of degradation in natural capital. They conclude that policy makers should focus on securing access to and continued availability of those ecosystem services most essential to poor citizens and ensure that policies and projects do not unintentionally degrade them.

Continuing to drill down to the local scale, in Chapter 4 Vira and Kontoleon examine the evidence of the extent to which poor people depend upon biodiversity and the nature of this dependence, whether for subsistence, income and/or insurance (risk reduction). Their exploration of what is understood by the terms *biodiversity* (or 'nature's resources') and *poverty* affirms their call for more careful and disaggregated identification of the pathways by which changes in biodiversity affect poor peoples' livelihood choices and strategies. They find a surprising lack of empirical data. Furthermore, where studies exist they show considerable variation in the contribution of nature's resources to household income. They find that when participation in biodiversity-based livelihood activities is broken down by wealth class, it is the poor who typically show higher levels of dependence. This is not, however, the case for high-value natural resources which tend to be appropriated by richer groups, often pushing poorer households into 'poverty traps'.

Besides subsistence and income dependency, the chapter highlights that agro-biodiversity provides the poor with a form of cost-effective and readily accessible insurance against the risk of food insecurity, but that the role of other aspects of biodiversity in risk reduction is poorly understood.

Biodiversity and poverty relationships in different ecological settings

Part II explores the linkages between biodiversity and poverty in different ecological settings – from coasts to drylands, and from forests to farms. The benefits of biodiversity as an insurance and risk reduction mechanism appear to cut across ecological settings, but so does the risk of overdependence on, and unequal access to, natural resources.

In Chapter 5, Belcher explores the interactions between forests and poverty. He notes that for many remote communities, forest-based livelihood contributions are more valuable than those from agriculture, particularly when fuel wood supplies are included. He explores what he calls the 'naïve' assumption that local people will conserve forests if they benefit from them, finding that the same factors that help to conserve forests may also contribute to perpetuating poverty, such as remoteness from markets, absence of infrastructure and poor soils. Competing claims to forest resources can lead to 'poverty traps' whereby high competition for open-access resources leads to the poorest people engaging in overharvesting, which in turn leads to price reductions. The poorest are the most dependent on forest resources for subsistence purposes, but the wealthiest are the main beneficiaries of forest-related income opportunities, notably from timber production.

In Chapter 6, Campbell and Townsley note the diverse biomes encompassed by the coastal environment which provide many opportunities for the poor. They note the difficulty of identifying the scale of linkages between poverty and coastal fisheries. The high diversity of coastal species provides for access by a wide spectrum of social groups (of different genders, ages and wealth groups), and a safety net in times of stress. Access to coastal biodiversity is, however, changing rapidly as a result of expanding population coupled with rapid coastal development, coastal degradation, resource overexploitation, climate change and other factors. This results in increased competition for resources – often with the effect of reduced access for the poor. To date there is a lack of evidence that improved coastal zone management can achieve both conservation and poverty reduction aims.

In Chapter 7, Mortimore describes how drylands are often considered as 'biodiversity deserts' – a strong contrast with forest and coastal ecosystems – where biodiversity loss is associated with overgrazing, overcultivation and deforestation. In drylands, biodiversity is often the product of co-evolving human and ecological systems. Genetic management (crop breeding, animal breeding, wild plant protection and harvesting) is the key 'bridge' between the human and ecological systems. The dependence of the poor is not so much on 'wild nature' but on agro-diversity (cultivars), useful plants (spontaneously regenerating), protected and spontaneous on-farm trees and domesticated livestock. The chapter corroborates Vira and Kontoleon's identification (in Chapter 4) of the importance of biodiversity as an insurance against risk, and concludes that the primary determinant of poverty in dryland systems is the state of health of the human and ecological systems.

In Chapter 8, Douma focuses on agricultural landscapes and confirms the role of agro-biodiversity in supporting the livelihoods of poor smallholder farmers – particularly in terms of improving their capacity to maintain productivity and cope with climate change. The chapter notes that 'sustainable' agricultural practices such as composting, crop rotation, zero tillage, application of green manure and abstention from chemical pesticides and fertilisers increase household incomes and reduce food security risks to poor farmers. The chapter shows that low-external-input agriculture, organic agriculture and smallholder systems have positive effects on both carbon sequestration and biodiversity and argues for further exploration of PES as a mechanism to reward poor farmers for their safeguarding of sustainability.

Poverty impacts of different conservation interventions

In the third section of the book, authors assess alternative types of conservation interventions for their impact on poverty, and find common ground in the relative lack of disaggregated poverty impact data. They also recognise that conservation interventions are generally, and understandably, targeted at delivering those livelihood benefits thought to be most likely to incentivise conservation action, rather than tackling poverty alleviation per se. The question of whether conservation interventions could and should do more to reach the poorest is an important one. Our authors appear to agree that conservation interventions are largely not reaching the poorest and argue that truly large-scale poverty alleviation requires government action at a level beyond the reach of the conservation sector alone.

Leisher *et al.* reviewed over 400 documents in order to explore the empirical evidence for biodiversity conservation as a mechanism for poverty alleviation. In Chapter 9, they find a relative lack of robust data, particularly with regard to poverty impacts other than income. They identify 10 types of intervention varying from nature tourism and local management of marine areas to improved agro-forestry and grassland management. They find that in many cases, it is *biomass* rather than *biodiversity* per se that determines the poverty alleviation potential of a conservation intervention – although it should be noted that biodiversity is often important for generating high biomass. They find some evidence of community-based forestry enterprises, nature-based tourism, fish spill-over from no-take zones in marine conservation areas, agro-forestry and grassland management interventions supporting routes out of poverty. They confirm that some interventions may lead to poverty traps, particularly where elites gain control of resources, noting that better-off households with greater physical and social capital are more likely to participate in conservation initiatives.

In Chapter 10, Holmes and Brockington evaluate the social impacts of protected areas (PAs), which remain a central strategy of conservation practice. Recognising that other chapters have focused on the potential positive links between biodiversity conservation and poverty alleviation, they review some of the more problematic aspects of this relationship. They find that social impacts of PAs tend to be concentrated at the local level in close proximity to the conservation area. Impacts include physical displacement, and the more common but less visible 'economic displacement' in which economic opportunities are curtailed. Social relations can also be changed by PAs, often negatively, although in some cases indigenous and community conserved areas (ICCAs) may protect the rights of vulnerable people. In terms of who is affected by PA impacts, the authors find that three factors are crucial: local existing micro-politics, PA governance regimes and wider political economies. As with so many other chapters in this book, the authors find that data on the impacts of PAs are seriously lacking. However, they also caution that the complexity of the relationship between PAs and local people may make it impossible to draw strong inferences from case studies, no matter how many are available.

In Chapter 11, Sandbrook and Roe explore the poverty impacts of the conservation of particular charismatic species, drawing on the example of great apes. They find that despite a high level of investment, positive impacts on poverty tend to be limited. Various strategies for linking species conservation and poverty alleviation have been implemented including tourism, community-based natural resource management (CBNRM) and integrated conservation and development. However, the focus of species-based conservation is on species, not poverty alleviation, and the benefits for local people are often too low or poorly distributed to make a significant difference to poverty levels. At the same time, the presence of species-based conservation can greatly increase the level of conservation law enforcement, which may exacerbate poverty for resource-dependent people in the short term.

In Chapter 12, Jones *et al.* explore the poverty impacts of community conservancies in Namibia, widely touted as a flagship example of a successful CBNRM programme. They evaluate whether benefits are fairly shared and reaching the household level, while at the same time pointing out that it would be misleading to frame CBNRM as a poverty alleviation strategy per se. They find that most conservancies have elected to provide benefits to members through social projects rather than make payments to members or households, and that those benefits that are paid at the household level in cash or kind are not being seized by elites. Jobs are the most significant benefit in terms of providing pathways out of poverty. With the exception of jobs, most benefits from CBNRM probably help alleviate poverty but do not transform people from poor to non-poor. Wildlife numbers are steadily growing in conservancy areas, creating new challenges as the livelihood costs of growing human–wildlife conflict fall more heavily on the poor.

In Chapter 13, Elliott and Sumba analyse whether biodiversity-based enterprises (e.g., tourism and the wild products trade) can deliver both conservation and poverty benefits. They find some positive examples in value chain interventions that affect a whole sector by improving incomes for primary producers. They find extensive evidence of income and employment benefits, but note that in many cases the beneficiaries appear to place significant value on non-economic benefits, notably improved livelihood security and empowerment, which are hard to quantify. The evidence suggests that the enterprise approach works best for high-value species and habitats but is not suitable for resources of low economic value – which tend to be the ones of importance to the poorest. They find that the very poorest and most marginalised members of society are hard to reach without the support of government welfare provisions. This finding is not limited to conservation enterprise and resonates with the experience of the development sector in supporting small enterprises.

Distributional and institutional issues

In Part IV authors explore in more depth some of the main cross-cutting distributional and institutional issues identified in earlier chapters. They shed further light on the pervasive challenges to fair and equitable distribution of benefits from conservation initiatives, and on the need for locally based actions to complement actions at national and international levels.

In Chapter 14, Wunder and Börner offer insights into the poverty impacts of PES schemes on three groups of poor people – those who supply the services, those who buy the services and those not included in the scheme. They note a lack of evidence of any 'poverty trap' effect for suppliers to PES schemes and conclude that in most cases PES has positive effects (though small impacts on poverty alleviation) as long as participation is voluntary and the poor have secure rights over environmental assets, usually land. PES schemes are often targeted at areas of low population density, which can increase the participation of the poor. They find that poor buyers or beneficiaries of PES also receive significant benefits, particularly in watershed protection schemes. However, in common with other conservation initiatives, the poor who are not included in a PES scheme tend to be losers, with the landless and marginalised most likely to lose. They conclude that the key going forward is to expand the coverage of PES schemes and to incorporate safeguards to prevent elite capture in larger schemes, notably in Reduced Emissions from Deforestation and Forest Degradation (REDD) schemes.

In Chapter 15, Homewood et al. identify serious issues in the creation and sharing of tourism benefits around protected areas in Kenya and Tanzania. They compare impacts of conservation on Maasai livelihoods in both countries and find only one site in five where significant benefits are delivered at the household level, with the bulk of (often significant) revenue levels typically captured by elites within the communities, by the private sector, government and NGOs. They find the situation in Tanzania to be significantly worse than in Kenya due to benefit capture by state agencies, community elites and global investors as well as a history of (and fear of future) conservation exclusions. Even within Kenya's Maasai Mara, where nature-based tourism income is significant, they find that the wealthiest group captures the majority of the income. At the same time the poorest are increasingly losing access to resources – with serious implications for livelihood security and for the future of conservation initiatives.

In Chapter 16, Thomas investigates the important role of local grassroots organisations in addressing conservation–poverty links, and specifically in addressing issues of sustainability, efficiency, legitimacy and fulfilment of rights. He concludes that their importance today is nowhere more apparent than in the opportunities emerging from PES schemes – including payments for REDD and from access and benefit sharing (ABS) schemes. Strengthening the capacity, tools, administrative skills and finance of grassroots organisations is important, but ultimately communities and their organisations need favourable conditions of rights, tenure and long-term security if they are to invest in local stewardship that delivers both conservation and sustainable livelihood benefits.

In Chapter 17, Berkes explores community incentives for conservation. Through analysis of case studies, he concludes that political, social and cultural objectives are equally as or more important than monetary objectives and that empowerment is almost always a key objective. He finds that for indigenous groups, political and empowerment objectives are usually central. The fact that objectives differ greatly between and within communities makes it impossible to design 'blueprint' solutions. He unravels the complexities of trying to link livelihood incentives to biodiversity conservation given the complex and multidimensional nature of poverty, and argues that the first step in doing a better job with the linked incentives model is to acknowledge the trade-offs. He finds that intra-community differences in needs and objectives are inevitable and hard to incorporate; conservation interventions tend not to reach the poorest directly, but better social organisation and increased social capital help communities to take care of their poor through sharing networks.

Biodiversity and poverty relationships in the context of global challenges

In Part V, the final section of the book, the linkages are placed in their global context. The implications of existing and anticipated rates of biodiversity loss are huge for all, but particularly for the world's poorest people. Will acting swiftly on global poverty help save biodiversity? Or will investing more in biodiversity conservation solve the poverty problem? Or are both part of a much bigger question – that of how we choose to live our lives, grow our economies and govern our sharing of the world's resources?

In Chapter 18, MacKinnon locates these questions within the context of the threat that climate change poses to sustainable development. She finds that climate change provides a unique opportunity to re-emphasize the multiple values of natural ecosystems and that biodiversity conservation can contribute to reducing the impacts of climate change, especially for the most vulnerable communities, notably by safeguarding water and food supplies and by reducing vulnerability to natural disasters. Greater investment in natural ecosystem management and the goods and services they provide is, she argues, a priority if we are to address the coming global challenges.

In Chapter 19, Adams emphasises that both biodiversity conservation and poverty alleviation are intensely political activities. Attempts to integrate them have proved difficult, expensive and often socially divisive. He argues that the focus on local 'win–wins' for both rural poverty and pristine biodiverse environments has side-tracked thinking about the broader linkages and resulted in failure to acknowledge or address overconsumption as the chief cause of biodiversity loss. He cites evidence that conservation has both positive and negative poverty impacts at the local scale, but finds evidence that fewer than 1% of the poor live in areas of 'intact ecosystems', which must limit any capacity for conservation in those places to contribute to poverty alleviation. Adams points out that 'frontier' conservation ignores 'transformed' urban and agricultural zones and therefore some of the biggest challenges to global biodiversity. Economic growth fuelled by resource and energy consumption has been the driving model for both poverty alleviation and wealth creation since the Second World War, and is now pushing us beyond planetary boundaries. Conservation interventions are largely framed within this neoliberal capitalist paradigm, he argues, instead of challenging it. He challenges conservation to move beyond concepts of wilderness and monetary value. If we are to survive the 'Anthropocene' era, we must find new paths to delink energy generation from carbon emissions and delink energy consumption from economic growth, while redistributing wealth and resource use.

Finally in Chapter 20, Roe et al. review the preceding chapters and highlight the need to be much more careful in making claims about biodiversity–poverty linkages. While some international policies and proclamations assert that conserving biodiversity and eliminating poverty are two sides of the same coin, it is clear from the chapters in this book that it is not a given that this will be a win–win relationship: all sorts of politics, power relationships, global environmental and economic pressures and governance issues mediate the relationship. Indeed, it is clear that biodiversity conservation is not a panacea for poverty alleviation. But neither is the current model of development. Biodiversity loss and persistent poverty share common driving forces:

overconsumption, poor governance and unsustainable development paths. It is in tackling these bigger issues that solutions to both problems are likely to be found.

References

Adams, W.M., Aveling, R., Brockington, D., Dickson, B., Elliott, J., Hutton, J., Roe, D., Vira, B. & Wolmer, W. (2004) Biodiversity conservation and the eradication of poverty. *Science*, 306, 1146–9.

Biodiversity Liaison Group (2005) *Biodiversity: Life Insurance for Our Changing World*, Secretariat of the Convention on Biological Diversity, Montreal, http://www.cbd.int/doc/statements/mdg-2005-en.pdf (accessed 2 May 2012).

Convention on Biological Biodiversity (CBD) (2002) *Decision VI/26: Strategic Plan for the Convention on Biological Diversity*. Secretariat of the Convention on Biological Diversity, Montreal.

Convention on Biological Biodiversity (CBD) (2010) *Decision X/2: Strategic Plan for Biodiversity 2011–2020*. Secretariat of the Convention on Biological Diversity, Montreal.

Fisher, B. & Christopher, T. (2007) Poverty and biodiversity: measuring the overlap of human poverty and the biodiversity hotspots. *Ecological Economics*, 62, 93–101.

King, K. & Palmer, R. (2007) *Skills Development and Poverty Reduction: A State of the Art Review*. European Training Forum, Turin.

Leader-Williams, N., Adams, W.M. & Smith, R.J. (2010) *Trade-offs in Conservation: Deciding What to Save*. Wiley-Blackwell in association with the Zoological Society of London, Chichester.

Millennium Ecosystem Assessment (MA) (2003) *Ecosystems and Human Well-being: A Framework for Assessment*. World Resources Institute, Washington, DC.

Millennium Ecosystem Assessment (MA) (2005) *Ecosystems and Human Well-being: Biodiversity Synthesis*. World Resources Institute, Washington, DC.

Nadkarni, M.V. (2000) Poverty, environment, development: a many-patterned nexus. *Economic and Political Weekly*, 35, 1184–90.

Redford, K.H., Levy, M.A., Sanderson, E.W. & de Sherbinin, A. (2008) What is the role for conservation organisations in poverty alleviation in the world's wild places? *Oryx*, 42, 516–28.

Roe, D., Walpole, M. & Elliott, J. (2010) Introduction: why link biodiversity conservation and poverty reduction? In *Linking Biodiversity Conservation and Poverty Alleviation: A State of Knowledge Review*, ed. D. Roe, pp. 9–12. Technical Series no. 55, Secretariat of the Convention on Biological Diversity, Montréal.

Sachs, J.D., Baillie, J.E., Sutherland, W.J., Armsworth, P.R., Ash, N., Beddington, J., Blackburn, T.M., Collen, B., Gardiner, B., Gaston, K.J., Godfray, H.C., Green, R.E., Harvey, P.H., House, B., Knapp, S., Kümpel, N.F., Macdonald, D.W., Mace, G.M., Mallet, J., Matthews, A., May, R.M., Petchey, O., Purvis, A., Roe, D., Safi, K., Turner, K., Walpole, M., Watson, R. & Jones,

K.E. (2009) Biodiversity conservation and the Millennium Development Goals. *Science*, 325, 1502–3.

United Nations (2006) *Report of the Secretary General on the Work of the Organization*. United Nations, New York.

United Nations (2008). *Official List of MDG Indicators*, http://mdgs.un.org/unsd/mdg/Host.aspx? Content = Indicators/OfficialList.htm (accessed 2 May 2012).

World Bank (2001) *World Development Report 2000/2001: Attacking Poverty*. World Bank, Washington, DC.

Part I
Biodiversity, Ecosystem Services and Poverty – The Potential for Synergies

The Potential, Realised and Essential Ecosystem Service Benefits of Biodiversity Conservation

Will R. Turner[1], Katrina Brandon[2], Thomas M. Brooks[3,4,5], Claude Gascon[6], Holly K. Gibbs[7], Keith Lawrence[1], Russell A. Mittermeier[1] and Elizabeth R. Selig[1]

[1]Conservation International, Arlington, VA, USA
[2]Independent Consultant, Beijing, China
[3]NatureServe, Arlington, VA, USA
[4]World Agroforestry Centre (ICRAF), University of the Philippines, Los Baños, the Philippines
[5]School of Geography and Environmental Studies, University of Tasmania, Tasmania, Australia
[6]National Fish and Wildlife Foundation, Washington, DC, USA
[7]Nelson Institute for Environmental Studies, Department of Geography, University of Wisconsin-Madison, Madison, WI, USA

Introduction

There are substantial debates regarding the relationships between biodiversity, ecosystem services and human well-being, even though the Convention on Biological Diversity (CBD), Millennium Development Goals (MDGs) and other international agreements explicitly connect biodiversity conservation to poverty alleviation (Sachs *et al.*, 2009). As noted in Chapter 1, for example, countries agreed that significantly reducing the rate of biodiversity loss at the global, regional and national levels

Biodiversity Conservation and Poverty Alleviation: Exploring the Evidence for a Link, First Edition.
Edited by Dilys Roe, Joanna Elliott, Chris Sandbrook and Matt Walpole.
© 2013 John Wiley & Sons, Ltd. Published 2013 by John Wiley & Sons, Ltd.

by 2010 was "a contribution to poverty alleviation and to the benefit of all life on Earth" (CBD, n.d.-a), and included this as part of Millennium Development Goal 7 (MDG7, Ensure Environmental Sustainability). In 2010, this agreement was extended to "take effective and urgent action to halt the loss of biodiversity in order to ensure that by 2020 ecosystems are resilient and continue to provide essential services, thereby securing the planet's variety of life, and contributing to human well-being, and poverty eradication" (CBD, n.d.-b). This goal is assessed by country-level indicators on species extinction risk, remaining forest cover and the percentage of lands and seas in protected areas, among others. While conservationists have long held that actions to protect biodiversity and ecosystem services support human well-being (e.g. Curry-Lindahl, 1972), until recently there has been little empirical, broad-scale evidence to support this. Conservation continues to be presented as both a constraint on development and a tool for achieving poverty reduction (Adams *et al.*, 2004; West *et al.*, 2006; Andam *et al.*, 2010; Barrett *et al.*, 2011).

As evidenced in this volume, debates continue over the relationships between biodiversity conservation and poverty reduction. However, there is strong evidence emerging that biodiverse systems show greater productivity and resilience (Flombaum & Sala, 2008; Isbell *et al.*, 2011), with links between conservation and resilience of human communities likely to become even more important given projected climate change impacts (Turner *et al.*, 2009). The links between biodiversity and human well-being can be expressed under the framework of ecosystem services (Millennium Ecosystem Assessment (MA), 2005). Yet the variety and values of these services remain greatly overlooked, by both beneficiaries and decision makers at virtually all scales, from local to global (The Economics of Ecosystems and Biodiversity, 2009).

Discussions of biodiversity conservation actions (what, where and how) and poverty have largely been reviewed at national or local scales, and, indeed, biodiversity and poverty often coincide (Fisher & Christopher, 2007). There is also a general expectation that conservation actions, such as creating protected areas, should benefit human well-being, help secure livelihoods and pose little risk to, if not directly benefit, the poor (World Commission on Environment and Development, 1987). Yet analyses to date have been insufficient to inform decision makers on the potential role of conservation for socio-economic development, especially regarding the linkages that conservation action has to poverty alleviation (Chapter 9, this volume).

Here, we undertake a global analysis by mapping ecosystem service flows from natural habitats to human communities in order to assess the distribution of services among countries and regions, to investigate the flow of these services to the poor in particular and to understand the connection between biodiversity conservation and these important services.

Approach and methods

Previous studies of ecosystem services, human well-being and/or biodiversity have conducted analyses across large, heterogeneous spatial units, such as biodiversity priority regions (Turner *et al.*, 2007; Naidoo *et al.*, 2008), drainage basins (Luck *et al.*, 2009), countries (Ebeling & Yasué, 2008) or the entire globe (Raudsepp-Hearne *et al.*, 2010; Duraiappah, 2011). One of the inherent challenges of this type of analysis is to assess the ecosystem services in a way that allows for accurate comparison among different sites. We resolve methodological problems[1] faced by previous studies by conducting all analyses on a grid of terrestrial equal-area hexagons[2] (Sahr *et al.*, 2003) and aggregate to larger units only for reporting.

We assess the flows of ecosystem services provided to people, especially to the poor, by priority habitats for terrestrial conservation, considering global distributions of biodiversity, physical factors and socio-economic context. Specific studies of a given area using local data may allow for a more in-depth understanding of that place; however, our data set offers the means to compare different places. The data for biodiversity (International Union for Conservation of Nature (IUCN), 2008), human population (LandScan, 2006) and poverty (Center for International Earth Science Information Network, 2005), as well as the base data for improving estimates for valuing climate regulation (Reusch & Gibbs, 2008) and other ecosystem services (Costanza *et al.*, 1997; Turner *et al.*, 2007), have been detailed elsewhere (Turner *et al.*, 2012).

We use four geographically explicit valuation alternatives to estimate ecosystem service value (ESV) delivered to different socio-economic contexts and to understand links between biodiversity conservation and the sources of these services:

1. **Potential services** generated by natural habitats, irrespective of whether people are close enough to receive benefits;
2. **Realised services**, which account for the human population that might capture the services;
3. **Essential services**, the services that flow directly to the poor and provide immediate benefits;
4. **Essential services with transfers**, essential services as in #3, plus amounts the poor could receive from payments for ecosystem services (PES) mechanisms.

In the analysis in this chapter, we use these methods to understand the value of ecosystem services delivered by conservation priority areas,[3] and how these values change when we incorporate human population and poverty into the calculation of

---

[1] Firstly, spatial variation relevant to ecosystem services is lost when aggregated to large regions. Secondly, unequal areas make it difficult to compare quantities such as those measuring ecosystem service value and biodiversity. Thirdly, boundaries coincide with features such as country and habitat borders that are correlated with multiple variables. Alternatives to these, such as rectangular geographic or equal-area grids, incur oversampling and shape distortion problems away from the equator (Potere & Schneider, 2007). Hexagonal grids avoid each of these problems.

[2] $2{,}592\,\mathrm{km}^2 \pm 11.6\,\mathrm{km}^2$ SD; N = 58,613.

[3] These areas represent high priorities for conserving terrestrial vertebrate species, and were based on mapped distributions of all threatened vertebrates in taxa comprehensively assessed by the IUCN Red List (IUCN, 2008). Two key criteria used to inform conservation priorities are irreplaceability and vulnerability, the former identifying where conservation options are most limited over space and the latter showing where they are most urgent (Margules & Pressey, 2000). We mapped endemic, threatened biodiversity as 'range-size rarity' or 1/(species range size), summed

ESV. We also explore the importance of ESV to the world's poor and geographic differences by region.

We explore 17 different classes of ecosystem services based on biome- and service-specific value estimates (Costanza *et al.*, 1997) refined by more recent studies using improved land cover and climate regulation data (Sutton & Costanza, 2002; Turner *et al.*, 2007, 2012). This approach does not account for within-biome variation; this and other known assumptions are only partially addressed by the refinements discussed in this chapter. Nevertheless, this approach is the only published, global mapping of values for a range of services and biomes, and has been used as a source for ESV estimates at regional (Viglizzo & Frank, 2006) and global scales (Balmford & Bond, 2005; Turner *et al.*, 2007). We report all monetary values in 2005 US dollars, converted where necessary according to published estimates of annual global consumer price inflation. Our study includes only terrestrial areas, but this does include high-value coastal services and areas. More comprehensive valuation of marine ecosystem services in relationship to biodiversity conservation and human well-being remains a critical research need.

Given our interest in the relationship with poverty, we present our findings comparing the world's regions as defined by the World Bank's *Country and Lending Groups* (World Bank, 2012). We grouped 70 high-income countries, and used the standard regional groupings for all developing countries: East Asia and Pacific (EAP) (24), developing Europe and Central Asia (23), Latin America and the Caribbean (LAC) (30), Middle East and North Africa (MNA) (13), South Asia (8) and Sub-Saharan Africa (47). All countries in these latter categories are considered to be developing, although many are middle-income developing countries.

Potential ecosystem service value: economic values irrespective of use

If all things were equal, the proportion of areas with high biodiversity and the total potential ESV for each region would be similar to its total proportion of land area. However, both biodiversity and ecosystem service value are distributed unequally over the globe. Figure 2.1a shows the variation in land area per region, and Figure 2.1b the biodiversity importance. Figure 2.1c then provides an overview of the proportion, by development status and region, of potential ESV in comparison to these variables.

The developing countries of Europe and Central Asia occupy a large land area (18% of the globe) and have potential ESV roughly proportionate to land area, but harbour only 1% of the total area of high biodiversity value. The high biodiversity and extensive

---

across all threatened species occurring in a cell (Williams *et al.*, 1996). We defined priority areas for biodiversity conservation as the upper quarter of cells according to this metric.

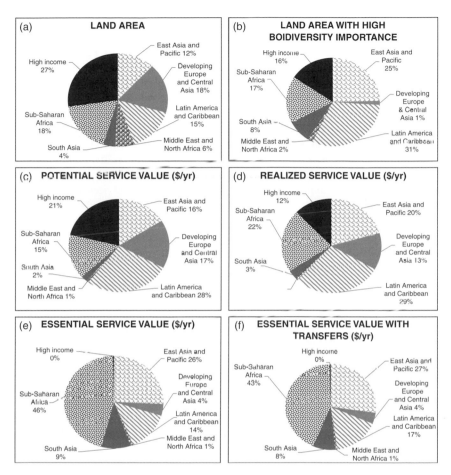

Figure 2.1   Percentage of global total, by development status and region, of (a) land area; (b) area of high biodiversity importance; (c) potential ecosystem service value (ESV); (d) realised ESV; (e) essential ESV and (f) essential ESV with transfers (payments for ecosystem services).

Note: Ecosystem service value estimates are for services originating from the habitats in a region, not necessarily those flowing to a region. In (e) and (f), developed countries generate some (<1% of global total) essential ecosystem services that originate within their borders but flow to developing countries in other regions.

Country groupings follow World Bank classifications, but the spatial area of regions and the number of countries (or areas) within vary greatly. There are 70 countries included in the high-income class. Groupings are East Asia and Pacific (n = 24 including China, Mongolia and Indonesia, as well as small island nations (e.g. Samoa and Fiji)), developing Europe and Central Asia (n = 23 with Europe's poor countries (e.g. Bulgaria), and Russia and former USSR), Latin America and the Caribbean (0), Middle East and North Africa (13), South Asia (8) and Sub-Saharan Africa (47).

tropical forests and watersheds in Latin America are reflected by the comparatively larger share of biodiversity and ESV reflected for that region – both around twice as high as expected on the basis of land area. The largely arid MNA region has lower concentrations of biodiversity relative to the area, and even less potential ESV, due in part to the near absence of forest ecosystem services in particular. The potential ESV and biodiversity conservation priority of Africa is roughly proportional to its land area, while EAP countries have ESV slightly higher than land area but twice the expected level of biodiversity importance. Developed nations have ESV roughly proportional to land area but with proportionately less high-biodiversity land area. These findings allow potential ESV to be mapped and hence the relative total potential ESV for different regions to be compared. Critically, however, they say little about the importance of services to people. Our analysis in the 'Realised services' section of this chapter explores this.

Realised services: valuing ecosystem services by considering their use

People, especially decision makers, assign higher value to ecosystem services that are directly captured by human beneficiaries. Some ecosystem service benefits may be captured locally, for example by people depending on firewood or wild food harvest. Yet for other services, those who benefit most live far from the area providing the service – this is especially the case for water supply and carbon storage, where key users may be in distant cities or countries. Our second valuation alternative estimates these *realised services* by classifying the 17 ecosystem services into three service flow models (Figure 2.2) including those services realised only within a given distance ('proximal' services such as pollination), services following river drainage patterns ('downstream' services such as water supply)[4] and services that benefit people everywhere ('global' services such as climate regulation).

While there is similarity in the places with high potential ESV and high realised ESV, the distribution of population density in particular drives major differences in the distributions of potential ESV (Figure 2.1c) and realised ESV (Figure 2.1d) among regions. Realised ESV is lower in regions where relatively few people can capture services, including tundra and boreal forests. The relative importance of realised services declines for developed countries given the low population density across places such as Canada and much of Australia. This is also the case for the developing countries of Europe (e.g. Russia) and Central Asia. Globally, accounting

---

[4] For those 'downstream' services following hydrological drainage patterns, we used 30-arc-second drainage direction data (Lehner *et al.*, 2008), where available, and Hydro1k drainage direction data (US Geological Survey, 2011) elsewhere to compute the set of cells downstream of each hexagon cell, up to 500 km, as the window for calculating beneficiary populations.

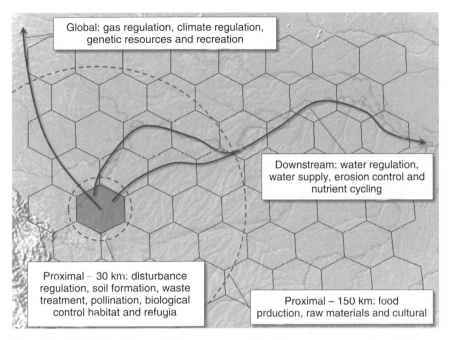

Figure 2.2 **Models for flows of 17 ecosystem services from source ecosystems to human beneficiaries.**

for capture by human beneficiaries results in an aggregate value of US$9.40 trillion in realised services, compared to US$14.03 trillion in potential services. While there are substantial differences in the potential and realised value within countries, in general the patterns for LAC, MNA and South Asia show a correspondence between potential and realised ESV. There is a modest increase for East Asia and the Pacific. The greatest increase from potential ESV to realised ESV – the value of services that people rely upon – is for Africa. Yet, although realised services reflect ecosystems' overall benefits to people, they do not identify how many poor depend on these services. This is explored in more detail in the 'Essential services' section of this chapter.

Essential services: valuing the ecosystem services the poor rely upon

People differ in their ability to access or pay for alternatives if ecosystem services are lost. While affluence frees people from direct dependence on local ecosystem services and can buffer them against some consequences of ecological change, the poor lack these buffers and substitutes (Chapter 4, this volume). Poor communities are thus often critically dependent on ecosystem services to sustain their lives and livelihoods (Luck *et al.*, 2009). Declines in wild resources, biodiversity or ecosystem health that provides food, fuel, clothing, medicines and shelter are linked to declines in rural health and welfare. For example, when water quality declines, the poor have neither the money nor access to buy clean water. *Essential services* are estimates of ESV benefits that flow directly to the poor and provide immediate benefits. Thus, our calculations of essential services resemble those for realised services, but only value benefits flowing directly to poor individuals, such as water supply, disturbance regulation or food production. Additionally, they exclude indirect or longer term benefits (i.e. climate regulation, gas regulation, nutrient cycling, genetic resources and recreation).

Developed countries, with fewer impoverished people, are nearly absent from the estimated maps of essential services (Figure 2.1e). Not surprisingly, Africa shows the highest relative increases in ESV (31% increase over potential ESV), when flows to poor people are taken into account. In these areas, ecosystem services are critical to human well-being. Africa comprises nearly half of the absolute value of essential ESV that the world's poor rely upon, followed by EAP, then LAC. However, the proportional decreases from realised to essential ESV for the LAC region and for developing Europe and Central Asia are offset by the modest increases in both South Asia and EAP.

The economic value of essential services flowing to the poor is high, equalling US$1.814 billion per year for all developing countries. For 36 of the 49 least developed countries, the value of essential ESV exceeds US$1 per person per day. This suggests that, at the national level, ecosystem services may provide a substantial subsidy for the world's poorest countries, representing a significant contribution towards meeting the MDGs.

Estimating all benefits to the poor: essential services and payments for environmental services

PES is based on the premise that market mechanisms should allow users who benefit from ecosystem services to provide equitable support to local residents for their stewardship of these resources (Chapter 14, this volume). There are numerous assessments of how PES systems work in practice, and reason for caution that the theory is easier than implementation (Wunder *et al.*, 2008). One might assume that there is tremendous potential to use PES to correct the failure of markets to value, and to capture the value of, environmental services, and transfer them to resource managers, providing substantial benefits for both nature and people. Yet this overall potential has rarely been addressed over broad spatial scales. For this analysis we assumed that we could identify the beneficiaries of ecosystem services based on their spatial relationship to service flows. We also assumed that non-poor beneficiaries could financially compensate residents who either manage or incur the opportunity cost for sustaining the source habitats (i.e. people within the same hexagonal cell as those habitats). This valuation alternative calculates total value as the value of essential services in a cell (those flowing directly to the poor), plus the value of PES transfers exclusively to the poor.[5] This assumption is important for watersheds, for example where non-poor residents downstream would pay resource managers upstream. In general, PES revenues are distributed without regard to economic status (Chapter 14, this volume). This implies that financial mechanisms targeting the poorest people in particular could deliver even greater PES value to the poor than that estimated here.

A significant finding is that the most important places for effective PES mechanisms are also the same places where the poor depend on essential ecosystem services (Figure 2.1f). If such PES transfers were uniformly realised, they would capture and channel an additional US$858 billion annually to poor resource stewards. This would be in addition to the essential ecosystem services that the poor receive and rely upon. While these numbers sound high, they represent the full value of the ESV if provided and transferred. If effective and equitable PES mechanisms were implemented comprehensively, ESV to the poor would exceed US$1 per person per day for 30% (331 million) of an estimated 1.1 billion people living in poverty. It will remain challenging to implement such transfers, for which a host of socio-economic, policy and institutional challenges must be addressed in many of these countries. But these results – specifically arising from the spatial distribution of population and poverty relative to the sources and flows of ecosystem services – suggest that progress on even a fraction of these possible PES systems could represent major income flows to the poor.

Conclusions: linking ecosystem service flows, people and biodiversity

Biodiversity is the foundation for ecosystem service provision (MA, 2005). We are only beginning to understand how, and how much, intact habitats and biodiversity can be altered without affecting ecosystem functioning and the lasting provision of ecosystem services. Biodiversity loss can change both the magnitude and the stability of ecosystem processes. Ecosystems retaining their original complexity are more resilient, with high numbers of species offering more buffers to change. For example, removing just one of many species of fish from a river can worsen freshwater quality (Taylor et al., 2006), and elimination of biodiversity from landscapes contributed to Lyme

---

[5] Because our interest is in poor communities in particular, we did not count PES flowing to the non-poor. Specifically, this means the value of PES flowing to the poor in a given cell is (Total PES Value Flowing to Cell) × (Cell Poverty Rate).

disease and hantavirus pulmonary syndrome becoming epidemic, opening pathways for human infection on a large scale (Keesing *et al.*, 2010).

Our analysis allows us to examine the distribution of ecosystem service value (Figure 2.3a) and biodiversity conservation priorities (Figure 2.3b). More importantly, we can identify the places with a high degree of overlap between threatened biodiversity, habitats that provide ecosystem services and poor residents who depend on these resources or services. Overall, there is a strong relationship between biodiversity conservation priorities and ecosystem service provision. For example, conserving the 25% highest priority cells for biodiversity would sustain a disproportionately large 39% of global potential ESV. The relationship between biodiversity conservation priorities and ecosystem service provision increases when realised services are considered: the 25% highest priority cells for biodiversity sustain a disproportionately large 50% of globally realised ESV services – the services actually used by people. And the

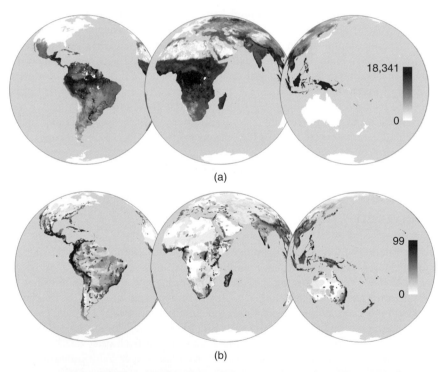

(a)

(b)

Figure 2.3 **Mapped global distributions of (a) ecosystem service value ($/ha/yr), specifically essential services with transfers (including payments for ecosystem services), and (b) biodiversity conservation priority (percentile rank) as measured by threatened species endemism.**

relationship becomes stronger still when poverty is considered: the highest 25% of land for biodiversity provides 56% of essential services benefitting the world's poorest, or 57% if PES transfers are additionally considered. Globally, these transfers could add US$858 billion per year in payments to poor communities over all land area; the top quarter of land for biodiversity alone accounts for US$509 billion (59.3%) of this annual increase.

Our findings are particularly critical in the poorest countries given the challenges that they face in achieving the MDGs. These challenges will become especially serious in developing countries as they see their populations swell dramatically by 2050 (Roberts, 2011), with much of this growth taking place in the world's poorest countries (United Nations Population Division, 2011). Our analysis shows that the loss of biodiversity is something that should be alarming to all. It is becoming increasingly evident that as biodiversity loss increases, a significant proportion of the planet's 7 billion are also being deprived of goods and services from biodiversity and ecosystems that they depend upon for their survival. While wealthier countries, or wealthier people within poor countries, have the economic means to replace lost services or pay for higher priced alternatives, poor countries and people cannot. This highlights the importance of conservation actions in supporting both biodiversity and human communities.

While the negative effects of conservation actions on human well-being have been emphasised, recent findings based on more sophisticated research techniques are testing whether people would be better off in the absence of conservation action. Quasi-experimental studies for Costa Rica and Thailand found that districts with protected areas had lower poverty: approximately 10% lower in Costa Rica and 30% in Thailand (Andam et al., 2010). Additionally, there is evidence from across Brazil that an initial economic boom from timber sales can be followed by a bust when the converted habitat loses agricultural productivity. In such boom-and-bust municipalities, the loss of biodiversity and ecosystem services becomes economically apparent, and higher rural poverty rates ensue (Rodrigues et al., 2009). These broad-scale findings suggest that for many countries, the ecosystem services provided by intact habitats outcompete alternative uses, especially in places with steep slopes and poor soils. A clear implication is that the international development community should place biodiversity conservation as a priority concern to avert the unintended deepening of poverty.

Our results suggest that widespread (though not universal) 'win–win' synergies are possible between poverty alleviation and conservation. There are many ways that multiple benefits for the poor, and for biodiversity, could be realised, including creating, expanding or improving management of formal protected area systems; expanding governance structures for land and resource management on community-conserved or indigenous lands (Lewis et al., 2011); decentralising rights over forest commons to forest dwellers (Chhatre & Agrawal, 2009; Ricketts et al., 2010) and introducing incentives for PES (Lipper *et al.*, 2009). The proposed international payment scheme for tackling forest-related carbon emissions – Reducing Emissions from Deforestation and Forest Degradation (REDD) – aims to capture and transfer global benefits to local resource stewards. Such efforts are in their early stages, but are beginning to widely introduce the idea of payment transfers, and may help establish the institutions and mechanisms that will allow other service values to be captured as well. There will undoubtedly be substantial challenges in implementing these programmes, but our ability to target is improving. Meanwhile, decision makers are increasingly looking for ways to accomplish multiple objectives with one investment, heightening discussions about the potential spatial concordance of poverty alleviation, biodiversity conservation and ecosystem service flow protection.

Along with poor governance, the failure of markets to value and capture the importance of natural areas has contributed to both biodiversity loss and persistent poverty when people in remote, rural areas view habitat conversion to marginal agriculture as their only option. Though widespread PES implementation faces challenges such as insufficient policy frameworks and, to date, poor spatial planning (Wunder *et al.*, 2008), our results suggest great potential if effective REDD mechanisms are designed and similar compensation mechanisms for other services are supported widely. Climate change – likely a dominant force impacting our planet in the decades ahead – will increase the value of the many ecosystem services that help ameliorate it and its negative impacts (Turner *et al.*, 2009). If some of the value of these environmental services was captured and efficiently and equitably returned to communities, there would be dual benefits: additional incentives for conservation and new sources of income for rural communities. While funding transfers alone may be insufficient to lift the poor from poverty, they offer better livelihood options and help buffer people facing the increasing challenges of an unpredictable climate. These findings provide important insight for policy makers, as human well-being and poverty alleviation depend critically on the substantial benefits provided by natural ecosystems. Sustaining the places of highest priority for biodiversity conservation will deliver benefits to human well-being and poverty alleviation that are large both in absolute terms and relative to costs and needs.

Acknowledgements

The authors thank the IUCN Species Survival Commission for access to the species distribution data from the IUCN Red List, and specifically BirdLife International for the bird data and John Iverson and Ross Kiester for the turtle and tortoise data.

References

Adams, W.M., Aveling, R., Brocking, D., Dickson, B., Elliott, J., Hutton, J., Roe, D., Vira, B. & Wolmer, B. (2004) Biodiversity conservation and the eradication of poverty. *Science*, 306, 1146–9.

Andam, K.S., Ferraro, P.J., Sims, K.R.E., Healy, A. and Holland, M.B. (2010) Protected areas reduced poverty in Costa Rica and Thailand. *Proceedings of the National Academy of Sciences of the USA*, 107, 9996–10001.

Balmford, A. & Bond, W. (2005) Trends in the state of nature and their implications for human well-being. *Ecology Letters*, 8, 1218–34.

Barrett, C.B., Travis, A.J. & Dasgupta, P. (2011) On biodiversity conservation and poverty traps. *Proceedings of the National Academy of Sciences of the USA*, 18, 13907–12.

Center for International Earth Science Information Network (CIESIN) (2005) *2005 Global Subnational Rates of Child Underweight Status*, Center for International Earth Science Information Network, Columbia University, Palisades, NY, http://www.ciesin.columbia.edu/povmap/ds_global.html (accessed 21 April 2012).

Chhatre, A. & Agrawal, A. (2009) Trade-offs and synergies between carbon storage and livelihood benefits from forest commons. *Proceedings of the National Academy of Sciences of the USA*, 106, 17667–70.

Convention on Biological Diversity (CBD) (n.d.-a) *COP Decision VI/26*, http://www.cbd.int/decision/cop/?id=7200 (accessed 21 April 2012).

Convention on Biological Diversity (CBD) (n.d.-b) *COP 10 Decision X/2*, http://www.cbd.int/decision/cop/?id=12268 (accessed 21 April 2012).

Costanza, R., d'Arge, R., deGroot, R., Farber, S., Grasso, M., Hannon, B., Limburg, K., Naeem, S., Oneill, R.V., Paruelo, J., Raskin, R.G., Sutton, P. & van den Belt, M. (1997) The value of the world's ecosystem services and natural capital. *Nature*, 387, 253–60.

Curry-Lindahl, K. (1972) *Conservation for Survival: An Ecological Strategy*. Morrow, New York.

Duraiappah, A.K. (2011) Ecosystem services and human well-being: do global findings make any sense? *BioScience*, 61, 7–8.

Ebeling, J. & Yasué, M. (2008) Generating carbon finance through avoided deforestation and its potential to create climatic, conservation and human development benefits. *Philosophical Transactions of the Royal Society of London*, 363, 1917–24.

Fisher, B. & Christopher, T. (2007) Poverty and biodiversity: measuring the overlap of human poverty and the biodiversity hotspots. *Ecological Economics*, 62, 93–101.

Flombaum, P. & Sala, O.E. (2008) Higher effect of plant species diversity in natural than artificial systems. *Proceedings of the National Academy of Sciences of the USA*, 105, 6087–90.

International Union for Conservation of Nature (IUCN) (2008) *2008 IUCN Red List of Threatened Species*, www.iucnredlist.org (accessed 21 April 2012).

Isbell, F., Calcagno, V., Hector, A., Connolly, J., Harpole, W.S., Reich, P.B., Scherer-Lorenzen, M., Schmid, B., Tilman, D., van Ruijven, J., Weigelt, A., Wilsey, B.J., Zavaleta, E.S. & Loreau, M. (2011) High plant diversity is needed to maintain ecosystem services. *Nature*, 477, 199–202.

Keesing, F., Belden, L.K., Daszak, P., Dobson, A., Harvell, C.D., Holt, R.D., Hudson, P., Jolles, A., Jones, K.E., Mitchell, C.E., Myers, S.S., Bogich, T. & Ostfeld, R.S. (2010) Impacts of biodiversity on the emergence and transmission of infectious diseases. *Nature*, 468, 647–52.

LandScan (2006) *Global Population Database (2006 Release)*, Oak Ridge National Laboratory, Oak Ridge, TN, http://www.ornl.gov/landscan/(accessed 21 April 2012).

Lehner, B., Verdin, K. & Jarvis, A. (2008) New global hydrography derived from spaceborne elevation data. *Eos, Transactions, American Geophysical Union*, 89, 93–94.

Lewis, D., Bell, S.D., Fay, J., Bothi, K.L., Gatere, L., Kabila, M., Mukamba, M., Matokwani, E., Mushimbalume, M., Moraru, C.I., Lehmann, J., Lassoie, J., Wolfe, D., Lee, D.R., Buck, L. & Travis, A.J. (2011) Community Markets for Conservation (COMACO) links biodiversity conservation with sustainable improvements in livelihoods and food production. *Proceedings of the National Academy of Sciences of the USA*, 108, 13957–62, http://www.pnas.org/content/108/34/13957 (accessed 21 April 2012).

Lipper, L., Sakuyama, T., Stringer, R. & Zilberman, D. (eds.) (2009) *Payment for Environmental Services in Agricultural Landscapes: Economic Policies and Poverty Reduction in Developing Countries*. Springer and Food and Agriculture Organization, London.

Luck, G.W., Chan, K.M.A. & Fay, J.P. (2009) Protecting ecosystem services and biodiversity in the world's watersheds. *Conservation Letters*, 2, 179–88.

Margules, C.R. & Pressey, R.L. (2000) Systematic conservation planning. *Nature*, 405, 243–53.

Millennium Ecosystem Assessment (MA) (2005) *Ecosystems and Human Well-Being: Synthesis*. Island Press, Washington, DC.

Naidoo, R., Balmford, A., Costanza, R., Fisher, B., Green, R.E., Lehner, B., Malcolm, T.R. & Ricketts, T.H. (2008) Global mapping of ecosystem services and conservation priorities. *Proceedings of the National Academy of Sciences of the USA*, 105, 9495–500.

Potere, D. & Schneider, A. (2007) A critical look at representation of urban areas in global maps. *GeoJournal*, 69, 55–80.

Raudsepp-Hearne, C., Peterson, G.D., Tengö, M., Bennett, E.M., Holland, T., Benessaiah, K., MacDonald, G.K. & Pfeifer, L. (2010) Untangling the environmentalist's paradox: why is human well-being increasing as ecosystem services degrade? *BioScience*, 60, 576–89.

Reusch, A. & Gibbs, H.K. (2008) *New IPCC Tier-1 Global Biomass Carbon Map for the Year 2000*, Oak Ridge National Laboratory, Oak Ridge, TN, http://cdiac.ornl.gov (accessed 21 April 2012).

Roberts, L. (2011) 9 billion? *Science*, 333, 540–43.

Rodrigues, A.S.L., Ewers, R.M., Parry, L., Souza, C. Jr., Veríssimo, A. & Balmford, A. (2009) Boom-and-bust development patterns across the Amazon deforestation frontier. *Science*, 324, 1435–7.

Sachs, J.D., Baillie, J.E.M., Sutherland, W.J., Armsworth, P.R., Ash, N., Beddington, J., Blackburn, T.M., Collen, B., Gardiner, B., Gaston, K.J., Godfray, H.C.J., Green, R.E., Harvey, P.H., House, B., Knapp, S., Kümpel, N.F., Macdonald, D.W., Mace, G.M., Mallet, J., Matthews, A., May, R.M., Petchey, O., Purvis, A., Roe, D., Safi, K., Turner, K., Walpole, M., Watson, R.T. & Jones, K.E. (2009) Biodiversity conservation and the Millennium Development Goals. *Science*, 325, 1502–3.

Sahr, K., White, D. & Kimerling, A.J. (2003) Geodesic discrete global grid systems. *Cartography and Geographic Information Science*, 30, 121–34.

Sutton, P.C. & Costanza, R. (2002) Global estimates of market and non-market values derived from nighttime satellite imagery, land cover, and ecosystem service valuation. *Ecological Economics*, 41, 509–27.

Taylor, B.W., Flecker, A.S. & Hall, R.O. Jr. (2006) Loss of a harvested fish species disrupts carbon flow in a diverse tropical river. *Science*, 313, 833–6.

The Economics of Ecosystems and Biodiversity (2009) *The Economics of Ecosystems and Biodiversity for National and International Policy Makers*. UNEP, Bonn.

Turner, W.R., Brandon, K., Brooks, T.M., Costanza, R., da Fonseca, G.A.B. & Portela, R. (2007) Global conservation of biodiversity and ecosystem services. *BioScience*, 57, 868.

Turner, W.R., Brandon, K., Brooks, T.M., Gascon, C., Gibbs, H.K., Lawrence, K., Mittermeier, R.A. & Selig, E.R. (2012) Global biodiversity conservation and the alleviation of poverty. *BioScience*, 62, 85.

Turner, W.R., Oppenheimer, M. & Wilcove, D.S. (2009) A force to fight global warming. *Nature*, 462, 278–9.

United Nations Population Division (2011) *World Population Prospects: The 2010 Revision*, United Nations, New York, http://esa.un.org/unpd/wpp/(accessed 21 April 2012).

US Geological Survey (2011) HYDRO1k Elevation Derivative Database, http://eros.usgs.gov/Find_Data/Products_and_Data_Available/gtopo30/hydro (accessed 21 April 2012).

Viglizzo, E.F. & Frank, F.C. (2006) Land-use options for Del Plata Basin in South America: tradeoffs analysis based on ecosystem service provision. *Ecological Economics*, 57, 140–51.

West, P., Igoe, J. & Brockington, D. (2006) Parks and peoples: the social impact of protected areas. *Annual Review of Anthropology*, 35, 251–77.

Williams, P.H., Prance, G.T., Humphries, C.J. & Edwards, K.S. (1996) Promise and problems in applying quantitative complementary areas for representing the diversity of some neotropical plants (families Dichapetalaceae, Lecythidaceae, Caryocaraceae, Chrysobalanaceae and Proteaceae). *Biological Journal of the Linnean Society*, 58, 125–57.

World Commission on Environment and Development (1987) *Our Common Future*. Oxford University Press, Oxford.

World Bank (2012) *Country and Lending Groups*, http://data.worldbank.org/about/country-classifications/country-and-lending-groups (accessed 21 April 2012).

Wunder, S., Engel, S. and Pagiola, S. (2008) Taking stock: a comparative analysis of payments for environmental services programs in developed and developing countries. *Ecological Economics*, 65, 834–52.

$$3$$

Poverty Reduction and Biodiversity Conservation: Using the Concept of Ecosystem Services to Understand the Linkages

Heidi Wittmer[1], Augustin Berghöfer[1] and Pavan Sukhdev[2]

[1]Helmholtz Centre for Environmental Research UFZ, Leipzig, Germany
[2]GIST Advisory Private Ltd, Haryana, India

Introduction

The international study on The Economics of Ecosystems and Biodiversity (TEEB) which was concluded in 2010 highlights the links between biodiversity, ecosystem services and human well-being in economic terms (2010a,b,c, 2011a). It has compiled a wide array of policy options to better take the values of biodiversity into account. One of TEEB's main messages is that poor people often depend significantly on nature, and thus economic values can best inform policy if distributive implications are considered. This chapter explores some of the implications that biodiversity loss has for human well-being and proposes a set of guiding questions for analysing the impacts on poor people.

We suggest assessing ecosystem services to understand the environmental and natural resource assets that are central to maintain and improve livelihoods. Furthermore, a focus on ecosystem services makes visible these linkages in an encompassing way, including most aspects of how human well-being depends on nature.

As discussed in Chapter 1 and throughout this volume, it is not easy to capture these linkages in a universal way. Both *biodiversity* and *poverty* are contested concepts because they refer to phenomena which escape straightforward definition

Biodiversity Conservation and Poverty Alleviation: Exploring the Evidence for a Link, First Edition.
Edited by Dilys Roe, Joanna Elliott, Chris Sandbrook and Matt Walpole.
© 2013 John Wiley & Sons, Ltd. Published 2013 by John Wiley & Sons, Ltd.

(Agrawal & Redford, 2006). While some causal chains have been repeatedly found – as, for example, the impact of road development on tropical forests – we cannot assume that such linkages follow universal patterns. The range, diversity and dynamics of influences on a local resource use regime and its underpinning social-ecological system are enormous (Agrawal, 2001; Ostrom, 2007).

So, a range of conceptualisations of biodiversity and poverty, respectively, seek to describe a highly complex issue. We argue here that a focus on *ecosystem services* can be usefully applied as a heuristic to capture some of the linkages between both realms – without assuming that a single valid conceptualisation needs to be found, or claiming that the entire set of variables intervening at the poverty–environment nexus needs to be understood.

We discuss the perspective on ecosystem services, focusing on two policy levels: (i) the 'GDP of the poor' for underprovided and poor regions or sectors of society as an alternative indicator for national economic welfare, and (ii) assessing ecosystem services at the local level for tailoring (environmental) policy decisions towards a better consideration of poverty reduction objectives.

Ecosystem services: essential to human well-being and a lifeline for the poor

Dependency on nature is sometimes directly visible, as with agriculture, fisheries and forestry. At other times, it is less visible; the water supply of urban areas, the food sold in supermarkets and clean air also rely on functioning ecosystems. In cities, urban parks and green spaces lower the summer temperature, improve air quality, reduce the amount of flooding after heavy rains and also significantly increase the recreational value of city life and the real estate value of adjacent property (Brack, 2002). In addition, ecosystems and biodiversity provide inspiration and are often an important basis of local culture. There is clear evidence for a central role of biodiversity in the delivery of many – but not all – services (TEEB, 2010a).

The TEEB reports (2010a, 2010b, 2010c, 2011a) take an economic perspective on these relationships and illustrate with practical examples how policy and management can improve human well-being by taking nature and its services into account in decision making. This chapter focuses on how ecosystem services could be examined in order to better structure and understand the complex relationships between biodiversity and poverty.

In 2005, of the 1.4 billion people globally estimated to live on less than US$1.25 per day, approximately 1 billion (~70%) lived in rural areas (United Nations, 2010). They largely depend on access to natural resources and ecosystem services, as Box 3.1 illustrates. Moreover, poorer households have very limited capacity to cope with losses of critical ecosystem services such as water purification or protection against natural hazards: there may be no alternatives, only degraded alternatives or alternatives that are much more costly or unaffordable (TEEB, 2010c). Limited coping capacity translates into higher reliance on stable, functioning ecosystems. According to the International Fund for Agricultural Development's (2011) rural poverty report, 'reducing and better managing risks and increasing resilience are critical for sustainable growth in the rural economies, and for growth to enable rural people to move out of poverty'. Regulatory and supporting ecosystem services are keys to such resilience.

Box 3.1 **The importance of nature's benefits to poor people**

Forest resources directly contribute to the livelihoods of 90% of the 1.2 billion people around the world living in extreme poverty (World Bank, 2004). About 80% of the population in developing countries relies on traditional medicine that is mainly derived from herbal plants (World Health Organisation, 2008).

At current rates of depletion, most fisheries will be exhausted by 2050, affecting approximately 30 million artisanal fishers and risking the main source of protein for around 1 billion people (United Nations Environment Programme (UNEP), 2011).

One billion city dwellers around the world live without clean water or adequate sanitation. Over 2 million children die each year as a result. Currently 700 million people globally live with insufficient access to water quantity. This is expected to increase to about 3 billion people by 2025 (United Nations Development Programme (UNDP), 2006). Healthy watersheds, water bodies and aquifers and access to these water sources by poor people are critical for their survival.

Protection from natural hazards is also critical: each hectare of mangrove forestland in India's Orissa State has been calculated to be worth more than US$8000 in protecting coastlines and minimising cyclone-related damage (UNDP-UNEP), 2008).

Getting aggregate indicators right – 'GDP of the poor'

Traditional measures of national income, such as GDP, do not systematically take natural capital into account. Not only do they inadequately reflect the flow of ecosystem services but also, due to the nature of aggregate macro-economic indicators, they do not recognise the importance of these flows to poverty reduction. These two shortcomings can be overcome by an adapted measure of the income of poor rural households – we name this the *GDP of the poor*.

By including ecosystem services more systematically and by attributing the income to different groups within the population, the relative share obtained directly from nature can be calculated. The first step consists in identifying what proportion of income from the national accounts of agriculture, forestry and fisheries can be attributed to poorer households. Then, various adjustments are made to include a broad range of ecosystem services: unrecorded timber and fuel wood from forestry, non-timber forest products, ecotourism (only 50% of locally accrued ecotourism revenues) and bioprospecting (25% thereof). Also indirect benefits such as nutrient and freshwater cycling can be considered as part of the estimate – not via market values but by means of a production function approach. To gain comparable percentage values, the contribution of these additional benefits to the overall economy needs to be estimated as well.

Such calculations show, for example, that for 480 million people in India (almost half of the population) ecosystem services account for 47% of goods and services consumed. In Brazil, the poor rural population (consisting of smallholders, forest-dwelling people and fishing communities) relies on ecosystem goods and services for up to 90% of its total consumption; in Indonesia this percentage equals about 75% (TEEB, 2011a). The respective percentages for ecosystem services included in the standard GDP measures (referring to agriculture, forestry and fisheries) were 16.5% for India, 6.1% for Brazil and 11.4% for Indonesia. Thus it becomes clear that if natural capital is degraded or lost, the rural poor are disproportionally affected, and this is not at all captured by the current economic metric.

Analysis of a GDP of the poor has provided estimates of the overall dependence of livelihoods on non marketed natural resources and other ecosystem services. It should be noted that the calculation of this hybrid indicator for poor rural households is not comparable to the computation of GDP according to the UN System of National Accounts, as it is not derived by summing up corporate profits, wages, interest income and so on. However, it does focus policy attention on the income aggregate that is probably the most relevant for sustainable development: the livelihood income of rural poor households. This can help to better focus national-level policies for rural development, poverty reduction and natural resource use.

Assessing ecosystem services reveals poverty implications

Assessing ecosystem services seems to us a convincing basis to design strategies for combined poverty–environment interventions. Especially when combining it with a property rights perspective, it allows for a systematic appraisal of all potential ways in which humans (individuals as well as social groups or strata), including future generations, depend on environmental benefits. It can thus provide necessary information for targeted policy design.

As shown in the other chapters of this book, the specific relationships between biodiversity conservation and poverty alleviation vary greatly from one setting to the next. It is thus often necessary to go to regional and local policy levels in order to target interventions in a mutually enhancing way. Here, rather than considering only income, an assessment of ecosystem services captures multiple aspects of livelihoods in relation to nature. Examining ecosystem services rather than 'ecosystems' or 'the environment' allows us to identify in much greater detail how exactly an environmental change affects different constituents of well-being.

The analysis should consider both the flow of ecosystem services and the property rights of different beneficiary groups. Why such a combined assessment? If we focus on the ecosystem service *flows* alone, we can identify changes in the availability of ecosystem services but remain unaware of changes in their distribution or access to them in case overall availability has remained the same. However, focusing only on *rights* to ecosystem services reveals issues of environmental justice but may neglect ecosystemic conditions for long-term supply. An example from Sri Lanka illustrates the combined assessment of the provision and distribution of different services (see Box 3.2).

Box 3.2 Multiple benefits from traditional water management in Sri Lanka

Early Sri Lankan society developed a system of irrigation tanks that retain river run-off mainly for the purpose of irrigated agriculture. Since the 1970s, the demand for water in upstream areas for modern, large-scale agriculture and hydropower has risen and traditional management practices have been lost. This led to increased sediment load and siltation with negative consequences for the livelihood of downstream users. In order to better understand these consequences, the International Union for Conservation of Nature (IUCN) together with the local authorities conducted an economic valuation of the goods and services that the traditional tank system provided for the livelihoods of local communities in the Kala Oya river basin. While officials considered the tanks only in terms of paddy yield, the analysis showed that, besides paddy, a large variety of goods provided by the tanks contribute to local livelihoods – including products such as fish, lotus flowers and roots – as well as water for domestic and animal use. In fact, the value of paddy was equivalent to only roughly 6% of the total value generated by the watertanks. With regard to distribution the study revealed that only 13% of households obtain benefits from paddy rice cultivation, while 93% benefit from access to domestic water.

As a result, the tanks were rehabilitated rather than investing in modern paddy irrigation infrastructure.

Source: Emerton (2005) and Vidanage *et al.* (2005).

An example from Indonesia illustrates how looking at ecosystem services reveals the impacts of different development scenarios (van Beukering *et al.*, 2003). Faced with rapid degradation of Leuser National Park and surrounding areas in Aceh Province, Indonesia, its scientific director commissioned a valuation study to compare the impact of different ecosystem management strategies on the province's potential for economic development through 2030. A detailed analysis of 11 different ecosystem services, including monetisation of expected benefits for three different land use scenarios, was calculated. The study estimated that conservation and selective use of the forest would provide the highest return for the region over the long term (US$9.1–9.5 billion). Continued deforestation would cause the degradation of ecosystem services, including water storage and erosion control, and generate a lower overall economic return for the province (US$7 billion). By analysing who would benefit and lose in each scenario, the valuation exercise further demonstrated that logging the tropical forest was the slightly better option for industry including logging companies but by far the worst option for rural forest communities. In fact, the latter would lose almost one-third of the benefits they could alternatively obtain in the conservation scenario (Figure 3.1).

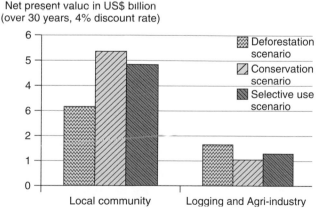

Figure 3.1   **Who benefits and who loses from forest conservation? Source: TEEB (2010c) and van Beukering *et al.* (2003).**

The most important services lost included water supply, hydropower, attractiveness for and thus income from nature tourism and fire prevention. It should be noted that in this example local communities were treated as a homogeneous unit. In practice, there are significant variations *within* and *between* communities. As Vira and Konteleon (Chapter 4, this volume) found in their review, often lower income strata depend more directly on natural resources collected from the wild, unless these are higher value products, where a certain crowding out of poorer households may occur.

Identifying property rights to nature

From a poverty perspective, it is the effective right to an ecosystem service that counts more than its physical availability. *Rights* often refer to the access, use, management and/or transfer of land or other resources (Ostrom & Schlager, 1996). Rights to different ecosystem services from the same piece of land vary. Timber grown on a private patch of land usually belongs to the land owner – yet many countries require permits for cutting trees, even on private land, and at the same time the collection of fuel wood, honey, wild fruits and other non-timber forest products is often not restricted.

Effective ownership and stipulation of rights depend on the characteristics of the service itself (e.g. can you delimit its borders? Is it quantifiable?). They depend on laws and environmental entitlements, that is, the capacity to have your right count (Leach *et al.*, 1999). They further depend on the ways in which they are being enjoyed, in particular whether cash flows are involved or not. Goods and services for which rights can be practically enforced often receive more attention (and protection) than the more elusive ones (Ruhl *et al.*, 2007).

In institutional economics, a prominent perspective on the nature of goods focuses on the characteristics of excludability and subtractability (Ostrom & Ostrom, 1977): *private goods* (e.g. the fruit in my garden), *public goods* (e.g. fresh air and protection from floods), *common-pool resources* (e.g. limited water for irrigation from a widely accessible source) and *toll or club goods* (e.g. a protected area with an entrance fee). These goods can be provided under different property regimes depending on who is in charge of managing the resources and providing the goods: private owners (private regime), the community (common property regime), government (public regime) or no one (open access regime). However, classifying the ownership of ecosystem services is more complicated (Tovey, 2008). One ecosystem can provide private, public and collective benefits. For human well-being – and decisively for poverty reduction – all types of ecosystem services are crucial, not only the provisioning ones.

Many regulating and cultural services are not managed privately but by the community or the state. For some benefits provided by ecosystems, private or collective ownership can be easily attributed (e.g. drinking water), whereas others remain essentially public goods (e.g. flood protection or reduction of waterborne disease). Currently, environmental policy instruments often focus on single ecosystem services. However, private and public ecosystem services tend to be 'bundled' rather than separate, meaning they need to be managed in bundles as well. Consequently the synergies and trade-offs between services need to be identified and taken into account.

As in this chapter's example from Sri Lanka, in many contexts the local population might use much more ecosystem services than most policy makers are aware of. Access to these benefits is often granted through informal or incomplete property rights. In many locations the (seasonal) collection of wild foods and raw materials or the access to sites of recreational value may be permitted on state land or even private lands. These benefits are thus accessible as public goods, or as a common-pool resource. The problem is that private goods usually have a much stronger lobby than public or common-pool goods. In other words, land use options that also ensure or increase public benefits instead of only maximising private ones, such as selective logging or forest conservation, usually have a weaker lobby.

In summary, poorer households have access to fewer resources, hold less formal and less secure rights and often lack the capacity to enforce their rights. In consequence they are more dependent on a reliable provision of public and common-pool goods for their well-being.

Through the lens of ecosystem services, it appears that rights to some (provisioning) benefit streams are mostly specified while other (mostly regulating) services often go unnoticed. For example, the provisioning service of growing food or fabrics is generally specified, whereas the pollinating service of insects does not seem to belong to anybody. Forest owners in many countries have their timber-harvesting rights regulated and have to follow provisions for water catchment and biodiversity protection. But rarely are their rights specified to benefits where causal connections are not easily quantifiable, or for whose enjoyment other users or beneficiaries are not easily excludable. Typical examples are the maintenance of genetic resources of crop wild relatives or the regulation of the regional climate. In consequence of this bias towards marketable provisioning services, the right to harvest or grow food and cash crops often outweighs to an important extent the responsibility for limiting damage on neighbouring ecosystems. The rights to intact regulating services that actually ensure functioning ecosystems in the long term are surprisingly weak. Land use regulation often struggles with adequately considering land use's negative side effects on other ecosystem services in the wider region. The polluting effects of shrimp farming on mangroves (Martinez-Alier, 2002) can be understood as rights to private goods (the shrimp ponds) affecting publicly owned ecosystem services (e.g. spawning grounds for coastal fish). Therefore, clarifying rights in particular to the range of regulating services seems a sound poverty reduction strategy. This clarification can take place in any property regime – as long as rights remain elusive, we have de facto an open access regime.

As Table 3.1 shows, different regimes have different strengths and weaknesses in providing services to poorer groups. If this was done for each ecosystem service, we would probably get a picture with marketable or scarce services in either private or public hands with restricted access to poor people. Rights to services are often specified at the moment that they become valuable in terms of either scarcity or market value. But specification of rights provides incentives for securing resource sustainability only if rights secure long-term access to a service and not only the good once it has been harvested (Hanna *et al.*, 1995).

Furthermore, such specification is necessary but insufficient: pursuits to clarify rights to a broader set of ecosystem services are no guarantee for pro-poor benefits. Services

Table 3.1 **Examples of poverty implications of different property regimes to two ecosystem services**

Ecosystem service	Private property regime	Public property regime	Common property regime	Open access regime
Wild medicinal plants from forest	If informal rights to medicinal plants are not recognised or specified, it is the forest owner who decides on admission, charges and harvest rules for wild plant collectors.	Often, de facto open access regime. If the state forest is a protected area, wild plant collection is typically forbidden. In other cases, formalisation of collectors' rights is sought as a basis for agreements (e.g. Table Mountain National Park, South Africa).	If collective property rights to plants are internally and externally recognised and members know each other, sustainable harvest rules and spatial arrangements are feasible. Such collective management capacity is often weakened if poverty-related need for ad hoc income increases.	As long as subsistence needs are met by abundant supply, medicinal plant collection is sustainable. If, for example, access to markets increases or market value rises, professional collectors tend to outcompete subsistence ones.

Table 3.1   (*Continued*)

Ecosystem service	Private property regime	Public property regime	Common property regime	Open access regime
Drinking water	The individual right holder to the well, water stream or piece of land where the water is found has the right to limit, exclude or sell access to water. Poor people have to pay for access (fees), for substitutes (bottled water), for poorer water quality (health risks) or for alternative sources (extra walking hours).	Government regulation ideally ensures sustainable provision of water (sufficient quantity and monitored quality) at affordable cost to all citizens. Poorly designed water fees to finance this can impose higher burden on poor people and/or lead to underprovisioning.	Collective dependence on a common water source has stimulated sophisticated rules for sustainable water use levels (as in traditional irrigation systems). It also facilitates joint action for maintenance of water infrastructure. But intra-community distribution of labour and watershed protection costs can (dis)advantage poor people.	Unregulated access to drinking water leads to medium-term degradation of water quality and quantity unless there is very low demand and abundant water supply. Absence of any regulation translates into rule of power – probably to the disadvantage of poor people.

can be valued and rights specified in a way that puts poor people at (further) disadvantage, for example when formalisation of rights ignores pre-existent customary rights. The on-going land seizure in many African countries for agro-industrial production (e.g. flowers and agro-fuels) is a notorious case of power asymmetries being reinforced by formalised new land rights (Meinzen-Dick *et al.*, 2010). Formalisation of rights can simply reconfirm and legitimize practices of exploitation (Global Forest Coalition, 2008). In order to avoid this, parties must agree *who* has *which* legitimate claim towards certain environmental benefit streams. There is no straightforward criterion for judging such legitimacy, for example in the case of competing claims from different population groups. The same problem applies to questions of intergenerational justice (how do we consider hypothetical claims of future users towards a limited benefit stream?). Thus, clarification of rights helps tackle problems of neglect and disregard of (in particular) regulating ecosystem services, but has difficulties of its own.

Working with rights to nature: considering property rights and regimes in policy design

Going beyond mere recognition and clarification of rights, we can take an important potential for poverty reduction into account by adapting management in light of rights and property *regimes*. Two examples can illustrate this. In the Keoladeo National Park near the city of Bharatpur in Rajasthan, India, property rights to public ecosystem benefits have been adapted at low cost, increasing access for all segments of society. The forests and wetlands of the park generate a beautiful setting and healthy microclimate amidst the urban landscape. Park authorities have designated a route for 'morning walkers' inside the park and have waived entrance fees to the park between 5:00 and 7:00 a.m. Thereby the recreational and health benefits have become widely accessible – with more than 1000 morning walkers every day (Stolton & Dudley, 2010). From a rights perspective, the park becomes a public good for 2 hours each day. At the same time the park's protective function is maintained and the park still generates income from entrance fees during the rest of the day.

In a second example, in the Drakensberg Mountains, South Africa, a world heritage site, management initially emphasised values of global interest and invested mainly in infrastructure for international tourism. This, however, did not increase local benefits; in fact, by restricting local use of the area, traditional and informal rights to its ecosystem services were challenged. By changing the management and public investment towards water, food and energy security, ecosystem services used by local communities and particularly poor people, their rights to these services were reconfirmed (Blignaut et al., 2011; TEEB, 2011b). Management, by determining which ecosystem services flows are reduced or increased, influences whose de facto rights are being respected and whose are not.

Considering different property regimes opens up room for finding adequate policy instruments for poverty reduction. In a private property regime, one can seek to enforce rights, incentivise provisioning of services or subsidise consumption in order to address poverty-related restrictions. In a common property regime, collective action at the level of the community or the local-level government would be a key focus for intervention. If essentially a public or open access regime, the state capacity needs strengthening for effectively maintaining the flow of the ecosystem service and regulating access to it.

How to structure a combined assessment of ecosystem services and related property rights

So far, we have argued that a combined assessment of ecosystem services and property rights is needed, with the former revealing the availability (and thereby management needs) of nature's benefits and the latter their distribution and management options.

How could one go about such an assessment? Firstly, it should respond to a prior established understanding of what kind of information is actually required in a given policy context. In what detail, for which time horizon and with what degree of certainty do we need to know about services and rights to them, in order to inform certain project or policy decisions? While this seems common sense, assessments are not often demand driven in their design.

Building on this understanding of information needs, the following set of questions is helpful in designing the combined assessment (Berghöfer, forthcoming):

1. **What ecosystem services** are in the area? This includes: which non-marketed provisioning services (e.g. wild foods, raw materials and non-timber forest products)? Which regulating services? What are their seasonal changes?
2. Which ecosystem aspects are – or, sometimes, which exact part of an ecosystem is – of critical importance to **system stability**? This includes: Can thresholds be anticipated? Do safe minimum standards apply?
3. Which population sub-groups **benefit** to what extent from the different ecosystem services?
4. What are the levels of **dependency** of different sub-groups on different services? This includes: Which services have substitutes and how accessible are these, especially to poorer households?
5. Who holds **what rights** to which services? This includes: How are these rights being recognised?
6. What **conflicts** over rights or over actual availability and use of services are manifest or imminent?
7. Under which **regime** (private, common property or public) can the service best be provided in the specific setting?
8. What options for **regime adaptation** or fine-tuning are at close reach?

Once specified and weighted in accordance with a concrete policy setting in a given region, these questions shape the analysis.[1]

If current resource use and expected impacts of new policies are assessed as suggested here, it becomes much more feasible to jointly address poverty reduction

---

[1] For further practical guidance on how to assess (and value) ecosystem services, please consult TEEB (2010a,c).

and biodiversity conservation. Targeted design of programmes becomes easier once ecosystem services and rights to them are understood. In the Sri Lankan case of Kala Oya (see Box 3.2), recognition of the collective dependence on the traditional water tank system paved the way for joint maintenance efforts and support from authorities to them. Sometimes, even slight modifications can go a long way, as with the Keoladeo National Park waiving entrance fees for early morning walkers.

Policy makers should aim for securing the access to and continued availability of the ecosystem services that are most essential to poor citizens. Efforts against poverty should certainly aim beyond maintaining people's sources of basic subsistence – but the issue for local policy makers is to ensure that policies and projects do not unintentionally degrade those ecosystem services which are currently essential for poor households.

Thus (environmental) policy interventions can be designed and implemented in a way much more sensitive to poverty reduction. The reverse strategy also merits more attention: investing in restoring or enhancing ecosystem services has important potentials for poverty reduction. In the wooded savannah areas of Shinyanga, Tanzania, for example, trees and woodland were previously cleared to eradicate tsetse fly, creating open (and private) lands for agriculture. This significantly reduced the public availability of the goods and services of trees and woodlands vital for the livelihoods of agro-pastoralists. Subsequently, traditional enclosures or fodder reserves, co-managed under a common property regime, were restored, which allowed trees to grow back and significantly increased water, fuel and fodder availability. Approximately 500,000 ha of woodlands have been restored by farmers, groups and villages across over 825 villages in Shinyanga region over a 25-year period, and the process has now spread to at least two other regions in Tanzania (Barrow & Shah, 2011). The emphasis on collective rights and management responsibilities was instrumental for this success.

Conclusions

We have argued that a focus on ecosystem services provides valuable support to poverty reduction interventions. Namely, we have emphasised the potential of (i) recognising the broad range of ecosystem services, also as a means to disaggregate benefits from biodiversity; (ii) clarifying the rights situation to them and (iii) exploring options for designing property regimes that better integrate poverty and biodiversity conservation concerns.

What is the relevance of these thoughts to policy making? We have a number of suggestions:

- For poverty reduction efforts we first need to get the knowledge base right. Certainly, change at the macro-systemic level is slow, but unless we reform key policy-guiding indicators such as the GDP, poverty interventions will be weakened by a misinformed policy context.

- The environmental basis for poverty reduction is a cross-cutting issue. Ecosystem services are (in)directly enhanced, ignored or jeopardised by a very broad range of policy areas. The issue is to ensure that policies and projects do not unintentionally degrade ecosystem services essential for poor households. Introducing needs-specific assessments of ecosystem services is key to respond to this cross-cutting issue.
- Investment in maintaining and restoring functional ecosystems can be a good poverty reduction strategy. Vulnerability can be reduced by increasing coping capacity and by protecting natural systems from destabilisation.
- Regulating ecosystem services, such as water purification or erosion control, need a lobby. Often poorly visible, these public goods tend to be neglected until their loss creates severe livelihood problems. Poverty reduction strategies need to actively engage with them before they have been lost.
- Ecosystem services are interconnected and should be dealt with as bundles. Consider their flows and the rights to them in a systemic perspective, this focus provides a high return in insight and orientation on investment in analysis.

These joint assessments of ecosystem services and related rights can help us to go beyond presumable trade-offs between biodiversity conservation and poverty reduction. As details matter, this type of analysis can help to fine-tune interventions in such a way that both policy goals can be achieved jointly.

References

Agrawal, A. (2001) Common property institutions and sustainable governance of resources. *World Development*, 29, 1649–72.

Agrawal, A. & Redford, K. (2006) *Poverty, Development and Biodiversity Conservation: Shooting in the Dark?* Working Paper 26. Wildlife Conservation Society, New York.

Barrow, E. & Shah, A. (2011) *TEEB Case: Traditional Forest Restoration in Tanzania*, http://www.eea.europa.eu/atlas/teeb/traditional-forest-restoration-in-tanzania-tanzania/view (accessed 24 April 2012).

Berghöfer, A. (forthcoming) *Building Governance Structures for Biosphere Reserves*. Unpublished PhD dissertation, NORAGRIC-UMB, Aas, Norway.

Blignaut, J., Zunckel, K. & Mander, M. (2011) *Assessing the Natural Assets of the uThukela District Municipality, South Africa, Specifically Considering a Range of Incentive Mechanisms to Secure a Buffer Zone around the uKhahlamba Drakensberg Park World Heritage Site*. Unpublished report based on a 2010 project conducted by Golder Associates.

Brack, C.L. (2002) Pollution mitigation and carbon sequestration by an urban forest. *Environmental Pollution*, 116, 195–200.

Emerton, L. (ed.) (2005) *Values and Rewards: Counting and Capturing Ecosystem Water Services for Sustainable Development*. IUCN Water, Nature and Economics Technical Paper No. 1.

International Union for Conservation of Nature, Ecosystems and Livelihoods Group Asia, Bangkok.

Global Forest Coalition (2008) *Life as Commerce: The Impact of Market-Based Conservation on Indigenous Peoples, Local Communities and Women*, Global Forest Coalition, CENSAT Agua Viva, COECOCEIBA, EQUATIONS, Alter Vida, the Timberwatch Coalition, http://www.globalforestcoalition.org/wp-content/uploads/2010/11/LIFE-AS-COMMERCE2008.pdf(accessed 24April 2012).

Hanna, S.S., Folke, C. & Mäler, K.G. (1995) Property rights and environmental resources. In *Property Rights and the Environment: Social and Ecological Issues*, eds. S.S. Hanna & M. Monasinghe, pp. 1–10. The Beijer International Institute of Ecological Economics and the World Bank, Washington, DC.

International Fund for Agricultural Development (2011) *Rural Poverty Report 2011*, IFAD, Rome, http://www.ifad.org/rpr2011(accessed 24April 2012).

Leach, M., Mearns, R. & Scoones, I. (1999) Environmental entitlements: dynamics and institutions in community-based natural resource management. *World Development*, 27, 225–47.

Martinez-Alier, J. (2002) *The Environmentalism of the Poor: A Study of Ecological Conflicts and Valuation*. Edward Elgar, Cheltenham.

Meinzen-Dick, R., Markelova, H. & Moore, K. (2010) *The Role of Collective Action and Property Rights in Climate Change Strategies*. Policy Brief No. 7. CAPRi–CGIAR, Washington, DC.

Ostrom, E. (2007) A diagnostic approach for going beyond panaceas. *Proceedings of the National Academy of Sciences*, 104, 15181–7.

Ostrom, V. & Ostrom, E. (1977) Public goods and public choices. In *Alternatives for Delivering Public Services: Toward Improved Performance*, ed. E.S. Savas, pp. 7–49. Westview Press, Boulder, CO.

Ostrom, E. & Schlager, E. (1996) The formation of property rights. In *Rights to Nature: Ecological, Economic, Cultural, and Political Principles of Institutions for the Environment*, eds. S.S. Hanna, C. Folke & K.G. Mäler, pp. 127–56. Island Press, Washington, DC.

Ruhl, J.B., Kraft, S.E. & Lant, C.L. (2007) *The Law and Policy of Ecosystem Services*. Island Press, Washington, DC.

Stolton, S. & Dudley, N. (2010). *Vital Sites: The Contribution of Protected Areas to Human Health*. The Arguments for Protection Series. WWF International, Gland, Switzerland.

The Economics of Ecosystem Services and Biodiversity (TEEB) (2010a) *The Economics of Ecosystems and Biodiversity: Ecological and Economic Foundations*. Earthscan, London.

The Economics of Ecosystem Services and Biodiversity (TEEB) (2010b) *The Economics of Ecosystems and Biodiversity for Business*. Earthscan, London.

The Economics of Ecosystem Services and Biodiversity (TEEB) (2010c) *The Economics of Ecosystems and Biodiversity for Local and Regional Policy Makers*. Earthscan, London.

The Economics of Ecosystem Services and Biodiversity (TEEB) (2011a) *The Economics of Ecosystems and Biodiversity in National and International Policy Making*. Earthscan, London.

The Economics of Ecosystem Services and Biodiversity (TEEB) (2011b) *The Economics of Ecosystems and Biodiversity: TEEB Manual for Cities: Ecosystem Services in Urban Management*. Earthscan, London.

Tovey, J.P. (2008) Whose rights and who's right? Valuing ecosystem services in Victoria, Australia. *Landscape Research*, 33, 197–209.

United Nations (2010) *The Millennium Development Goals Report*. United Nations, New York.

United Nations Development Programme (UNDP) (2006) *Human Development Report: Beyond Scarcity – Power, Poverty and the Global Water Crisis*, UNDP, New York, http://hdr.undp.org/en/media/HDR06-complete.pdf (accessed 2 May 2012).

United Nations Development Programme and United Nations Environment Programme (UNDP-UNEP) (2008) *Making The Economic Case: A Primer on the Economic Arguments for Mainstreaming Poverty-Environment Linkages into National Development Planning*, UNDP-UNEP Poverty-Environment Initiative, Nairobi, http://www.protos.be/library/plomino_documents/fd91eea2a279c37940941af526f6e2e1 (accessed 2 May 2012).

United Nations Environment Programme (UNEP) (2011) *Towards Green Economy: Pathways to Sustainable Development and Poverty Eradication A Synthesis for Policy Makers*, United Nations Environment Programme, Nairobi, www.unep.org/greeneconomy (accessed 2 May 2012).

van Beukering, P.J.H., Cesar, H.S.J. & Janssen, M.A. (2003) Economic valuation of the Leuser National Park on Sumatra, Indonesia. *Ecological Economics*, 44, 43–62.

Vidanage, S., Perera, S. & Kallesoe, M. (2005) *The Value of Traditional Water Schemes: Small Tanks in the Kala Oya Basin, Sri Lanka*. IUCN Water, Nature and Economics Technical Paper No. 6. International Union for Conservation of Nature, Ecosystems and Livelihoods Group Asia, Bangkok.

World Bank (2004) *Sustaining Forests: A Development Strategy*. World Bank, Washington, DC. http://siteresources.worldbank.org/INTFORESTS/Resources/SustainingForests.pdf (accessed 2 May 2012).

World Health Organisation (2008) *Traditional Medicine*, Fact sheet 134, World Health Organisation, Geneva, http://www.who.int/mediacentre/factsheets/fs134/en (accessed 2 May 2012).

(4)

Dependence of the Poor on Biodiversity: Which Poor, What Biodiversity?

Bhaskar Vira[1] and Andreas Kontoleon[2]

[1]Department of Geography, University of Cambridge, Cambridge, UK
[2]Department of Land Economy, University of Cambridge, Cambridge, UK

Introduction

Although the links between the biodiversity conservation and poverty alleviation agendas have been widely accepted, the specific nature of the relationship between biodiversity and poverty is still not well understood. At their broadest level, two types of links can be identified: (i) biodiversity as a means of subsistence or income, providing inputs into poor peoples' livelihoods, and (ii) biodiversity as insurance, providing a buffer against risks and shocks, and helping smooth livelihoods and consumption patterns. While these relationships have been empirically documented in a wide variety of circumstances, there is still a considerable need to investigate these linkages more critically, and with greater analytical clarity. Crucially, recognising that both biodiversity and poverty can manifest themselves in different guises, it is important to interrogate a more nuanced question: which groups of the (differentiated) poor depend, in which types of ways, on different elements of biological diversity?

This chapter represents a first attempt to address this question. It is based on a systematic review of literature on the ways in which poor people depend on biodiversity as a direct contribution to their subsistence, income and other livelihood needs, and as a source of risk coping and insurance. It examines the published literature, in order to document broad trends emerging from existing knowledge about these relationships, and to identify key areas where there are knowledge gaps.

Biodiversity Conservation and Poverty Alleviation: Exploring the Evidence for a Link, First Edition.
Edited by Dilys Roe, Joanna Elliott, Chris Sandbrook and Matt Walpole.
© 2013 John Wiley & Sons, Ltd. Published 2013 by John Wiley & Sons, Ltd.

Definitional issues

Biodiversity

Biodiversity is widely understood to refer to three dimensions within which variability occurs: *genetic*, meaning the variation of genes within a species, sub-species or population; *population* or *species*, meaning the variation between living species and their component populations at different spatial scales (local, regional or global); and *community* or *ecosystem*, meaning the variation within ecological complexes of which species are a part.

Defined in this way, relatively few of the studies we reviewed explicitly focussed on 'biodiversity'. The term *nature's resources* better captures the generic categories of resources that have been studied. These include forests, in terms of both wood-based and non-timber forest products (NTFPs); mangroves; fish; wild animals (bushmeat) and wild plants (including herbs) and common pool resources (CPRs) more generally.

Many studies focus on tropical natural environments, so it is possible to make inferences about the importance of biodiversity, but we would suggest that these links need to be established more carefully. Keeping this in mind, we need to be cautious about how we interpret the material that has been the subject of this review. While nature's resources are clearly very central to the livelihood strategies of the poor, we cannot make the assumption that these activities require or depend upon the existence of biodiversity. Indeed, lack of diversity may not harm certain types of uses (such as the harvesting of particular NTFPs), as long as the specific resource that is being exploited remains relatively abundant. Monoculture plantations of the most valuable species may provide sustainable inputs into household livelihoods, but may not be related to biological diversity in any recognisable sense.

Moreover, as several studies document, wild plants that are in increasing demand (e.g. for medicinal purposes) are frequently being domesticated for cultivation as their values increase, bringing their use patterns closer to those of farming systems (and thereby breaking the link with 'biodiverse nature'). The dependence of poor (and rich) rural populations on these species for their livelihoods does not necessarily change, but they are being managed in conditions that are very different to their origins in the wild. This raises important challenges to the ways in which we conceptualise these use patterns, although these distinctions (between 'wild' and 'farmed') may in some cases be less meaningful to the communities who are actually engaged in the management and exploitation of these species. In the agricultural context, studies show that *in situ* conservation of agro-biodiversity and the protection of wild species may have additional insurance value (see Chapters 7 and 8, this volume).

Furthermore, the different components of biodiversity (genetic, species and ecosystem) are not necessarily equally important in order to maintain the flows of resources on which the livelihoods of the poor depend. The literature on resource dependence does not trace the links between these components, the resources that emerge from nature and the livelihoods of the poor. In terms of contributions to livelihoods, what is often valuable is the volume (in terms of extent and abundance) of a resource, rather than diversity. While an assumption can be made that wild resources, harvested from nature, do depend on the existence of biological diversity in a general sense, there is an urgent need to document more clearly the specific parameters of this relationship. Diversity at the ecosystem level is likely to be important for enhancing resilience, but the precise nature of this relationship needs to be explored in greater detail (one recent example is Isbell *et al.*, 2011). Thus, while biodiversity–poverty links are often asserted the specific pathways through which changes in biodiversity affect poor people's livelihood choices and strategies need to be more carefully identified (Ash & Jenkins, 2007).

Poverty

In their review of the links between poverty, development and biodiversity conservation, Agrawal and Redford (2006) propose a useful way of "parsing poverty" in two ways: first, 'aspects' of poverty, in terms of incidence, intensity, inequality, temporality and spatiality and, second, 'dimensions' of poverty, such as income and wealth, education, health, nutrition, food security, political autonomy, empowerment and social equality. These two concepts were used to interrogate the detailed case studies that form the focus of this review. However, the reviewed material does not engage with poverty in its multiple aspects and dimensions (Appendix 4.1), and tends to focus almost exclusively on the incidence of poverty defined in material wealth, or income terms (although the concept is usually expanded to include the values of non-market goods and services derived from nature). Some studies also pay attention to issues of inequality, though this is still measured in terms of income inequality.

What is missing from most analyses are several of the more interesting dimensions that have been highlighted in recent poverty research, many of which have potential impacts on the ways in which rural people interact with nature's resources. For instance, while studies point to the seasonality of resource use, and the importance of natural resources for meeting both consumption needs and employment and income needs during lean seasons (e.g. de Merode *et al.*, 2004; Béné *et al.*, 2009), these analyses do not address the volatility of poverty, and the extent to which some rural populations cycle in and out of poverty, while others remain chronically poor. If nature's resources help to temporally smooth consumption and incomes, their poverty impacts may be better captured through an explicit focus on this temporality as part of our poverty measure, instead of restricting our understanding to annualised income or consumption (in which these temporary contributions from nature do not always feature as significant). Similarly, trying to incorporate wider issues of empowerment, social exclusion and autonomy may be very important in understanding the context within which particular groups experience material deprivation, and it may help frame our understanding of the potential for resource-based interventions to offer potential pathways out of poverty. So, for instance, increasing the value of nature-based goods and services may result in their capture by politically powerful local actors, thereby excluding the very poor from access to potential benefits. Unless issues of political decision making and social inclusion are tackled at the same time, such resource-based interventions may do little to help the resource-dependent rural populations who are their intended targets.

Approach and methods

This review is based on a careful analysis of the current state of knowledge, and differentiates between robust evidence, on one hand, and claims that are less well founded on empirical experience, on the other. Furthermore, it highlights uncertainties and differences of opinion in the available literature. The specific steps that were undertaken as part of this process were:

1. An examination of the peer-reviewed literature, as published in journals and books; and
2. An examination of websites and portals of major organisations and forums working on biodiversity conservation and poverty alleviation.

   The literature that was identified in these two steps was then further short-listed to focus on studies that provided direct evidence that was relevant to this review. A final set of 200 studies were examined in detail (as listed in the References). From these studies, 27 specifically provided empirical evidence for the dependence of the poor on biodiversity for their income and subsistence needs (see Appendix 4.2 for details), while a further 22 dealt with evidence on risk coping and insurance.

Dependence on biodiversity: direct livelihood linkages

This review focussed on the question: which groups of the (differentiated) poor depend, in which types of ways, on different elements of biological diversity? As has been discussed in this chapter, the extent to which the different case studies provided evidence on these issues varied greatly. Appendix 4.2 summarises the evidence from these case studies, which provides the basis for the discussion in this section. Given the variation in the extent to which each study has covered issues such as differences between different income classes in the population, impacts on economic and social inequality and the extent of dependence on identifiable elements of biological diversity, this section will summarise key trends that are emergent from this literature, but will not attempt any overall summary. Furthermore, all meta-analyses suffer from an important limitation, which is that the lack of consistency in case study methods precludes any easy aggregation of results, so this will not be attempted here.

Figure 4.1 provides details of the geographical regions that were represented in the 27 studies that were the subject of more detailed analysis, while Figure 4.2 breaks down the studies by resource type.

This section provides an overview of the evidence on different types of dependence on biodiversity-based resources, primarily derived from micro studies at specific local sites. Unsubstantiated claims, such as those that are often found in larger macro or sectoral studies, have not been included here, since the evidence for these claims is not available. What emerges from this review is a complex picture, with site- and resource-specific patterns of access, use and dependence, often reflecting very divergent patterns, although some regularities can also be observed from these studies.

Evidence on dependence

Most studies use income from biodiversity-based resources as a percentage of total household income as their indicator of the extent of dependence. Table 4.1 summarises the findings from these studies.

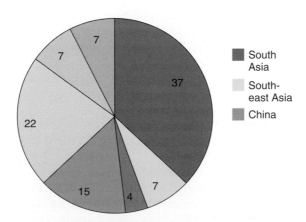

Figure 4.1   Case studies by geographical region (numbers are percentages).

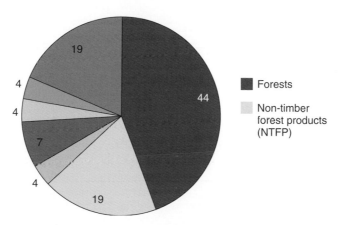

Figure 4.2   **Case studies by resource type (numbers are percentages).**

As Table 4.1 demonstrates, there is considerable variation reported in the extent of household income that is contributed by biodiversity-based resources. Some of this dependence is very specific to particular groups (such as the poorest fish-dependent groups in Béné *et al.* (2009), for whom fishing represents 90% of household income). Moreover, some multi-sited studies demonstrate variability across different sites, reflecting both the availability of alternative income sources (such as in the case of Fu *et al.*'s 2009 study of two sites in China), as well as access issues and previous resource use patterns (such as the variation across three proximate sites reported in Shaanker *et al.*, 2004).

A number of studies additionally report the proportion of households engaged in particular types of activities, which will be used here as an indicator of the 'depth' of dependence on biodiversity resources. These findings are summarised in Table 4.2.

Table 4.2 shows that the depth of dependence reported in these studies is high, although there is some variation when this is broken down by wealth class, with the poor typically showing higher levels of dependence.

A further set of studies also focus on biodiversity-based resources as part of household consumption and production strategies (without monetising these values as a proportion of income). These studies are summarised in Table 4.3.

Overall, these data suggest reasonably high levels of dependence on biodiversity-based resources, in terms of contributions to household incomes, as well as production and consumption strategies. Levels of participation in biodiversity-based livelihood activities are also high, suggesting that the depth of dependence on these resources is significant.

Table 4.1   **Evidence on dependence on biodiversity for income**

Source	Region	Evidence	Resource type
Bahuguna (2000)	South Asia	48.7% of household income.	Forests: fuel, fodder and employment
Béné *et al.* (2009)	West Africa	From 90% (poorest) to 29.7% (richest)	Fish
Cavendish (2000)	Southern Africa	35.4% of household income in 1993–4, and 36.9% in 1996–7	Wild foods, wood, grasses and other environmental resources
Coomes *et al.* (2004)	Latin America	20% of household income	Fish, palm products, timber and hunting
de Merode *et al.* (2004)	West Africa	24% of cash sales	Wild foods
Fisher (2004)	Southern Africa	30% of household income	Forests
Fu *et al.* (2009)	China	1.7% of household income in Site 1, and 12.2% in Site 2	Non-timber forest products (NTFP)
Jodha (1995)	South Asia	14–23% of total household income	Common pool resources
Kamanga *et al.* (2009)	Southern Africa	15% of total household income	Forests
Levang *et al.* (2005)	South-east Asia	30% of total household income	Forests
Mamo *et al.* (2007)	East Africa	39% of total household income	Forests
Narain *et al.* (2008a)	South Asia	Income quartile 1: 9%; quartile 2: 7.2%; quartile 3: 7.9% and quartile 4: 8% of permanent income	Fuel wood, dung for fuel, manure, fodder and construction wood
Shaanker *et al.* (2004)	South Asia	Site 1: 16%; Site 2: 24% and Site 3: 59% of household income	NTFP
Viet Quang and Anh (2006)	South-east Asia	For 30% of households, over 50% of total income; further 15%, 25–50% of total income	NTFP

Table 4.2   **Evidence on depth of dependence on biodiversity resources**

Source	Region	Evidence	Resource type
Coomes *et al.* (2004)	Latin America	66% of households depend on resource extraction.	Fish, palm products, timber and hunting
Dovie *et al.* (2007)	Southern Africa	98% of households use non-timber forest products (NTFP).	NTFP
Dovie *et al.* (2007)	Southern Africa	91% of households use wild herbs.	Wild herbs
Glaser (2003)	Latin America	68% of households depend on mangroves,	Mangrove resources, especially crabs and fish
Jha (2009)	South Asia	70% of households depend on beedi making or firewood.	Forests
Jodha (1995)	South Asia	84–100% of poor depend on common pool resources (CPR).	CPR
Jodha (1995)	South Asia	10–19% of rich depend on CPR.	CPR
Levang *et al.* (2005)	South-east Asia	72% of households depend on forest products.	Forests
Mamo *et al.* (2007)	East Africa	42% of households depend on forest for grazing.	Forests
Narain *et al.* (2008a)	South Asia	Income quartile 1: 77.5%, Q2: 81.5%; Q3: 72.8% and Q4: 61.4% of households collect.	Fuel wood, dung for fuel, manure, fodder and construction wood
Shackleton and Shackleton (2006)	Southern Africa	96–100% of households purchase NTFP.	NTFP
Shackleton and Shackleton (2006)	Southern Africa	8% (rich), 15% (middle) and 36% (poor) households sell NTFP.	NTFP
Sharma *et al.* (2009)	South Asia	75% of household fuel and fodder needs from forests.	Forests

Table 4.3   Other evidence on dependence on biodiversity resources

Source	Region	Type of data	Evidence	Resource type
Béné *et al.* (2009)	West Africa	Consumption	Varies from 33% (poorest) to 20% (richest)	Fish
de Merode *et al.* (2004)	West Africa	Consumption	10% of household consumption	Wild foods
de Merode *et al.* (2004)	West Africa	Production	31% of household production	Wild foods

Relative dependence on biodiversity resources: are the poor disproportionately dependent?

The evidence on the relative dependence of rich and poor groups on biodiversity-based resources has been the subject of considerable interest, and is somewhat mixed. An early set of studies seemed to suggest unambiguously that the poor were disproportionately dependent on such resources (e.g. Jodha, 1995; Cavendish, 2000). This became accepted wisdom, with overviews of the field consistently suggesting that the poor depend proportionately more on nature for their resource needs (Millennium Ecosystem Assessment, 2005). However, more recent work has started to question this accepted view, and the studies reviewed here present a more mixed picture.

One factor that has been highlighted in recent work is the complementarity between asset ownership (especially land and cattle) and the use of certain types of biological resources (see e.g. Adhikari *et al.*, 2004; Coomes *et al.*, 2004; Fisher, 2004; Narain *et al.*, 2008a; Coulibaly-Lingani *et al.*, 2009). In these circumstances, asset-rich households tend to depend more on nature's resources. If this difference in asset ownership is further reflected in greater political power at the local level, rich households are also able to use their dominance to secure access to resources, and to exclude the relatively poor.

These differences in political power suggest another reason why resource use may be skewed in favour of the rich. While a biodiversity-based resource remains relatively low value, rich users tend not to feel the need to restrict access. However, with returns to certain types of resources increasing as they become more valuable and in greater demand, rich and powerful groups may try to capture these resources, and may exclude the poor from access (see e.g. Fisher, 2004).

Table 4.4 summarises the evidence of the reviewed studies on the extent to which resource access increases or decreases with increases in household wealth.

Table 4.4   **Evidence on relative dependence of rich and poor on biodiversity resources**

Reference	Region	Resource	Relative dependence
Adhikari *et al.* (2004)	South Asia	Fodder	Increases with wealth
Adhikari *et al.* (2004)	South Asia	Leaf litter	Increases with wealth
Babulo *et al.* (2008)	East Africa	Forests	Decreases with wealth
Béné *et al.* (2009)	West Africa	Fish	Decreases with wealth
Cavendish (2000)	Southern Africa	Multiple	Decreases with wealth
Coomes *et al.* (2004)	Latin America	Hunting	Increases with land ownership
Coomes *et al.* (2004)	Latin America	Palm fruit	U-shaped: first decreases, then increases with wealth
Coulibaly-Lingani *et al.* (2009)	West Africa	Forests	Increases with wealth
De Merode *et al.* (2004)	West Africa	Bush meat	Consumption and sale increase with wealth
De Merode *et al.* (2004)	West Africa	Fish	Consumption and sale increase with wealth
De Merode *et al.* (2004)	West Africa	Wild plants	Consumption and sale decrease with wealth
Fisher (2004)	Southern Africa	Low-return forest activities	Decreases with wealth
Fisher (2004)	Southern Africa	High-return forest activities	Increases with wealth
Fu *et al.* (2009)	China	Non-timber forest products (NTFP)	Decreases with wealth
Jha (2009)	South Asia	Firewood	Decreases with wealth
Jha (2009)	South Asia	Beedi making	Increases with wealth
Jodha (1995)	South Asia	Common pool resources (CPR)	Decreases with wealth
Kamanga *et al.* (2009)	Southern Africa	Forests	Decreases with wealth
Levang *et al.* (2005)	South-east Asia	Forests	Decreases with wealth
Mamo *et al.* (2007)	East Africa	Forests	Decreases with wealth
Narain *et al.* (2008a)	South Asia	Fodder and construction wood	Increases with wealth
Narain *et al.* (2008a)	South Asia	Fuel, dung fuel and dung manure	Decreases with wealth
Paumgarten and Shackleton (2009)	Southern Africa	NTFP	Sale decreases with wealth

(continued overleaf)

Table 4.4   (*Continued*)

Reference	Region	Resource	Relative dependence
Reddy and Chakravarty (1999)	South Asia	Forests	Decreases with wealth
Sapkota and Oden (2008)	South Asia	Forests	Decreases with wealth
Shaanker *et al.* (2004)	South Asia	NTFP	Decreases with wealth
Shackleton and Shackleton (2006)	Southern Africa	NTFP	Sale decreases with wealth
Shackleton and Shackleton (2006)	Southern Africa	Fuel wood	Consumption decreases with wealth
Shackleton and Shackleton (2006)	Southern Africa	Edible herbs	Consumption decreases with wealth
Sharma *et al.* (2009)	South Asia	Forests	Decreases with wealth
Viet Quang and Anh (2006)	South-east Asia	NTFP	Decreases with wealth

A number of interesting patterns emerge from the data that are presented in Table 4.4. 'Inferior', or low-value, goods and services tend to be the ones that reflect the conventional wisdom, in that the poor tend to depend disproportionately more on these resources. In contrast, where commercial production and sales are involved, or if resources complement existing assets such as land and livestock, this relationship may be reversed. This has important implications for the potential of biodiversity-based resources to be used as part of poverty alleviation strategies. If the rich capture resources once they become more valuable, increasing the value of biodiversity-based resources may not be a feasible strategy, since the poor may eventually lose out in such a scenario (Chapter 5, this volume; Angelsen & Wunder, 2002).

A second implication of this material is that there appears to be some confirmation of what Angelsen and Wunder (2002) have referred to as a 'poverty trap'. The poor appear to be linked with low-value nature-based resource use, but these low values may serve to perpetuate poverty. Here, poverty is endogenous, in the sense that biodiversity resource dependence is a symptom of poverty, and it is only by 'leaving the forest' that the poor can hope to escape poverty (Levang *et al.*, 2005). This, of course, is an important issue for discussion, since it has been widely believed that biodiversity-based resources provide an essential safety net for the poor, preventing them from destitution. But, if this dependence is reproducing or reinforcing existing patterns of poverty, it may be important to examine alternative livelihood strategies in order to benefit these economically marginalised groups.

Impacts on inequality: does the inclusion of biodiversity-based resources improve distributional outcomes?

If the poor do depend disproportionately on biodiversity-based resources for their livelihoods, one outcome that emerges is that the inclusion of such resources in estimates of household income is likely to improve equity. A number of studies explicitly focus on this issue (Table 4.5).

While the data in Table 4.5 clearly demonstrate that these studies find improvements in distributional outcomes due to the inclusion of biodiversity-dependent livelihood strategies in their analyses, there is reason to be cautious in light of the discussion here about the dependence of wealthier groups on certain high-value resources. Clearly, in these circumstances, distributional outcomes would be worse if we were to include biodiversity resources in our analysis, and a biodiversity-based strategy would not necessarily improve equity.

Apart from their focus on measures of income inequality, a couple of studies also looked at other indicators of social inequality. Thus, Adhikari *et al.* (2004) report a lower level of resource dependence amongst both female-headed households and lower social castes. This is explained due to restrictions on access, as well as the lack of ownership of complementary assets such as land and livestock. The influence of caste is also remarked upon by Sapkota and Oden (2008), although they show a greater degree of dependence among lower castes. In their study, Béné *et al.* (2009) report that while only 69% of women sell fish, 98.6% of men do so, thereby suggesting that women are

Table 4.5   **Equity implications of biodiversity resource dependence**

Reference	Region	Resource	Impact on inequality
Fisher (2004)	Southern Africa	Forests	Reduces by 12%
Jodha (1995)	South Asia	Common pool resources (CPRs)	Lowers Gini coefficient
Kamanga *et al.* (2009)	Southern Africa	Forests	Lowers Gini coefficient
Mamo *et al.* (2007)	East Africa	Forests	Lowers Gini coefficient
Reddy and Chakravarty (1999)	South Asia	Forests	Reduces inequality
Shaanker *et al.* (2004)	South Asia	Non-timber forest products (NTFPs)	Lowers Gini coefficient

less likely to engage in commercial or market-driven development opportunities in this context. Glaser (2003) remarks that 'pure subistence products' are most important for the weakest sections, especially women and children. Levang *et al.* (2005) reflect on the importance of geographical isolation, and suggest that resource dependence is highest in remote areas due to the lack of alternatives (suggesting an interesting spatiality to the issue of biodiversity resource dependence). This is also reflected in Fu *et al.*'s (2009) study which shows greater dependence in the less developed (remote) village as compared to the more developed (better connected) village.

Dependence on biodiversity: insurance and risk coping

This section focusses on reviewing the empirical evidence that evaluates the 'insurance value' of biodiversity to the poor based on how biodiversity is defined or conceptualised in each study. The empirical literature on biodiversity as a means for risk coping is *considerably smaller* than that on biodiversity as a source of livelihood. Moreover, the methods used to assess 'dependency' are more varied than those on biodiversity as a source of livelihood. This makes tabulation of results (as in the 'Impacts on Inequality' section of this chapter) less informative.

The literature discusses three main categories of risks in relation to biodiversity – food security risks (e.g. high variability in crop production), environmental and weather hazards (e.g. storms, floods and mudslides) and health risks (e.g. risk from infectious diseases, or risk of illness due to lack of wild medicinal herbs) – with the bulk of the literature focussing on food security. A fourth type of risk that is discussed is that associated with degrading ecosystem resilience. This refers to a stability property of ecosystems that reflects the capacity of a system to absorb shocks (Perrings, 1995), or the rate at which a system variable returns to the reference state after perturbation (Schläpfer *et al.*, 2002). This is a more complex risk to isolate as it permeates the three other type of risks mentioned here.

Agro-biodiversity for food security insurance

There is considerable evidence from the ecology and agronomy literatures on the relationship between agro-biodiversity and crop productivity and variability and yield shocks (see also Chapter 8, this volume). There are various complex channels that give rise to these effects. For example, diverse crop species are shown to adapt better to environmental changes since the larger pool of different metabolic traits and metabolic pathways enables them to more effectively use resources (such as water and soil nutrients) over a broad range of environmental conditions (Schläpfer *et al.*, 2002).

Also, biodiversity has been shown to improve ecosystem resilience which provides insurance against crop failure due to shocks (Perrings, 1995; Tilman, 1996).

At an economic policy level, the benefits of agro-biodiversity for the poor have been acknowledged within the environment and development literature (e.g. see the review by Smale & Drucker, 2008). Further the theoretical economics literature has developed bio-economic models that have clarified and defined in economic terms the insurance benefits or value that agro-biodiversity entails (e.g. Weitzman, 1993; Polasky & Solow, 1995; Evenson et al., 1998; Schläpfer et al., 2002; Chavas, 2009; Baumgartner & Quaas, 2010). These biodiversity benefits are then conjectured to be more important for the poor as they provide a cost-effective insurance policy against the risk of food insecurity to segments of the populations that do not have alternative risk-coping mechanisms (Heal, 2000).

Our review suggests that the empirical evidence documenting the degree and nature of this dependence (i.e. how valuable this insurance value is to the poor) is a dynamic body of work with several notable advancements but with significant shortcomings. The most informative body of work concerns studies that have explored the impact of on-farm crop genetic diversity on output (mean of yields) and variability (variance of yields). Only a few studies have focussed on assessing variability of income (e.g. Di Falco & Perrings, 2005). The main examples of these studies are Just and Candler (1985), Smale et al. (1998, 2008), Widawsky and Rozelle (1998), Meng et al. (2003), Di Falco and Perrings (2005), Di Falco and Chavas (2008b) and Heisey et al. (1997).

A very strong and consistent finding across all these studies is that the coefficient of on-farm crop genetic diversity has a strong positive effect on the mean of crop yields and a negative effect on the variance of crop yields. This implies that on-farm agro-biodiversity reduces food risk insecurity, something which is particularly important for the poor. It is important to note that this result is robust against different production function specifications, different types of crops, different scales of data (regional vs. plot specific) and different measures of crop genetic diversity. The policy implication of this body of work is that though environmental changes such as rainfall reductions have adverse effects on agro-ecosystem productivity, these adverse effects can be buffered in the short term and possibly reversed in the longer term under increased agro-biodiversity. In other words, this body of evidence suggests that agro-biodiversity can buffer and insure the poor against negative environmental effects and support the resilience of the system under adverse weather conditions associated with climate change (Di Falco & Perrings, 2005).

A few recent studies (Di Falco & Chavas, 2006, 2008a, 2009) have expanded upon the work noted here and have also tried to examine the value of biodiversity not only as insurance against yield variability but also against the risks of total crop failure due to exogenous shocks (e.g. storms or new invasive pests). For example, Di Falco and Chavas (2009) use plot-level barley production data from Ethiopia and show that agro-biodiversity (in this case, a more diverse portfolio of barley landraces) increases farm productivity of barley in Ethiopia farms. Further they show that agro-biodiversity decreases downside risk exposure. This effect is shown to dominate other confounding effects so that higher biodiversity tends to reduce the cost of risk (as measured by their estimated risk premiums). Finally, they also find that the risk benefit of biodiversity becomes larger under less fertile soils which offers empirical evidence that biodiversity can aid farmers to cope with harsh climatic conditions, especially in degraded lands. This last finding has implications for poorer segments of the population that tend to use and occupy degraded, marginal and less fertile lands.

Some studies have tried to compare more directly the impact of adopting modern crop varieties (often provided by aid agencies) as opposed to using traditional landraces on agricultural production decisions and outcomes. The rationale in this empirical work is that widespread adoption of modern varieties erodes genetic diversity, and this may have implications for coping with food security risk in marginal low-productivity lands. For example, Lipper et al. (2008) study subsistence-level sorghum production in Ethiopia. They show that the likelihood of crop failure due to drought increases as the likelihood of adopting modern crop varieties (over traditional landraces) increases. This effect is found to be worse for marginal low-production farms occupied by the poorer segments of the population. This provides further support regarding the insurance value of traditional landraces (which have a higher degree of genetic diversity as compared to modern varieties).

Other evidence from the literature assesses the factors that impact decisions to conserve in situ crop and animal diversity (see Van Dusen & Taylor, 2005; Van Dusen et al., 2007). Lastly, there are a few stated preference studies relying on choice experiment methods. These methods directly elicit farm household preferences for different levels of agro-biodiversity with the aim of calculating the welfare (consumer surplus) associated with conserving on-farm genetic diversity (e.g. Birol et al., 2009).

With respect to the role of livestock as insurance for the poor, discussion and evidence in the development economics and development studies literature suggest that this is significant (e.g. Dercon, 1998; Fafchamps et al., 1998, Kinsey, 1998). Further there is some recent literature discussing the special role of animal genetic resources (as a distinct concept from livestock), characterised by the properties of flexibility, resilience and diversity. It is suggested that such animal genetic resources provide an enhanced form of insurance, as they are vital assets for the livelihoods of the poor (Anderson, 2003; Wollny, 2003). However, there are no empirical studies which detail the insurance value of access to diverse animal genetic resources. Likewise, there has been little empirical exploration of the benefits to the poor of landscape-level biodiversity as an insurance mechanism against high levels of crop variability (e.g. via its function as enhancing resilience), although this link is scientifically plausible and explainable in both ecological and economic terms.

Wild food products, biodiversity and food security

The literature on the dependence of the poor on biodiversity-related income is also useful in providing insights on the role of wild food products, in particular for coping with the risk of food insecurity. For example, Vedeld *et al.* (2007) suggest that together with fuel wood, wild food products are the main source of forest-related income and consumption. This sort of evidence lends support to the idea of forests as 'safety nets' (see also Chapter 5, this volume). Yet (and what concerns this chapter), the value of diverse ecosystems and of the 'diversity' of wild food items as providing better qualities and quantities of such wild foods is not well documented.

A few studies assess the role of tropical forests (and hence biodiversity-rich ecosystems) as an insurance against food security (and income) variability. Pattanayak and Sills' (2001) study on the Peruvian rainforest and Takasaki *et al.*'s (2004) work on the Brazilian Amazon suggest that poor households in these tropical areas use the forest to cope with ex-ante risks and ex-post shocks. Pattanayak and Sills (2001) found that time spent collecting forest products was correlated with agricultural yield risks (an income-smoothing response) and unforeseen production shocks (a consumption-smoothing response). One of the main findings of the Takasaki *et al.* (2004) study was that the insurance value of the forest (as a source of wild non-timber forest resources during unforeseen shocks) was much more significant for the poorest segments of their sample. The micro-econometric study by Fisher and Shively (2003) on communities living at the margins of tropical forests of Malawi corroborates and complements these earlier findings. They find that rural households rely on tropical forests (for wild foods) for coping with income and consumption shocks and that asset-poor households are even more dependent on forests for dealing with such shocks. Similar findings are reported in Akinnifesi *et al.* (2006), World Bank (2007), McSweeney (2003) and Sunderlin *et al.* (2000). Hence, at least for the case of poor communities living close to tropical forests, there appears to be support for the conjecture that forests act as a safety net against food insecurity.

Biodiversity and natural hazards

Rural poor communities face serious risk from natural hazards, the most common being floods, fires, hurricanes and storms, landslides and dust storms. The lack of market or government insurance of the poor against such hazards has been shown to lead to an exacerbation of poverty (Dercon & Krishnan, 2000; Zimmerman & Carter, 2003; Dercon, 1996, 2004, 2005, 2006). Natural ecosystems can play an important role in mitigating these risks as they provide cost-effective insurance. This is achieved through complex interrelationships between local geo-morphological traits, weather conditions as well as soil and land cover characteristics. Of these, vegetation and soil conditions are more susceptible to human interference (at least at the local level and in the short run). Ash and Jenkins (2007) summarise the links between genetic diversity (be it at the soil, vegetation or landscape level) and the mitigation of flood and fire risks. There is less evidence on the possible links between biodiversity and mitigating against other hazards, such as landslides, hurricanes and dust storms.

The risk of flooding is directly related to the water-retentive capacity of the soils. This is related to soil and forest land traits. Diversity in soils and forests (e.g. type of plant coverage) is important for regulating water flows, though the exact mechanisms are not entirely clear. Still, the literature does seem to conclude that *natural forests* (a term that to some degree implies higher levels of biodiversity compared to plantation forests and agricultural landscapes) are associated with higher degrees of flood protection. Higher levels of biodiversity are associated with improved ecosystem capacity to regulate fire patterns, their frequency and their severity. Eroding plant biodiversity and introducing invasive tree species alter fire patterns by reducing their frequency but severely increasing their intensity and extent. Natural resources such as mangroves have been shown to help local communities deal with the risk of hurricanes and storms (Das & Vincent, 2009). Yet, case study evidence linking diversity per se and protection against such harsh climatic events is scant. Both landslides and dust storms are impacted by the extent and nature of vegetation coverage across the relevant ecosystem landscape. With respect to the nature or type of vegetation, there is scientific evidence that shows that endemic vegetation outperforms invasive or introduced tree or bush varieties. Again, links with biodiversity per se seem to be weak or at least not well understood (Ash & Jenkins, 2007).

Biodiversity and health risks

There is a sizable body of research on the links between environmental conditions and health from the medical, epidemiological and social science literatures (see World Bank, 2007, for a review). A small subset of this literature tries to investigate the links between biodiversity and health vulnerability (e.g. Daily & Ehrlich, 1996; Chivian & Bernstein, 2004, 2008). Recent literature has identified two main avenues through which biodiversity provides a means for mediating health risk for the poor. The first has to do with the impact that biodiversity has on reducing the risk of infectious diseases. The second has to do with biodiversity as a source of accessible medicinal regimens which are not only curative but also preventive, thereby reducing health risks (Grifo & Rosenthal, 1997; Burlingame, 2000; Chivian & Bernstein, 2004, 2008; Huynen *et al.*, 2004; Frison *et al.*, 2005; Johns, 2006; Johns & Eyzaguirre, 2006).

Biodiversity and risk of infectious disease

There is a growing literature documenting how biodiversity reduces risk of exposure to several types of infectious diseases. Biodiversity at the ecosystem level produces the appropriate balance between predators and prey, hosts, vectors and parasites which allows for appropriate controls and checks for both the spread of 'endemic' infectious disease as well as resistance towards invasive pathogens (from humans, animals or insects). Ash and Jenkins (2007) identify a large list of diseases as being particularly dependent on changes in ecosystem biodiversity. Most of these diseases are of particular relevance to the poor and include malaria (in all ecosystem types); schistosomiasis, lymphatic filariasis and Japanese encephalitis (particularly in cultivated and inland water systems in the tropics); dengue fever (particularly in tropical urban centres); leishmaniasis and Chagas disease (in forest and dryland systems); meningitis (in the Sahel); cholera (in coastal, freshwater and urban systems) and rabies transmission (in tropical forest lands).

It is evident that the regions in the world with the highest levels of poverty are most vulnerable to such diseases. Biodiversity not only plays a role in reducing the risk of such diseases spreading within an ecosystem and the human population living within it, but also in many cases reduces the risk of allowing invasive diseases to enter a particular system. For example, there is evidence to show that cholera, kala-azar and schistosomiasis have not become established in the (biodiversity-rich) Amazonian forest ecosystem despite the risk of this happening from human migration and settlements (Ash & Jenkins, 2007).

Biodiversity and preventive wild medicines

Biodiversity has been recognised as an important source of traditional medicines (such as herbal medicines) for people in developing countries, especially where they have little (if any at all) access to formal healthcare. This lack of access to medicines and health services is even more acute in the remote and normally more poverty-stricken areas of the developing world. It is estimated that approximately 75% of the world's population depends primarily on traditional medicines gathered from the wild. For example, the 'sweet wormwood' plant produces traditional medicines that are increasingly important in combating drug-resistant strains of malaria, particularly in Africa (Department for International Development, 2001). Though traditional medicines may not be as effective as compared to scientifically tested drugs, they do provide a cost-effective and accessible option in poverty-stricken communities. Ash and Jenkins (2007) make reference to the importance of biodiversity as a medicinal source by providing some evidence of the number of wild plant species used as a source for curative and preventive drugs. For example, this includes well over 50,000 wild Chinese species (around 20% of all Chinese flora), over 7000 wild species in India, and 10% of Indonesia's flora. Though these facts are interesting, they do not provide a robust empirical indication of the degree of dependency. For this, we would need an assessment of the lives saved or illness incidents avoided as a result of higher levels of biodiversity and then use the appropriate value of statistical life estimates to assess the magnitude of these impacts on human welfare. Such studies are non-existent.

Biodiversity and resilience

The term *resilience* has been used to denote an ecosystem's ability to maintain its basic functions and controls under disturbances (Holling, 1973; Carpenter *et al.*, 2001; Baumgartner & Strunz, 2009). A higher degree of resilience is found in ecosystems that exhibit higher degrees of biodiversity (see Swift *et al.*, 2004, for a review of these studies from an ecological perspective). The economic relevance of ecosystem resilience is obvious, as a system flip may entail huge welfare losses since the continued provision of several key ecosystem services would be at risk of total collapse. For example, a combination of drought, fire and ill-adapted livestock grazing management in Sub-Saharan Africa, central Asia and Australia have led to severe degradation and desertification of semi-arid rangelands that provide subsistence livelihoods for more than one billion people. Once these grasslands are degraded, they can no longer be used as pasture (Baumgartner & Strunz, 2009). Hence, resilience is related to the threat of 'irreversible' ecosystem damage.

There is a growing literature that discusses the value of resilience as insurance against irreversible damage. These papers do make specific reference to this value in relation to the world's poor. Yet, the discussion is mainly conceptual and model based (e.g. Perrings & Stern, 2000; Perrings, 2006; Baumgartner & Quaas, 2008, 2010; Maler, 2008; Baumgartner & Strunz, 2009). There are no empirical studies on the degree of dependency on biodiversity as a source of protection against the risk of declining resilience. This is to be expected, given that resilience is associated with a form of 'wider insurance' against potentially multiple risks. It is also a concept related to 'entire systems' and not specific aspects of biodiversity, and as such is not easy to isolate or to trace back to specific behavioural decisions.

Conclusions

While there are many papers that refer to the existence of a relationship between biodiversity and poverty, there are surprisingly few studies that subject this relationship to critical empirical scrutiny. Indeed there is a real paucity of grounded empirical information about the particular ways in which people (especially the poor) use and benefit from the existence of biological diversity. However, it is not necessarily clear that adding to the stock of case study material would necessarily result in the discovery of a hitherto-unknown dimension of the poverty–biodiversity nexus; many of the current known patterns are likely to be repeated in additional studies.

The limited use of the poverty concept in the existing literature was a particular source of disappointment. Here, we would suggest that there are significant missed opportunities, and a more expanded notion of poverty is likely to result in much greater analytical traction for an understanding of the biodiversity–poverty link.

The biodiversity–poverty studies that do exist point to some interesting patterns which are worthy of generalisation. On income and subsistence, our review suggests that there is some evidence supporting the hypothesis that the poor do depend significantly on biodiversity, but this needs to be looked at with some caution. In some cases, there is clear evidence that the poor make extensive use of their natural resources, as long as these remain relatively low value and subsistence oriented, but there is also evidence that these same groups either lose access to, or are actively excluded from, more highly valued resource uses. This suggests that there is some evidence of a possible 'poverty trap', with poorer users stuck in low-value extractive uses but unable to make the transition out of this resource-dependent mode.

There is relatively robust information to show that the poor rely on farm agro-biodiversity to insure against food (in-)security and risk. However, we still know very little empirically about the significance of other forms of biodiversity in terms of risk insurance, protection against natural hazards, health and ecosystem resilience. This is an important knowledge gap, and our review points to the need for more systematic and robust studies which examine these linkages in greater depth.

Appendix 4.1   List of detailed case studies providing evidence on the dependence of the poor

Reference	Region	Resource	Activity	Aspect	Dimension
Adhikari et al. (2004)	South Asia	Forests	Multiple	Incidence	Income
Babulo et al. (2008)	East Africa	Forests	Multiple	Incidence	Income
Bahuguna (2000)	South Asia	Forests	Multiple	Not specified	Not specified
Béné et al. (2009)	West Africa	Fish	Multiple	Incidence	Income
Cavendish (2000)	Southern Africa	Multiple	Multiple	Incidence	Income
Coomes et al. (2004)	Latin America	Multiple	Multiple	Incidence	Income
Coulibaly-Lingani et al. (2009)	West Africa	Forests	Multiple	Not specified	Not specified
de Merode et al. (2004)	West Africa	Wild animals	Multiple	Incidence	Income
Dovie et al. (2007)	Southern Africa	Wild plants	Subsistence consumption	Incidence	Income
Fisher (2004)	Southern Africa	Forests	Multiple	Incidence	Income
Fu et al. (2009)	China	Other non-timber forest products (NTFP)	Multiple	Incidence	Income
Glaser (2003)	Latin America	Mangroves	Multiple	Incidence	Income
Jha (2009)	South Asia	Forests	Multiple	Incidence	Income
Jodha (1995)	South Asia	Multiple	Multiple	Incidence	Income
Kamanga et al. (2009)	Southern Africa	Forests	Multiple	Incidence	Income

Levang et al. (2005)	South-east Asia	Forests	Multiple	Incidence	Income
Mamo et al. (2007)	East Africa	Forests	Multiple	Incidence	Income
Narain et al. (2008a)	South Asia	Multiple	Multiple	Equality	Income
Narain et al. (2008b)	South Asia	Multiple	Multiple	Incidence	Income
Osemeobo (2005)	West Africa	Wild plants	Multiple	Not specified	Not specified
Paumgarten and Shackleton (2009)	Southern Africa	Other NTFP	Multiple	Incidence	Income
Reddy and Chakravarty (1999)	South Asia	Forests	Multiple	Equality	Income
Sapkota and Oden (2008)	South Asia	Forests	Multiple	Incidence	Income
Shaanker et al. (2004)	South Asia	Other NTFP	Multiple	Incidence	Income
Shackleton and Shackleton (2006)	Southern Africa	Other NTFP	Multiple	Incidence	Income
Sharma et al. (2009)	South Asia	Forests	Multiple	Incidence	Income
Viet Quang and Anh (2006)	South-east Asia	Other NTFP	Multiple	Incidence	Income

Appendix 4.2   Findings from detailed case studies

Reference	Region	Explanatory factors	Extent of dependence	Relative dependence	Inequality
Adhikari et al. (2004)	South Asia	Education reduces dependence; complementary productive assets are livestock and land increase use.		Tree and grass fodder leaf litter increases with wealth; fuel wood does not vary (slight increase with wealth, not statistically significant).	Female-headed and lower caste groups collect less.
Babulo et al. (2008)	East Africa	Lack of education, female-headed groups, lack of land, access to credit, roads and livestock increase dependence.		Poverty trap: forest dependence is inferior.	
Bahuguna (2000)	South Asia		48.7% of total income from forests, mainly fuel, fodder and employment.		
Béné et al. (2009)	West Africa	Periodic use: fish as 'bank' in the water.	Consumption: Q1 33%, Q2/Q3 23% and Q4 20%; and income: Q1 90%, Q2 67%, Q3 64% and Q4 63%, and for 29.7% fish is their only source of cash income.	The poor are more dependent on fish for income and consumption.	98.6% men and 69% women sell fish.

Cavendish (2000)	Southern Africa	Environment income: 35.4% in 1993–4 and 36.9% in 1996–7.	Poor more dependent than rich; quantity consumed increases with income; cash income falls with wealth (93–4: lowest quintile 50%, middle 60% =30% and riches 25%; 96–7: 34% poorest and 6% richest).
Coomes et al. (2004)	Latin America	Those with younger households, more nets and more members fish more; and those with younger households, land, equipment and experience hunt more; there are also species-level explanations.	66% of households depend on resource extraction, with a value of 20% of total income.	Resource draw not related to poor households, reliance for fishing not linked to wealth, reliance on hunting more for land-rich households and palm fruit reliance declines with wealth, then increases.
Coulibaly-Lingani et al. (2009)	West Africa	Age impact depends on product, household size, ethnicity, land and livestock.		Higher incomes increase dependence.

(continued overleaf)

Appendix 4.2 (Continued)

Reference	Region	Explanatory factors	Extent of dependence	Relative dependence	Inequality
de Merode et al. (2004)	West Africa	Seasonality of use.	Wild foods: 31% of household production, 10% of self-consumption and 24% of sale.	Bushmeat and fish consumed and sold less by the poor, and wild plants are sold more by the poor.	
Dovie et al. (2007)	Southern Africa		98% of households use NTFPs (value US$559 per household) and 91% use wild herbs (value US$167 per household).	No association with wealth – all depend.	
Fisher (2004)	Southern Africa	Lack of land, education and goats leads to reliance on low-return forest activities (LRFAs); lack of education, goats and available male labour as well as location lead to reliance on high-return forest activities (HRFAs).	Forest income is 30% of household income.	As income increases, reliance declines for LRFAs and increases for HRFAs.	Addition of forest income reduces inequality by 12%.
Fu et al. (2009)	China		Income from NTFP, % of income and dependence more in less developed village.	Poor more dependent on NTFP income.	

Source	Region			
Glaser (2003)	Latin America		Over 50% depend on crabs, 30% on commercial fishing and 80% on mangroves overall; 68% earn income from mangroves.		Pure subsistence products most important for voiceless – women and children.
Jha (2009)	South Asia		70% depend on beedis or firewood.	Dependence decreases with increased income, except beedis which are made by rich.	
Jodha (1995)	South Asia		Poor: 84–100%; rich: 10–19%; overall CPR income: 14–23% of total.	Poor more dependent.	Inclusion of CPR income lowers Gini coefficient.
Kamanga et al. (2009)	Southern Africa	Those who are younger, are less educated and have more members have greater forest income. Cash income less in remote areas.	Cash income from forests is low; fuel wood is 66–78% of forest income.	Poor lowest forest income, but poor and medium poor depend more (22%) compared to less poor (9%).	Gini coefficient increases with removal of forest income.
Levang et al. (2005)	South-east Asia		72% of households depend on forest products; forest products are 30% of total income, and main cash activity for 16%.	Poverty trap – get out of the forest.	Dependence high in remote areas because of few options.

(continued overleaf)

Appendix 4.2   (*Continued*)

Reference	Region	Explanatory factors	Extent of dependence	Relative dependence	Inequality
Mamo *et al.* (2007)	East Africa		Forest products are 39% of household income, with firewood the most important (59%); 42% depend on forest for grazing.	Poor rely more (59%, vs. 30% of rich) but extract less (US$95 vs. US$191).	Gini coefficient increases with removal of forest income
Narain *et al.* (2008a)	South Asia	Private land holdings reduce need; higher availability of biomass increases dependence.	Q1:11.6%; Q2: 8.9%; Q3: 10.9% and Q4: 13%.	Use of fuel, dung fuel and dung manure declines with income; use of fodder and construction wood increases with income.	
Narain *et al.* (2008b)	South Asia		Middle-income households most likely to collect.	Conditional on collection, rich use more common pool resources (CPRs).	At income extremes, bimodal use and dependence.
Osemeobo (2005)	West Africa		Average value of wild plants per household is US$11,957; net income is US$6743.		

Paumgarten and Shackleton (2009)	Southern Africa		Negligible effect of wealth on NTFP use; all use.	Poor sell more NTFPs, and rich sell high-value curios.	Inequality increases if forestry income not included.
Reddy and Chakravarty (1999)	South Asia		Poverty increases if forestry income set to zero.	Poorest of the poor disproportionately dependent.	Lower castes are more dependent.
Sapkota and Oden (2008)	South Asia	Proximity to source		Share of fuel wood from community forests higher for poor.	Inclusion of NTFP income lowers Gini coefficient.
Shaanker et al. (2004)	South Asia		Three sites: 16%, 24% and 59% of cash income.	Decreased with increase in wealth index.	
Shackleton and Shackleton (2006)	Southern Africa			Poor are more involved in sale of NTFPs; per capita consumption of fuel wood and edible herbs is greater for poor, but there is no difference for grass brushes.	
Sharma et al. (2009)	South Asia		75% of fuel and fodder are from forest.	Poor are more dependent.	
Viet Quang and Anh (2006)	South-east Asia	Urban proximity	30% of households: NTFP income is over 50% of total; further 15%, NTFP income is 25–50% (all poorest).	Poor are more dependent.	

References

Adams, W.M., Aveling, R., Brockington, D., Dickson, B., Elliott, J., Hutton, J., Roe, D., Vira, B. & Wolmer, W. (2004) Biodiversity conservation and the eradication of poverty. *Science*, 306, 1146–9.

Addison, T., Hulme, D. & Kanbur, R. (2009) *Poverty Dynamics: Measurement and Understanding from an Interdisciplinary Perspective*. Oxford University Press, Oxford.

Adhikari, B., Di Falco, S. & Lovett, J.C. (2004) Household characteristics and forest dependency: evidence from common property forest management in Nepal. *Ecological Economics*, 48, 245–7.

Agrawal, A. & Redford, K. (2006) *Poverty, Development and Biodiversity Conservation: Shooting in the Dark*. Working Paper No. 26. Wildlife Conservation Society, New York.

Akinnifesi, F., Waibel, H. & Mithöfer, D. (2006) *The Role of Food from Natural Resources in Reducing Vulnerability to Poverty: A Case Study from Zimbabwe*. Proceedings of the German Development Economics Conference, Research Committee Development Economics, Berlin, August.

Angelsen, A. & Wunder, S. (2002) *Exploring the Forest-Poverty Link*. CIFOR Occasional Paper 40. CIFOR, Bogor, Indonesia.

Ash, N. & Jenkins, M. (2007) *Biodiversity and Poverty Reduction: The Importance of Biodiversity for Ecosystem Services*. UNEP-World Conservation Monitoring Centre, Cambridge.

Babulo, B., Muys, B., Nega, F., Tollens, E, Nyssen, J., Deckers, J. & Mathijs, E. (2008) Household livelihood strategies and forest dependence in the highlands of Tigray, Northern Ethiopia. *Agricultural Systems*, 98, 147–55.

Bahuguna, V.K. (2000) Forests in the economy of the rural poor: an estimation of the dependency level. *Ambio*, 29, 126–9.

Baumgartner, S. & Quaas, M.F. (2008) Agro-biodiversity as natural insurance and the development of financial markets. In *Agrobiodiversity, Conservation and Economic Development*, ed. A. Kontoleon, U. Pascual & M. Smale, pp. 293–317. Routledge, London.

Baumgartner, S. & Quaas, M.F. (2010) Managing increasing environmental risks through agro-biodiversity and agri-environmental policies. *Agricultural Economics*, 41, 483–93.

Baumgartner S. & Strunz, S. (2009) The economic insurance value of ecosystem resilience. Paper presented at the European Society for Ecological Economics Conference, Ljubljana, Slovenia, 29 June–2 July.

Béné, C., Steel, E., Luadia, B.K. & Gordon, A. (2009) Fish as the 'bank in the water' – evidence from chronic-poor communities in Congo. *Food Policy*, 34, 108–18.

Birol, E., Villalba, E. & Smale, M. (2009) Farmer preferences for milpa diversity and genetically modified maize in Mexico: a latent class approach. *Environment and Development Economics*, 14, 521–40.

Burlingame, B. (2000) Wild nutrition. *Journal of Food Composition and Analysis*, 13, 99–100.

Cavendish, W. (2000) Empirical regularities in the poverty-environment relationship of rural households: evidence from Zimbabwe. *World Development*, 28, 1979–2003.

Chivian, E. & Bernstein, A. (2004) Embedded in nature: human health and biodiversity. *Environmental Health Perspectives*, 112, A12–A13.

Chivian, E. & Bernstein, A. (eds.) (2008) *Sustaining Life: How Human Health Depends on Biodiversity*. Oxford University Press, Oxford.

Coomes, O.T., Barham, B.L. & Takasakic, Y. (2004) Targeting conservation-development initiatives in tropical forests: insights from analyses of rain forest use and economic reliance among Amazonian peasants. *Ecological Economics*, 51, 47–64.

Coulibaly-Lingani, P., Tigabu, M., Savadogo, P., Oden, P.-C. & Ouadba, J.M. (2009) Determinants of access to forest products in southern Burkina Faso. *Forest Policy and Economics*, 11, 516–24.

Daily, G.C. & Ehrlich, P.R. (1996) Impacts of development and global change on the epidemiological environment. *Environment and Development Economics*, 1, 311–46.

Das, S. & Vincent, J.R. (2009) Mangroves protected villages and reduced death toll during Indian super cyclone. *Proceedings of the National Academy of Sciences of the USA*, 106, 7357–60.

de Merode, E., Homewood, K. & Cowlishaw, G. (2004) The value of bushmeat and other wild foods to rural households living in extreme poverty in Democratic Republic of Congo. *Biological Conservation*, 118, 573–81.

Delang, C.O. (2006) The role of wild food plants in poverty alleviation and biodiversity conservation in tropical countries. *Progress in Development Studies*, 6, 275–86.

DFID (Department for International Development (DFID) (2001) *Biodiversity – A Crucial Issue for the World's Poorest*, Department for International Development, London, https://www.dfid.gov.uk/Documents/publications/biodiversity.pdf (accessed 7 May 2012).

Di Falco, S. & Chavas, J.-P. (2006) Crop genetic diversity, farm productivity and the management of environmental risk in rainfed agriculture. *European Review of Agricultural Economics*, 33, 289–314.

Di Falco, S. & Chavas, J.-P. (2008a) Diversity, productivity and resilience in agro-ecosystems: an example from cereal production in Southern Italy. In *Agrobiodiversity, Conservation and Economic Development*, ed. A. Kontoleon, U. Pascual & M. Smale. Routledge, London.

Di Falco, S. & Chavas, J.-P. (2008b) Rainfall shocks, resilience and the dynamic effects of crop biodiversity on the productivity of the agroecosystem. *Land Economics*, 84, 83–96.

Di Falco, S. & Chavas, J.-P. (2009) On crop biodiversity, risk exposure and food security in the Highlands of Ethiopia. *American Journal of Agricultural Economics*, 91, 599–611.

Di Falco, S. & Perrings, C. (2003) Crop genetic diversity, productivity and stability of agroecosystems: a theoretical and empirical investigation. *Scottish Journal of Political Economy*, 50, 207–16.

Di Falco S. & Perrings, C. (2005) Crop biodiversity, risk management and the implications of agricultural assistance. *Ecological Economics*, 55, 459–66.

Dovie, D.B.K., Shackleton, C.M. & Witkowski, E.T.F. (2007) Conceptualizing the human use of wild edible herbs for conservation in South African communal areas. *Journal of Environmental Management*, 84, 146–56.

Duraiappah, A.K. (1998) Poverty and environmental degradation: a review and analysis of the nexus. *World Development*, 26, 2169–79.

Evenson, R.E., Gollin, D. & Santaniello, V (eds.) (1998) *Agricultural Values of Plant Genetic Resources*. CABI Publishing, Wallingford.

Fafchamps, M., Udry, C. & Czukas, K. (1998) Drought and saving in West Africa: are livestock a buffer stock? *Journal of Development Economics*, 55, 273–305.

Fisher, M. (2004) Household welfare and forest dependence in Southern Malawi. *Environment and Development Economics*, 9, 135–54.

Fisher, M. & Shively, G. (2003) Do tropical forests provide a safety net? Income shocks and forest extraction in Malawi. Paper presented at the Agricultural Economics Association Meeting, Montreal, July.

Frison, E., Smith, I.F., Johns, T., Cherfas, J. & Eyzaguirre, P. (2005) Using biodiversity for food, dietary diversity, better nutrition and health. *South African Journal of Clinical Nutrition*, 18, 112–14.

Fu, Y.N., Chen. J., Guo, H.J., Chen, A.G., Cui, J.Y. & Hu, H.B. (2009) The role of non-timber forest products during agroecosystem shift in Xishuangbanna, southwestern China. *Forest Policy and Economics*, 11, 18–25.

Glaser, M. (2003) Interrelations between mangrove ecosystem, local economy and social sustainability in Caete Estuary, North Brazil. *Wetlands Ecology and Management*, 11, 265–72.

Grifo, F. & Rosenthal, J. (eds.) (1997) *Biodiversity and Human Health*. Island Press, Washington, DC.

Heal, G. (2000) Biodiversity in the marketplace. *World Economics*, 1, 149–77.

Heisey, P., Smale, M., Byerlee, D. & Souza, E. (1997) Wheat rusts and the costs of genetic diversity in the Punjab of Pakistan. *American Journal of Agricultural Economics*, 79, 726–37.

Huynen, M., Martens, P. & Groot, R.D. (2004) Linkages between biodiversity loss and human health: a global indicator analysis. *International Journal of Environmental Health Research*, 14, 13–30.

Jackson, L.E., Pascual, U. & Hodking, T. (2007) Utilising and conserving agrobiodiversity in agricultural landscapes. *Agriculture, Ecosystems and Environment*, 121, 196–210.

Jha, S. (2009) Household-specific variables and forest dependency in an Indian hotspot of biodiversity: challenges for sustainable livelihoods. *Environment, Development and Sustainability*, 11, 1–9.

Jodha, N.S. (1995) Common property resources and the dynamics of rural poverty in India's dry regions. *Unasylva*, 180, 23–9.

Johns, T. (2006) Agrobiodiversity, diet and human health. In *Managing Biodiversity in Agricultural Ecosystems*, ed. D.I. Jarvis, C. Padoch & D. Cooper, pp. 382–406. Columbia University Press, New York.

Johns, T. & Eyzaguirre, P.B. (2006) Symposium on wild-gathered plants: basic nutrition, health and survival: linking biodiversity, diet and health in policy and practice. *Proceedings of the Nutrition Society*, 65, 182–189.

Just, R.E. & Candler, W. (1985) Production functions and rationality of mixed cropping. *European Review of Agricultural Economics*, 12, 207–31.

Kamanga, P., Vedeld, P. & Sjaastad, E. (2009) Forest incomes and rural livelihoods in Chiradzulu District, Malawi. *Ecological Economics*, 68, 613–24.

Kinsey, B., Burger, K. & Gunning, J.W. (1998) Coping with drought in Zimbabwe: survey evidence on responses of rural households to risk. *World Development*, 26, 89–110.

Lenne, J. & Wood, D. (1999) Optimizing biodiversity for productive agriculture. In *Agrobiodiversity: Characterization, Utilization and Management*, ed. D. Wood & J. Lenne. CABI Publishing, Wallingford.

Levang, P., Dounias, E. & Sitorus, S. (2005). Out of the forest, out of poverty? *Forests, Trees and Livelihoods*, 15, 211–35.

Lipper, L., Cavatassi, R. & Hopkins, J. (2008) The role of crop genetic diversity in coping with drought: insights from eastern Ethiopia. In *Agrobiodiversity, Conservation and Economic Development*, ed. A. Kontoleon, U. Pascual & M. Smale. Routledge, London.

Mamo, G., Sjaastad, E. & Vedeld, P. (2007) Economic dependence on forest resources: a case from Dendi District, Ethiopia. *Forest Policy and Economics*, 9, 916–27.

McSweeney, K. (2003) Tropical forests as safety nets? The relative importance of forest product sale as smallholder insurance, eastern Honduras. Paper presented at the International Conference on Rural Livelihoods, Forests and Biodiversity, Bonn, May.

Millennium Ecosystem Assessment (MA) (2005) *Ecosystems and Human Well-being: Biodiversity Synthesis*. World Resources Institute, Washington, DC.

Narain, U., S. Gupta, S. & Veld, K. van't. (2008a) Poverty and resource dependence in rural India. *Ecological Economics*, 66, 161–76.

Narain, U., Gupta S. & Veld, K. van't. (2008b) Poverty and the environment: exploring the relationship between household incomes, private assets, and natural assets. *Land Economics*, 84, 148–67.

Osemeobo, G.J. (2005) Living on wild plants: evaluation of the rural household economy in Nigeria. *Environmental Practice*, 7, 246–56.

Pattanayak, S.K. & Sills, E. (2001) Do tropical forests provide natural insurance? The microeconomics of non-timber forest products collection in the Brazilian Amazon. *Land Economics*, 77, 595–612.

Paumgarten, F. & Shackleton, C.M. (2009) Wealth differentiation in household use and trade in non-timber forest products in South Africa. *Ecological Economics*, 68, 2950–9.

Perrings, C. (1995) Biodiversity conservation as insurance. In *The Economics and Ecology of Biodiversity Decline*, ed. T.M. Swanson. Cambridge University Press, Cambridge.

Perrings, C. (2006) Resilience and sustainable development. *Environment and Development Economics*, 11, 417–27.

Perrings, C. & Stern, D.I. (2000) Modelling loss of resilience in agroecosystems: rangelands in Botswana. *Environmental and Resource Economics*, 16, 185–210.

Polasky, S. & Solow, A. (1995) On the value of a collection of species. *Journal of Environmental Economics and Management*, 29, 298–303.

Reddy, S.R.C. & Chakravarty, S.P. (1999) Forest dependence and income distribution in a subsistence economy: evidence from India. *World Development*, 27, 1141–9.

Roe, D. & Elliott, J. (2005) *Poverty-Conservation Linkages: A Conceptual Framework*. Poverty and Conservation Learning Group, London.

Sapkota, I. & Oden, P. (2008) Household characteristics and dependency on community forests in Terai of Nepal. *International Journal of Social Forestry*, 1, 123–44.

Schlapfer, F., Tucker, M. & Seidl, I. (2002) Returns from hay cultivation in fertilized low diversity and non-fertilized high diversity grassland. *Environmental and Resource Economics*, 21, 89–100.

Shaanker, R.U., Ganeshaiah, K.N., Krishnan, S., Ramya, R., Meera, C., Aravind, N.A., Kumar, A., Rao, D., Vanraj, G., Ramachandra, J., Gauthier, R., Ghazoul, J., Poole, N. & Chinnappa Reddy, B.V. (2004) Livelihood gains and ecological costs of non-timber forest product dependence: assessing the roles of dependence, ecological knowledge and market structure in three contrasting human and ecological settings in south India. *Environmental Conservation*, 31, 242–53.

Shackleton, C.M. & Shackleton, S.E. (2006) Household wealth status and natural resource use in the Kat River valley, South Africa. *Ecological Economics*, 57, 306–17.

Sharma, C.M., Gairola, S., Ghildiyal, S.K. & Suyal, S. (2009) Forest resource use patterns in relation to socioeconomic status. *Mountain Research and Development*, 29, 308–19.

Smale, M. (2006) *Valuing Crop Biodiversity: On-Farm Genetic Resources and Economic Change*. CABI Publishing, Wallingford.

Smale, M. & Drucker, A. (2008) Agricultural development and the diversity of crop and livestock genetic resources: a review of the economics literature. In *Agrobiodiversity, Conservation and Economic Development*, ed. A, Kontoleon, U. Pascual & M. Smale. Routledge, London.

Smale, M., Hartell, J., Heisey, P.W. & Senauer, B. (1998) The contribution of genetic resources and diversity to wheat production in the Punjab of Pakistan. *American Journal of Agricultural Economics*, 80, 482–93.

Smale, M., Singh, J., Di Falco, S. & Zambrano, P. (2008) Wheat diversity and productivity in Indian Punjab after the Green Revolution. *Australian Journal of Agricultural and Resource Economics*, 52, 419–32.

Sunderlin, W.D., Resosudarmo, I.A.P., Rianto, E. & Angelsen, A. (2000) *The Effect of Indonesia's Economic Crisis on Small Farmers and Natural Forest Cover in the Outer Islands*. CIFOR Occasional Paper 28. CIFOR, Bogor, Indonesia.

Swift, M.J., Izac, A.M.N. & Noordwijk, M. van (2004) Biodiversity and ecosystem services in agricultural landscapes – are we asking the right questions ? *Agriculture Ecosystems and the Environment*, 104, 113–24.

Takasaki, Y., Barham, B.L. & Coomes, O.T. (2004) Coping strategies in tropical forests: floods, illnesses, and resource extraction. *Environment and Development Economics*, 9, 203–24.

Tilman, D. (1996) Biodiversity: population versus ecosystem stability. *Ecology*, 77, 350–63.

Twyman, C. (2001) Natural resource use and livelihoods in Botswana's Wildlife Management Areas. *Applied Geography*, 21, 45–68.

Van Dusen E., Gauchan, D. & Smale, M. (2007) Farm conservation of rice biodiversity in Nepal: a simultaneous equation approach. *Journal of Agricultural Economics*, 58, 242–59.

Van Dusen, E. & Taylor, J.E. (2005) Missing markets and crop diversity: evidence from Mexico. *Environment and Development Economics*, 10, 513–31.

Van Staden, J. (1999) Medicinal plants in southern Africa: utilization, sustainability, conservation – can we change the mindsets ? *Outlook on Agriculture*, 28, 75–6.

Viet Quang, D. & Nam Anh, T. (2006) Commercial collection of NTFPs and households living in or near the forests: case study in Que, Con Cuong and Ma, Tuong Duong, Nghe An, Vietnam. *Ecological Economics*, 60, 65–74.

Villamor, G.B. & Lasco, R.D. (2009) Rewarding upland people for forest conservation: experience and lessons learned from case studies in the Philippines. *Journal of Sustainable Forestry*, 28, 304–21.

Walpole, M. & Wilder, L. (2008) Disentangling the links between conservation and poverty reduction in practice. *Oryx*, 42, 539–47.

Weitzman, M.L. (1993) What to preserve? An application of diversity theory to crane conservation. *Quarterly Journal of Economics*, 108, 157–83.

Widawsky, D. & Rozelle, S. (1998) Varietal diversity and yield variability in Chinese rice production. In *Farmers, Gene Banks, and Crop Breeding*, ed. M. Smale. Kluwer, Boston.

Wollny, C.B.A. (2003) The need to conserve farm animal genetic resources in Africa: should policy makers be concerned? *Ecological Economics*, 45, 341–51.

World Bank (2007) *Poverty and Environment: Understanding Linkages at the Household Level*. World Bank, Washington, DC.

Wunder, S. (2001) Poverty alleviation and tropical forests-what scope for synergies? *World Development*, 29, 1817–33.

Zimmerman, F.J. & Carter, M.R. (2003) Asset smoothing, consumption smoothing and the reproduction of inequality under risk and subsistence constraints. *Journal of Development Economics*, 71, 233–60.

Part II
Biodiversity and Poverty Relationships in Different Ecological Settings

Forests, Poverty and Conservation: An Overview of the Issues

Brian Belcher[1,2]

[1]Centre for Livelihoods and Ecology, Royal Roads University, Victoria, British Columbia
[2]Centre for International Forestry Research, Bogor, Indonesia

Introduction

This book considers whether and how biodiversity conservation and poverty interact. Are conservation objectives compatible with rural development objectives, or are they mutually exclusive? Are there practical ways to achieve gains on both fronts? If so, in what context? Understanding the interactions of forests and poverty is critical in this context. Forests, and specifically tropical forests, contain the highest levels of terrestrial biodiversity and 735 million rural people live in or near forests and woodlands in the tropics, 70 million of those within closed forest areas (Chomitz *et al.*, 2006). These people use forests and the underlying lands to support their livelihoods in a variety of ways, and their behaviour affects the quality and quantity of forests and, by extension, the biodiversity of the area.

The Brundtland Commission (World Commission on Environment and Development, 1987) popularised the idea that conservation and development are compatible objectives. The Millennium Development Goals, agreed in 2000, re-emphasised poverty alleviation and the concept of sustainable development (United Nations, 2011). However, despite high expectations (Scherr *et al.*, 2004) and substantial effort over decades to support *forest-based poverty alleviation* (FBPA), there are few success stories to support the sustainable development hypothesis. Now, with heightened concern about climate change and clear recognition of the importance of forests and forestry in

Biodiversity Conservation and Poverty Alleviation: Exploring the Evidence for a Link, First Edition.
Edited by Dilys Roe, Joanna Elliott, Chris Sandbrook and Matt Walpole.
© 2013 John Wiley & Sons, Ltd. Published 2013 by John Wiley & Sons, Ltd.

carbon management, there is revitalised interest in and attention to forests. That, in combination with unacceptably high levels of both poverty and biodiversity loss, gives new urgency to understand the real and potential relationships between forests and poverty.

This is not a new issue. Many authors have contributed excellent analyses on the topic (Dove, 1993; Pimentel *et al.*, 1997; Byron & Arnold, 1999; Neumann & Hirsch, 2000; Wunder, 2001; Angelsen & Wunder, 2003; Sunderlin *et al.*, 2005). This chapter aims to review and summarise the key issues and considerations from the perspective of combining biodiversity and conservation objectives. Therefore, it will focus at the micro scale. Wunder (2001) argued convincingly that forestry has limited potential for poverty alleviation at the economy-wide scale, with limited producer benefits, consumer benefits and labour absorption. At the community or forest level, there are four aspects to consider: the nature of 'forest dependency'; forests, poverty and remoteness; the influence of forest dwellers on biodiversity conservation and the influence of conservation activities on poverty. We can then consider opportunities for combining biodiversity conservation and poverty alleviation objectives.

The nature of 'forest dependency'

Many poorly substantiated claims exist about the number of people who 'depend' on forests. The World Bank's *Sustaining Forests* notes that:

> More than 1.6 billion people depend to varying degrees on forests for their livelihoods. About 60 million indigenous people are almost wholly dependent on forests. Some 350 million people who live within or adjacent to dense forests depend on them to a high degree for subsistence and income. (World Bank, 2004: 16)

International Network for Bamboo and Rattan (2011) claims that bamboo and rattan alone are "integral to the lives of up to 1.5 billion people".

Byron and Arnold (1999) give careful consideration to the idea of 'forest dependency'. They refer to a dictionary definition of *dependency* as "to be unable to do without". They suggest that large numbers of people use forest products as a matter of choice; far fewer are truly dependent in the sense that forest products are integral to their livelihood systems and that their condition would worsen if they no longer had access to those resources. Byron and Arnold (1999) also recognised the important fact that substantial quantities of 'forest products' are in fact produced from bush fallow and trees in agricultural fields rather than from true forests. Still, there are large numbers of people living near forests and, according to Chomitz *et al.* (2006), 70 million of these people are living within closed forest areas.

Defining forest products and types of use is important; measuring that use has also been problematic. Vedeld *et al.* (2007) reviewed 51 case studies from 17 countries in a meta-analysis of the contribution of forest environmental income to household income. Over the range of studies they reviewed, an average of 22% of total income came from environmental sources, with fuel wood, wild foods and fodder as the most important products. They noted that forest environmental income has a strong equalising effect on local income distribution. And they helpfully observed that the literature contains substantial flaws and inconsistencies in how income is defined and estimated, including within individual studies.

Studies that have used clear definitions and measures such as Cavendish (2000), Belcher *et al.* (2011) and the 33 case studies of the on-going Poverty and Environment Network project (Angelsen, 2011) corroborate these findings, documenting similar overall average forest-based components in total household incomes. They also highlight fuel wood as the single most important component of forest income. But these average figures disguise tremendous variation between households within villages, between villages within studies and between studies. For example, in a sample of 1215 households from 27 villages in Jharkhand, India, average household income (subsistence and sales) from forest sources at the village scale ranged from a low of around 2000 INR (US$47) up to nearly 27,000 INR (US$630), with different products predominating in different villages (Belcher *et al.* 2011). There is a pattern of specialisation in certain households and certain villages that is determined by factors such as resource availability, resource tenure, market access and institutions and of course the opportunity costs of land and labour (Belcher *et al.*, 2005).

Conditions in one place that allow one or many families to generate substantial income and employment from forest products may be absent even in nearby villages. Early efforts (and many current projects) at integrated conservation and development naïvely interpreted current non-timber forest products (NTFP) use, or even the simple presence of NTFPs in the forest environment, as an indication of potential for development and a way to encourage local conservation interest (Belcher & Schreckenberg, 2007). This approach is exemplified by the UN Food and Agriculture Organization's *Community-Based Tree and Forest Product Enterprises: Market Analysis and Development* field manual (Lecup & Nicholson 2000). The manual aims to provide facilitators with a framework for "identifying, planning and developing tree and forest product enterprises that will provide income and benefits without degrading the resource base" (Lecup & Nicholson, 2000: Booklet A:5) and includes the explicit assumption that "Community members will conserve and protect forest resources if they receive the economic benefits from sustainable use" (Booklet A:4). In practice, many of the necessary conditions for NTFP development, or any kind of development, are weak or absent (Belcher & Schreckenberg, 2007). This is captured well in the concept of 'remoteness'.

Forests, poverty and remoteness

As Turner *et al.* discuss in Chapter 2 of this volume, there are high poverty rates (i.e. a high proportion of people who are poor) in forested areas of the tropics. This highlights the relevance of appreciating and understanding the relationship between poverty and conservation; these are many of the same areas where high levels of biodiversity remain.

The same factors that help to conserve forests also contribute to perpetuating poverty. These areas tend to be remote, with poor transportation and other communications infrastructure. They often have steep topography, poor soils, hostile climates and low agricultural potential. The forest remains, at least in part, because it has been difficult and uneconomical to exploit the timber and because there has been relatively low pressure to convert the land to other uses. Markets for inputs and for outputs are distant and costly to access, so people living in the area must rely on mixed, low-intensity, subsistence-oriented systems. Commercial trade of agricultural and forest products often requires long journeys to markets, with high transaction costs in the form of transportation expenses; perished products; official and unofficial tariffs, taxes and bribes and uncertain demand and prices when they reach market. Alternatively, producers must rely on itinerant traders who typically have a stronger bargaining position than local producers. These 'middlemen' face the same high costs and risks of transporting goods from remote areas, so farm-gate prices remain low relative to retail prices.

Remoteness also limits opportunities for alternative employment or income. Where jobs are available, off-farm income is often a key avenue out of poverty (Haggblade *et al.*, 2002; Ruiz-Pérez *et al.*, 2004), but there are few waged jobs locally in remote forested areas. People must migrate to urban areas for work, with all of the risks and social disruptions that implies. And when the young, the strong and the better educated migrate, it further reduces capacity in their home villages.

The people who live in forested areas also tend to be marginalised people. Many are indigenous or culturally conservative, with long-established subsistence-based livelihood traditions (Sunderlin *et al.*, 2005). Others are migrants who have left poor conditions seeking new economic opportunities. Most are relatively powerless, with limited political and social capital. For these reasons, and as another consequence of their remoteness, communities in forested areas tend to be overlooked and underserved by government. They suffer poor education and primary health care services and generally low levels of human capital in another self-reinforcing cycle typical of poverty.

The key resources available to people in these remote areas – the 'natural capital' – are often contested by a range of stakeholders in a way that inhibits or prevents people from escaping poverty. Traditional societies often have well-developed systems of common property that work well for subsistence-level production and with low population pressure. Increasing population pressure and commercial production can stress these systems. More importantly, the state often does not recognise traditional ownership over forest areas and lays claim to the land and resources (McKean & Ostrom, 1995). It is not uncommon to have several overlapping claims to an area, with traditional claimants competing with government-allocated logging, mining or plantation concessions, as well as with new migrants, parks and protected areas (Peluso, 1995; White & Martin, 2002).

Under these conditions, forest products are treated as open access resources and there are low barriers to entry, at least for collection and trade, resulting in high competition which further depresses already low prices. This can result in a kind of 'poverty trap' in which the poorest, with the most limited options, are driven by need and declining prices to harvest and sell more, causing even further price declines in a self-reinforcing cycle.

Influence of forest dwellers on biodiversity

When we consider the influence of 'poverty' on biodiversity conservation, we need to keep in mind that the poverty of people who live in areas of high conservation value is defined by the complex suite of conditions that we have discussed in this chapter. They are not just poor in financial income and assets – they face high barriers that inhibit or prevent them from accumulating capital in any form.

People use forest and land resources in a wide variety of ways. Swidden (shifting) cultivation is the most common agricultural practice in forest-based cultures. It is sustainable under low-population conditions but unsustainable when population increases, available land area decreases and fallow periods become too short for sufficient regeneration. Swidden systems have been widely perceived as a threat to forests and incompatible with biodiversity conservation, and there has been strong government and other pressure to reduce shifting cultivation and to replace it with permanent agriculture (Fox et al., 2000; Rerkasem et al., 2009). However, as Reskasem et al. (2009) point out, this has ironically caused substantial agro-biodiversity losses as these relatively diverse agricultural systems are displaced and as traditional networks of crop genetic resource (seed) conservation and exchange collapse.

Many of the enormous variety of forest products used and traded by people in forest communities are harvested from wild, unmanaged populations (Neumann & Hirsch, 2000). This includes fruits, nuts, vegetables, mushroom, medicinal plants, fragrances and essences, leaves, vines and stems for fibre and structural use as well as animals and animal products for food, skins, trophies and medicines. With relatively low demand for local subsistence use or even small-scale commercial trade, such harvesting may have only a minor impact on local populations. The actual impact depends on the kind of harvesting, local abundance, reproduction and growth rates of the target species or product. As demand increases, there is a risk of local overexploitation of valuable species, especially those that are slower growing or less fecund or where harvesting requires killing the plant or animal. There are major concerns with overharvesting for the bushmeat trade in Africa (Nasi *et al.*, 2008), and there are many examples of serial overexploitation of NTFPs, such as gaharu (Momberg *et al.*, 2000), rattan and tree species used for wood carving (Cunningham *et al.*, 2005).

Timber markets tend to be out of reach of poor, forest-based people. Industrial-scale timber production requires a high level of investment. The poor typically lack the authority and resources for large-scale forest management. There are also strong political-economic biases against the poor in the timber sector; tropical forests and the underlying lands have high economic rent and have proven useful as political capital and sources of extra-budgetary income in corrupt, crony-capitalist systems. Thus poor villagers may be involved as labourers for legal timber operations, or as 'illegal loggers', using labour-intensive means to harvest and deliver logs to middlemen or directly to processors, for low returns. In some cases, where forest communities were granted rights to timber, as in the rapid decentralisation of resource management in Indonesia in the late 1990s, they were quick to sell the resources for short-term gains (Barr *et al.*, 2001). And, notably, some community-managed forest enterprises have succeeded in timber markets (Antinori & Bray, 2005). But for the most part, the industrial timber sector has been inaccessible to the poor (Wunder, 2001).

More useful species may also be cultivated and/or managed using enrichment planting, improving fallows and protecting and/or cultivating within agricultural fields. There is a rich variety of these intermediate systems (Belcher *et al.*, 2005a). Such approaches permit increased production and give more control to producers. As demand and prices increase, producers may further intensify management towards fully domesticated systems, as predicted by Homma (1992), with negative impacts on wild biodiversity, similar to any other agricultural land use.

Forest products production in China provides a good example of intensification to meet commercial demand. While there is still wild harvesting where natural forests are available, there are also well-developed horticultural production and market systems producing a range of forest products. This is facilitated by the official classification of many horticultural crops (e.g. tung oil, tea tree oil, star anise and even tea) as 'forestry' species, and by widely implemented policies to promote reforestation and afforestation on sloping lands (Hogarth *et al.*, unpublished). While this has increased forest cover and forest-based incomes, there have been limited biodiversity gains. This illustrates the danger of assuming that forest-based enterprises will necessarily create incentives for conservation.

Still some of the main demands on forests are for the underlying land for permanent agriculture and, increasingly, for industrial plantations.

Influence of conservation activities on poverty

There are three basic approaches to forest or biodiversity conservation: (i) strict protected areas, which aim to limit human activity in the conservation areas (so-called *fortress conservation*) (Hutton *et al.*, 2005); (ii) combined goals of local development and conservation through integrated conservation and development projects (ICDPs) (McShane & Wells, 2004; Blom *et al.*, 2010) or broader community-based conservation, which emphasizes community involvement and (iii) payments for environmental services, in which resource owners or managers are paid to protect a source of environmental services (Wunder, 2005).

These various approaches have been well described and analysed in other chapters of this book and elsewhere. Here we will briefly consider the implications from a livelihoods perspective. Information about the social impacts of protected areas is still very poor (Brockington *et al.*, 2006; Chapter 10, this volume). However, considering evidence about the importance of 'environmental income' to people living in remote areas (discussed in this chapter), and the fact that many of these protected areas are inhabited, it is easy to anticipate that restrictions on resource use and access will have negative impacts unless there is appropriate compensation or mitigation. Environmental income contributes an important component of livelihoods for people in these areas, and relocations and restrictions on use in protected areas directly reduce the resources available to people.

There is also an important cultural dimension that must be considered. Pretty *et al.* (2008), Berkes (Chapter 17, this volume) and others have explored the close interrelationship between biodiversity and culture, each shaping the other. When people who have strong cultural ties to an area are excluded or restricted from using that area, damage can be done to the culture and to the environment.

ICDP approaches including NTFP commercialisation, ecotourism development and other enterprise development activities aim, at least, to provide compensation to local people for benefits forgone in conservation activities or, more ambitiously, to promote conservation by making it economically attractive. These approaches have been notoriously unsuccessful (Sunderland & Murdiyarso, 2010), although Elliott and Sumba (Chapter 13, this volume) usefully highlight the conditions that increase the chances of success.

NTFPs have been a main focus in discussions of livelihoods and rural development because there has been a perception that forest communities have high levels of current use and that poor households tend to be disproportionately dependent on forest resources. NTFPs are seen as resources that are accessible to the poor, with simple (often traditional) harvesting and processing technologies, relatively high labour requirements, low barriers to entry and important opportunities for women (Shackleton & Shackleton, 2004; Marshall *et al.*, 2006). There was also a prevailing idea that NTFP production is somehow more benign environmentally than alternative resources uses. In combination, these factors seem to indicate that NTFP commercialisation should offer pro-poor development opportunities (Chapter 9, this volume). But current use does not necessarily indicate that there is good development potential, and studies of NTFP commercialisation have showed limited development potential (Belcher *et al.*, 2005b; Belcher & Schreckenberg, 2007) and real trade-offs between conservation and development objectives (Kusters *et al.*, 2006).

Krüger (2005) reviewed 251 cases of ecotourism described in the literature and found some successes in terms of conservation benefits. Key success factors included detailed a priori planning, local involvement and control measures. But again there are limits on the degree of engagement and the amount of benefits that can be captured by local people, because of inherent lack of capacity.

Community-based conservation efforts may put more emphasis on community engagement, institution building and the devolution of authority and responsibility to local people. But there is still doubt about the feasibility of combining conservation and community development objectives (Berkes, 2004). Campbell and Vainio-Mattila (2003: 434) note a tendency to focus on "getting people on-side with conservation", at the cost of genuine development benefits. Kellert *et al.* (2000) found, in the cases they reviewed, more success in human development but also that conservation goals were compromised. This is consistent with McDermott and Schreckenberg (2009), who reviewed community forestry programmes and found benefits at community and higher levels, but less at the household level.

PES schemes, and the widespread current popularity of Reduced Emissions from Deforestation and Forest Degradation (REDD) schemes in particular, may offer new opportunities to generate benefits to local people (Chapter 14, this volume). In addition to the value of direct payments, REDD approaches may change the economics enough to permit low-intensity management of forests. However, as various PES mechanisms gain momentum, indigenous peoples, civil society activities and forest policy experts have expressed concern about the risk of further marginalising forest peoples – for example, through stricter enforcement of forest protected areas – and still failing to tackle the underlying causes of deforestation (Griffiths, 2008).

Combining biodiversity conservation and poverty alleviation in forest communities

When thinking about the contribution of forests to tackling poverty, it is important to distinguish between poverty mitigation (alleviating and preventing poverty) and poverty reduction (lifting people out of poverty).

In order for conservation activities to 'do no harm', they need to recognise and protect the role that forests play in alleviating – or preventing – poverty (e.g. through the use of forest products to help meet basic needs or to act as a safety net in times of emergency) and fully compensate losses experienced by local people. We now have better information that shows that, in many households in forested areas, forest-based income rivals agriculture as the most important income source, and more so in more remote communities (i.e. in high conservation value areas). Fuel wood is the most important forest product universally in terms of market value, but there is an extremely large range of products used for subsistence, for sale and as inputs for processed products. We also should not discount the importance of products or species used in smaller quantities. They may have high value in supplying micro-nutrients, in filling gaps between harvests or in supplying medicines, or they may have traditional ritual or spiritual value. Overall, more and better effort is required to account for that component of livelihoods, to understand how planned interventions will affect it and to appropriately protect existing values and mitigate and/or compensate losses.

Poverty reduction is a much more difficult objective on its own, especially when combined with conservation objectives. In some situations and for some people, the key to livelihood improvement is to leave the forest environment (Levang et al., 2005). Many attempts at improving livelihoods opportunities in situ through ICDPs, NTFP commercialisation and ecotourism have failed, for many different reasons. There are changes taking place that seem to open new opportunities for small-scale producers (Scherr et al., 2004), though not necessarily in a way that is compatible with conservation objectives. There is a trend toward increased local ownership and control of forest resources (Sunderlin et al., 2008), growing demand for forest products, technical and market developments that have increased demand for a wider range of forest products, increased demand for environmental services (REDD) and improved communications capabilities (the cell phone can facilitate unprecedented access to market information by village-scale producers). There are growing markets for a wide range of products – medicinal and cosmetic products, essential oils and resins, fruits, essences and flavours as well as specialty woods – that can be produced in biodiverse production systems and in which small-scale producers may be able to compete. There are also growing niche markets for particular types of production and marketing arrangements; green markets and fair trade enterprises sell positive images of the ecological and/or social impacts of their products along with the inherent characteristics of the products themselves. But again, increased demand can only be met with increased production through intensification and/or extensification, with negative implications for biodiversity conservation. Improved tenure is good for landowners but may lead to deforestation (Chomitz et al., 2006). Improved agricultural technology promotes growth but also may increase demand to open new land (Angelsen & Kaimowitz, 1999). Roads promote rural development but also forest clearing (Chomitz et al., 2006). More importantly, the basic conditions required for successful commercial development and for sustainable livelihoods are not met in poor remote areas, by definition. The basic infrastructure (roads and communications) and institutions (including input and output markets and property rights over the resources) are typically weak or absent, and human and social capital are also weak. So, genuine efforts to alleviate poverty will need to address these weaknesses.

There is scope for poverty alleviation through improved governance and strengthening of local institutions. For example, McDermott and Schreckenberg (2009) considered issues of equity in community forestry. They found that community forestry programmes can reduce inequality when they explicitly target the poor and marginalised (i.e. it needs to be an explicit goal). Community forestry expands decision-making spaces, enabling change and benefit capture, with a range of benefits through changes at community and higher levels. This is an area that needs more and better support in the context of conservation projects.

Successful poverty reducing interventions will need to be broad based and aimed to achieve general improvements in these aspects of rural life. Realising development potential will require investments beyond the natural resources sector and into other areas, such as improved primary healthcare, education and transportation infrastructure, but again with potential conflicts with conservation aims.

References

Angelsen, A. & Kaimowitz, D. (1999) Rethinking the causes of deforestation: lessons from economic models. *World Bank Research Observer*, 14, 73–98.

Angelsen, A. & Wunder, S. (2003) *Exploring the Forest – Poverty Link: Key Concepts, Issues and Research Implications*. CIFOR Occasional Paper No. 40. Center for International Forestry Research (CIFOR), Bogor, Indonesia.

Angelsen, A., Wunder, S., Babigumira, R., Belcher, B., Börner, J. & Smith-Hall, C. (2011) Environmental incomes and rural livelihoods: a global-comparative assessment. Paper presented at the 4th Wye Global Conference, Rio de Janeiro, 9–11 November.

Antinori, C. & Bray, D. (2005) Community forest enterprises as entrepreneurial firms: economic and institutional perspectives from Mexico. *World Development*, 33, 1529–43.

Barr, C., Wolleberg, E., Limberg, G., Anau, N., Iwan, R., Sudana, I.M., Moeliono, M. & Djogo, T. (2001) *Case Study 3: The Impacts of Decentralization on Forest and Forest-Dependent Communities in Malinau District, East Kalimantan*. Center for International Forestry Research, Bogor, Indonesia.

Belcher, B., Dewi, S. & Achdiawan, R. (2011) Livelihoods strategies under different access and natural asset levels in Jharkhand, India. Paper presented to the Poverty and Environment Network (PEN) Symposium, University of East Anglia, Norwich.

Belcher, B.M., Michon, G., Angelsen, A., Ruiz-Pérez, M. & Asbjornsen, H. (2005a) The socioeconomic conditions determining the development, persistence and decline of forest garden systems. *Economic Botany*, 59, 245–53.

Belcher, B.M., Ruiz Pérez, M. & Achdiawan, R. (2005b) Global patterns and trends in the use and management of commercial NTFPs: implications for livelihoods and conservation. *World Development*, 33, 1435–52.

Belcher, B.M. & Schreckenberg, K. (2007) NTFP commercialisation: a reality check. *Development Policy Review*, 25, 355–7.

Berkes, F. (2004) Rethinking community-based conservation. *Conservation Biology*, 18, 621–30.

Blom, B., Sunderland, T. & Murdiyarso, D. (2010) Getting REDD to work locally: lessons learned from integrated conservation and development projects. *Environmental Science & Policy*, 13, 164–72.

Brockington, D., Igoe, J. & Schmidt-Soltau, K. (2006) Conservation, human rights, and poverty reduction. *Conservation Biology*, 20, 250–2.

Byron, N. & Arnold, M. (1999) What futures for the people of the tropical forests? *World Development*, 27, 789–805.

Campbell, L. & Vainio-Mattila, A. (2003) Participatory development and community-based conservation: opportunities missed for lessons learned? *Human Ecology*, 31 (3), 417–37.

Cavendish, W. (2000) Empirical regularities in the poverty-environment relationship of rural households: evidence from Zimbabwe. *World Development*, 28, 1979–2003.

Chomitz, K.M., Buys, P., de Luca, G., Thomas, T.S. & Wertz-Kanounnikoff, S. (2006) *At Loggerheads? Agricultural Expansion, Poverty Reduction and Environment in the Tropical Forests*. International Bank for Reconstruction and Development, Washington, DC.

Cunningham, A., Belcher, B. & Campbell, B. (2005) *Carving out a Future: Forests, Livelihoods and the International Woodcarving Trade*. Earthscan, London.

Dove, M.R. (1993) A revisionist view of tropical deforestation and development. *Environmental Conservation*, 20, 17–24.

Fox, J., Truong, D.M., Rambo, A.T., Tuyen, N.P., Cuc, L.T. & Leisz, S. (2000) Shifting cultivation: a new old paradigm for managing tropical forests. *Bioscience*, 50, 521–28.

Griffiths, T. (2008) *Seeing 'REDD'? Forests, Climate Change Mitigation and the Rights of Indigenous Peoples*. Forest Peoples Programme, Moreton-in-Marsh, UK.

Haggblade, S., Hazell, P. & Reardon, T. (2002) *Strategies for Stimulating Poverty-Alleviating Growth in the Rural Nonfarm Economy in Developing Countries*. EPTD Discussion Paper No. 92. World Bank, Washington, DC.

Hogarth, N., Belcher, B., Campbell, B. & Stacey. N. (Unpublished) *The Role of Forest Income in Rural Livelihoods and Poverty Alleviation in the Border-Region of Southern China*. Draft manuscript.

Homma, A.K.O. (1992) The dynamics of extraction in Amazonia: a historical perspective. In *Advances in Economic Botany 9: Non-Timber Products from Tropical Forests: Evaluation of a Conservation and Development Strategy*, ed. D.C. Nepstad & S. Schwartzman, pp. 23–32. Advances in Economic Botany 9. New York Botanical Garden Press, New York.

Hutton, J., Adams, W. & Murombedzi, J.C. (2005) Back to the barriers? Changing narratives in biodiversity conservation. *Forum for Development Studies*, 32, 341–57.

International Network for Bamboo and Rattan (2011) *Vision Statement*, http://www.inbar.int/Board.asp?Boardid=51 (accessed 23 April 2012).

Kellert, S., Mehta, J., Ebbin, S. & Lichenfeld, L. (2000) Community natural resource management: promise, rhetoric and reality. *Society and Natural Resources*, 13, 705–15.

Krüger, O. (2005) The role of ecotourism in conservation: panacea or Pandora's box? *Biodiversity and Conservation*, 14, 579–600.

Kusters, K., Achdiawan, R., Belcher, B. & Ruiz Pérez, M. (2006) Balancing development and conservation? An assessment of livelihood and environmental outcomes of nontimber forest product trade in Asia, Africa, and Latin America. *Ecology and Society*, 11, 20, http://www.ecologyandsociety.org/vol11/iss2/art20/(accessed 23 April 2012).

Lecup, I. & Nicholson, K. (2000) *Community-Based Tree and Forest Product Enterprises: Market Analysis and Development*. Food and Agriculture Organization of the United Nations (FAO), Rome.

Levang, P., Dounias, E. & Sitorus, S. (2005) Out of the forest, out of poverty? *Forests, Trees and Livelihoods*, 15, 211–35.

Marshall, E., Schreckenberg, K. & Newton, A.C. (eds.) (2006) *Commercialisation of Non-timber Forest Products: Factors Influencing Success: Lessons Learned from Mexico and Bolivia and Policy Implications for Decision-makers*. UNEP World Conservation Monitoring Centre, Cambridge.

McDermott, M. & Schreckenberg, K. (2009) Equity in community forestry: insights from North and South. *International Forestry Review*, 11, 157–70.

McKean, M. & Ostrom, E. (1995) Common property regimes in the forest: just a relic from the past? *Unasylva*, 46, 3–15.

McShane, T.O. & Wells, M.P. (2004) Integrated conservation and development? In *Getting Biodiversity Projects to Work*, ed. T.O. McShane & M.P. Wells, pp. 3–9. Columbia University Press, New York.

Momberg, F., Puri, R. & Jessup, T. (2000) Exploitation of gaharu, and forest conservation efforts in the Kayan Mentarang national park, East-Kalimantan, Indonesia. In *People, Plants and Justice: The Politics of Nature Conservation*, ed. C. Zerner, pp. 259–84. Columbia University Press, New York.

Nasi, R., Brown, D., Wilkie, D., Bennett, E., Tutin, C., van Tol, G. & Christophersen, T. (2008) *Conservation and Use of Wildlife-Based Resources: The Bushmeat Crisis*. Technical Series No. 33. Secretariat of the Convention on Biological Diversity, Montreal, and Center for International Forestry Research, Bogor, Indonesia.

Neumann, R.P. & Hirsch, E. (2000) *Commercialisation of Non-Timber Forest Products: Review and Analysis of Research*. Center for International Forestry Research (CIFOR), Bogor, Indonesia, and Food and Agriculture Organization of the United Nations (FAO), Rome.

Peluso, N.L. (1995) Whose woods are these? Counter-mapping forest territories in Kalimantan, Indonesia. *Antipode*, 27, 383–406.

Pimentel, D., McNair, M., Buck, L., Pimentel, M. & Kamil, J. (1997) The value of forests to world food security. *Human Ecology*, 25, 91–120.

Pretty, J., Adams, B., Berkes, F., Ferreira de Athayde, S., Dudley, N., Hunn, E., Maffi, L., Milton, K., Rapport, D., Robbins, P., Samson, C., Sterling, E., Stolton, S., Takeuchi, K., Tsing, A., Vintinner, E. & Pilgrim, S. (2008) *How Do Biodiversity and Culture Intersect?* Plenary paper for Sustaining Cultural and Biological Diversity in a Rapidly Changing World: Lessons for Global Policy conference, 2–5 April, http://www.greenexercise. org/pdf/How%20do%20biodiversity%20and%20culture%20intersect.pdf (accessed 23 April 2012).

Rerkasem, K., Lawrence, D., Padoch, C., Schmidt-Vogt, D., Ziegler, A.D. & Bech Bruun, T. (2009) Consequences of swidden transitions for crop and fallow biodiversity in Southeast Asia. *Hum Ecology*, 37, 347–60.

Ruiz-Pérez, M., Belcher, B., Maoyi, F. & Xiaosheng, Y. (2004) Looking through the bamboo curtain: an analysis of the changing role of forest and farm income in rural livelihoods in China. *International Forestry Review*, 61, 306–16.

Scherr, S., White, A. & Kaimowitz, D. (2004) *A New Agenda for Forest Conservation and Poverty Reduction: Making Markets Work for Low Income Producers*. Forest Trends, Washington, DC.

Shackleton, C.M. & Shackleton, S.E. (2004) The importance of non-timber forest products in rural livelihood security and as safety nets: a review of evidence from South Africa. *South African Journal of Science*, 100, 658–64.

Sunderlin, W.D., Angelsen, A., Belcher, B., Burgers, P., Nasi, R., Santoso, L. & Wunder, S. (2005) Livelihoods, forests, and conservation in developing countries: an overview. *World Development*, 33, 1383–402.

Sunderlin, W., Hatcher, J. and Liddle M. (2008) *From Exclusion to Ownership? Challenges and Opportunities in Forest Tenure Reform*. Rights and Resources Initiative (RRI), Washington, DC.

United Nations (2011) *Millennium Development Goals*, http://www.un.org/millenniumgoals/bkgd.shtml (accessed 23 April 2012).

Vedeld, P., Angelsen, A., Bojö, J., Sjaastada, E. & Kobugabe Berga, G. (2007) Forest environmental incomes and the rural poor. *Forest Policy and Economics*, 9, 869–79.

White, A. & Martin, A. (2002) *Who Owns the World's Forests? Forest Tenure and Public Forests in Transition*. Forest Trends, Washington, DC.

World Bank (2004) *Sustaining Forests: A Development Strategy*. World Bank, Washington, DC.

World Commission on Environment and Development (1987) *Our Common Future*. Oxford University Press, Oxford.

Wunder, S. (2001) Poverty alleviation and tropical forests – what scope for synergies? *World Development*, 29, 1817–33.

Wunder, S. (2005) *Payments for Environmental Services: Some Nuts and Bolts*. CIFOR Occasional Paper No. 42. CIFOR, Bogor, Indonesia.

(6)

Biodiversity and Poverty in Coastal Environments

Jock Campbell and Philip Townsley

Integrated Marine Management Ltd, UK

Introduction: the nature and scale of coastal diversity

The coast is a dynamic, complex interface between land and water that includes marine, estuarine, wetland and terrestrial habitats and species. According to the Millennium Ecosystem Assessment (MA, 2005a), the coast is the interface between ocean and land, extending seawards to about the middle of the continental shelf and inland to include all areas strongly influenced by the proximity to the ocean.

This interface between land and sea is one of the most diverse environments on the planet. It combines land-based ecosystems with those of the coastal waters but is also significantly affected by inland processes through rivers and estuaries, and by offshore processes through water body mixing. The very fluidity of the sea and the mobile nature of many of the life stages of different species also mean that ecosystem interactions are common. The diversity which exists in the coast is biological – including genetic, species and ecosystem diversity – and also relates to the livelihoods of people.

The sea covers some 71% of the Earth's surface, but the degree of biological diversity is not even across all of this area. Both species and ecosystem diversity are greatest in the coast. There is also generally an increase in biodiversity from the poles to the tropics. The interaction of coastal waters with coastal land creates an interface which is particularly important in this regard (Silvestri & Kershaw, 2010) and this creates strongly interlinked biomes which are often very dynamic, such as estuaries.

Biodiversity Conservation and Poverty Alleviation: Exploring the Evidence for a Link, First Edition. Edited by Dilys Roe, Joanna Elliott, Chris Sandbrook and Matt Walpole. © 2013 John Wiley & Sons, Ltd. Published 2013 by John Wiley & Sons, Ltd.

Other coastal ecosystems include coral reefs, mangroves, mudflats, sea grass beds, kelp forests, lagoons, beaches, dunes, salt marshes, coastal forests and farmland. Some of the ecosystems with the most species diversity are coral, mangrove and sea grass beds. Fish species diversity on coral reefs is thought to be the highest species diversity of any vertebrate assemblage on a local level (Ormond & Roberts, 1997), and at least 25% of all marine fish species are associated with coral reefs which are generally located in coastal waters (McAllister, 1991).

The coast is also an area of high human livelihood diversity, in part because of the concentration of people but also in response to the biodiversity. Linkages with the coast are sometimes expressed as geographical proximity – that is, coastal communities are defined as those living within 100 km. Estimates of the people dependent on the coast vary from a quarter to a half of the world's population (Small & Nicholls, 2003; United Nations Environment Programme (UNEP), 2006; Wilkinson, 2008). But these figures mean little unless dependency or other causal links between livelihoods and coastal ecosystems (such as direct or indirect resource use or exposure to risks from coastal flooding) can also be measured. For instance, McGranahan et al. (2006, 2007) estimated that 10% of the world's population, more that 600 million people, lived in low-elevation coastal zones of less than 10 m above sea level. These areas are prone to flooding, particularly when high tides combine with storm surges and/or high river flows (McGranahan et al., 2007). In reality, the number of people who depend upon the coast for their livelihoods is unknown, partly because of varying ways of measuring dependency and partly because the coast is extremely dynamic, often with strong seasonal dimensions.

Estimates of coastal poverty are equally difficult to make as the poor often exist outside of official statistics. Brown et al. (2008) estimate the global coastal poor to be in the order of 252 million people of whom 27% are in India, 13% in Indonesia and 9% in Bangladesh. Overall 80% of the coastal poor are estimated to be located in just 15 countries. Wilkinson (2008) estimates that some 500 million people depend upon coral reefs for food, coastal protection, building materials and income from tourism and 30 million are virtually totally dependent upon coral reefs for their livelihoods.

Global fisheries statistics can give some indication of coastal dependence as a large proportion of fishing activity is concentrated in productive coastal waters. According to the Food and Agriculture Organization (FAO, 2008), 43.5 million people depend directly on fishing (80%) and fish farming (20%) – mainly coastal – for a living. Of these 43.5 million, 86% are in Asia. Fishers and ancillary workers (e.g. processors, traders, net menders and boat builders) are estimated to be around 170 million people and, with their dependents, are thought to total nearly 520 million people, the vast majority of whom are based in the coast. Globally, fish provide more than 1.5 billion people with almost 20% of their average per capita intake of animal protein, and nearly 3 billion people with 15% of such protein (FAO, 2008).

Linking coastal biodiversity and poverty

The linkages between biodiversity and poverty are not always obvious, particularly as coastal areas are sometimes centres of development and often appear relatively wealthy. But detailed study has described and analysed the ways in which coastal biodiversity and poverty are linked (Campbell & Beardmore, 2001; Integrated Marine Management (IMM) *et al.*, 2003).

The characteristics of inshore coastal habitats, such as rocky shores, highly dynamic and unpredictable coastal waters, shallow seas and poor communications and market access, often provide livelihood niches for the poor but make exploitation by commercial operators unattractive. Likewise the poor often have limited access to the sort of capital which allows them to exploit resources further away from the coast. This commercial marginalisation of some highly diverse coastal areas has also meant that formal institutions have played less of a role in regulating resource access. This has resulted in access to resources being effectively open or characterised by poorly defined tenure arrangements leaving them for the poor to exploit without the associated transaction costs (IMM *et al.*, 2003).

In addition, the high diversity of ecological 'niches' often found in coastal areas, particularly in the tropics, provides poor people with a diverse range of livelihood opportunities that can help them to deal with seasonal changes and variations. The poor often make use of these livelihood characteristics to create complex household strategies. Johannes (1981) details the extent of Micronesian fishers' knowledge of the seasonal rhythms of coastal fish behaviour and adaptations of their fishing practices so as to provide some degree of livelihood stability through the year. This rich biodiversity can also help to create alternative market opportunities which can offset adverse changes in prices for some species and allow the poor to modify their livelihood strategies and substitute for products that are less in demand (Whittingham *et al.*, 2003). Similarly, the availability of a wide diversity of natural resources often helps to mitigate the potentially negative effects of competition, an important factor for poor people who are inevitably less able to deal with competition because of their lack of influence and limited investment potential.

Households also go through livelihood strategy cycles where their emphasis on exploitation changes to meet their changing personal attributes (e.g. age, physical ability and attitudes to risk), responsibilities (e.g. parenthood or carer), assets (physical, financial, natural, social and human) and other employment opportunities (IMM *et al.*, 2005). The richness of coastal biodiversity allows this to occur with relative ease. Diversity also allows for unforeseen adverse changes in livelihood strategies, for example a sudden loss of productive assets does not necessarily mean exclusion from resource exploitation because, often, alternative harvestable species are available.

However, that situation is rapidly changing as technological innovation allows more intensive exploitation of inshore and coastal resources. The same lack of access regulation and clearly defined tenure that has, in the past, encouraged poor people to seek a livelihood in the coast is also – with appropriate technology in place – increasingly providing opportunities for wealthier, larger scale interests to exploit these areas or find alternative uses for them. Areas that may previously have been regarded as 'underutilised' or 'wasteland', such as saline land, sand dunes and beaches, and coastal swamps and wetlands, are increasingly attractive for tourism, real estate, industry and aquaculture development (MA, 2005b). However, the extent to which the poor benefit from these changes is highly questionable, and in many cases the effects on them are negative.

Ecosystem services available to the poor

Coastal biodiversity gives rise to a wide diversity of ecosystem services that are used by the poor. The most obvious is the assurance of food security through fish used for both food and trade. However, particularly for the poor involved in small-scale production, fish diversity is as important as its biomass. A diversity of species, which are sometimes only available for short periods of time in relatively limited habitats, allows people of different ages and sex, with different skills, different levels of wealth and different physical abilities to participate in harvesting resources in ways that best suit their needs and capacities. The importance of this ecosystem service provided by coastal diversity also extends to people involved in fish processing and trade (often women), for whom the diversity of species allows a diversity of products to be produced which can be used to access different markets at different times of the year.

The poor also use a wide range of plants and animals for food. Worsley (1997) identified 119 different kinds of fish, 42 kinds of shellfish, five turtles, four crustaceans, two sea mammals and squid regularly used by aboriginals in the Torres Straits. In the Solomon Islands, most of the 400 locally named reef fish species are used (Ruddle et al., 1992).

Other examples of services provided by coastal ecosystems include the coral reef material used as a building material and for cement production. In some communities, reefs also underpin important cultural functions, such as the space for women's interaction, cooperation, knowledge sharing and mutual support activities provided by collaborative reef-gleaning activities (Whittingham et al., 2003). Reef species have been used as currency (e.g. shell money in Papua New Guinea and the Maldives). In many coastal communities, the different coastal ecosystems, species or physical structures also have spiritual meaning and significance which may reinforce community identity and social cohesion (Lokani, 1995; Innes, 1996; McClanahan et al., 1998; Rengasamy et al., 2003).

Regulating ecosystem services, such as water purification through coastal wetlands, coastal protection from reefs and mangroves and coastal micro-climates that can encourage precipitation that ensures the availability of drinking water in underground lenses, is also key in many areas. Whilst these ecosystem services are available to all, they are disproportionately important to the poor because ecosystem service alternatives are often prohibitively expensive for them or simply not available (Whittingham *et al.*, 2003).

Benefit flows to the poor

Most of the benefits from coastal biodiversity flow directly to the poor through their proximity to, and use of, coastal resources. Where tenure arrangements are poorly defined or not formalised, or where open access systems are predominant, the poor may choose to take the lead in directly exploiting coastal resources, particularly when it involves work which better off people are unwilling to undertake. The collection of shrimp fry in coastal waters to service coastal aquaculture is a good example of this.

In other cases, notably coastal small-scale fisheries, the concentration of economic activity in coastal fishing communities can act as a growth pole which stimulates a wide range of other economic activities, micro-enterprises and services, increasing the diversity of income-earning opportunities. These growth poles will often create spin-off opportunities for the poor. Where higher value resources are involved, much of the surplus value from these resources may be captured by better off sections of communities, but this can still leave important opportunities in service provision (food vending and fish handling) and maintenance work (net mending) that are particularly appropriate for the poor (IMM *et al.*, 2005; Allison & Horemans, 2006).

While the flow of benefits to the poor from the use of coastal biodiversity and ecosystem services is often relatively small in absolute terms, it is important to recognise how critical it can be for supporting the livelihoods of large numbers of poor people. This factor is often not recognised in wider economic development discourse in which coastal areas used exclusively by poor people may often be perceived as being 'underutilised' because they generate little surplus value for the wider economy and what value they produce tends to be 'diffused' among many beneficiaries. However, for these poorer beneficiaries, these limited benefits may be disproportionately important for their livelihoods and extremely difficult to replace should their access to them be denied, the resources degraded or the benefit flows redirected (Whittingham *et al.*, 2003).

In some situations the existence of high-value coastal resources, such as beach-front land for hotel or housing construction, can lead to ecosystem services being captured by wealthier or more powerful groups. The allocation of such rights may be used politically for maintaining political stability by the allocation of use rights to elites and may be considered as a necessary price to pay for stability which allows the poor to access less attractive resources. Another means of channelling the benefits of ecosystem services to the poor is through the reallocation of resource rents by the state to the poor. However, whether or not this reallocation of resource rents actually reaches the poor or benefits them often depends on a complex set of factors relating to the governance arrangements, power relations and institutional structures that constitute the political economy context (Robinson *et al.*, 2006; Konstantinidis, 2009).

Threats to poverty–biodiversity linkages

The Millennium Ecosystem Assessment estimates that ecosystem services from marine and coastal areas are deteriorating at a faster rate than other ecosystems (UNEP, 2006) and coastal areas are at greatest risk of large-scale unfavourable ecological changes (MA, 2005b). Major threats to coastal biodiversity and thus to ecosystem service flows to the poor include wider environmental change, habitat degradation, pollution and inappropriate resource use. However, the interconnectedness of the land–water interface in the coast creates challenges when attempting to attribute ecosystem service reduction to one driver (Silvestri & Kershaw, 2010).

Wider environmental changes include climate change and ocean acidification which are likely to be more severe in tropical coastal environments than in many other environments. Sea temperature changes are resulting in a change of composition of coastal species with some species migrating to remain within their temperature ranges (Cheung, 2009). This will remove some species and change the distribution patterns and seasonal availability of others (Silvestri & Kershaw, 2010) and not only will remove many harvesting opportunities from the poor but also may, in some situations, call into question the continued value of traditional ecological knowledge which has been so important to their livelihoods.

Surface temperature changes are also believed to be closely linked to coral-bleaching effects and these have been occurring with increased frequency since the 1980s (Parry *et al.*, 2007). Whilst ocean acidification changes are relatively predictable the effects on living organisms are less so (Royal Society, 2005). Coral reefs live within narrow ranges of temperature, light and aragonite (a form of calcium carbonate) saturation, outside of which reef formation becomes difficult. The earth has lost 19% of the original area of coral reefs; 15% are seriously threatened with loss within the next 10–20 years and 20% are under threat of loss in 20–40 years (Wilkinson, 2008). Phytoplankton and zooplankton are likely to be adversely affected by acidification and these are major sources of food for many species of fish and other animals (Royal Society, 2005) that coastal people depend upon.

Coral reefs have experienced significant ecological damage due to bleaching events, coral mining, increased coastal sediment loads from farming and deforestation, coastal pollution and damage from inshore vessels. Given the high levels of biodiversity associated with reefs, this has had a significant effect on coastal biodiversity in some areas. In addition, coastal deforestation and conversion of wild areas into agricultural land have significantly affected the composition of coastal biodiversity (Burke *et al.*, 2001).

Climate change–induced increased storm occurrence, raised sea levels, on-shore droughts, increased glacial melt, changed river flows and flooding patterns are all likely to affect both biodiversity and access to biodiversity-related ecosystem services by the poor. Inundation by saltwater into coastal areas is likely to affect large areas of coastal wetlands and salt intrusion will affect the productivity of coastal farmlands (Intergovernmental Panel on Climate Change Secretariat (IPCC), 2007). Tropical cyclone hazards are on the increase and these affect around 120 million people in coastal and low-lying areas annually (IPCC, 2007). Raised sea levels are likely to have major effects on low-lying coastal areas (Nicholls, 2002; McGranahan, 2006, 2007) which are often occupied by the poor (Blakie *et al.*, 1994). Increased coastal erosion and flood damage to poor people's land and homes, and transport limitation and poor access to flood warnings increase threats to the poor (Blakie *et al.*, 1994). They also have fewer reserves and often limited social capital to help rebuild their lives after a disaster.

There has also been an increase in the effects of human exploitation of coastal resources. Mangrove forests have declined by 35% in the last two decades (MA, 2005b), largely because of aquaculture development and harvesting for fuel wood and building materials (MA, 2005b).

The FAO (2008) estimates that in 2007, 52% of monitored fish stocks globally were fully exploited and a further 28% were overexploited, depleted or recovering from depletion leaving little room for expansion of wild-caught fish. This has major implications for the food security and nutrition, employment opportunities and incomes of coastal fishers and fish processors. The movement of fishing effort further offshore in search of unaffected resources is changing where fish are landed, what species are landed, the ownership of harvesting capital and access to those fish by fish processors and traders who are often women (IMM & Integrated Coastal Management (ICM), 2003).

Pollution, especially increased nitrogen inflows from agriculture, has had significant effects on coastal eutrophication in some areas resulting in high levels of algal growth and reducing light penetration. Most of the world's major cities (greater than 500,000 people) are located within 50 km of the coast and have a significant effect on pollution loads. Abstraction of freshwater for human consumption, industrial use and agriculture, and changes in flood patterns because of dam construction, are also affecting coastal environments.

These are only some of the impacts that changes in the ecosystem services generated by coastal biodiversity will have on the poor. The poor often have the least capacity to adapt to changes in their environment and are most severely affected.

Conservation and its impact on the poor

The dual pressure applied to both biodiversity and the livelihoods of dependent communities has prompted an array of conservation initiatives which claim to address both. Such conservation measures can be broadly summarised as: limiting access rights, enhancing resource-dependent livelihoods to become more sustainable over time, compensating for livelihood loss and seeking livelihood alternatives.

Access limitations include measures such as restricted-entry fisheries, and marine and forest reserves. Fisheries management measures, where the poor have been fully included in the process, have the potential to provide sustainable benefit flows to those communities. However, the combination of complexity and urgency to find fisheries management solutions has created a market for quick and technical solutions to management problems which all too often lead to tunnel vision (Degnbol & McCay, 2006). Given the increases in population pressure in coastal areas, these management measures need to be balanced by the creation of wider alternative livelihood options if they are to be successful in the long term (Campbell *et al.*, 2006). If access favours the wealthier operators, then distributional issues need to be addressed. Marine protected areas are now widely recognised as having considerable potential to provide conservation benefits (e.g. Gell & Roberts, 2002) but the livelihood benefits for the poor remain questionable (Sunde & Isaacs, 2008). All too often, fishing communities are seen as homogeneous groups and little effort is made to differentiate the benefit flows to specific groups of people within those communities, especially to the poor. Likewise, a lack of understanding of how biodiversity generates ecosystem services for different groups of poor people, at different times of the year or at different stages of their lives, can result in using oversimplistic indicators of success.

Other approaches have included engaging communities in integrated coastal area management, community management systems or addressing specific exploitation methods such as improved fishing gear, seasonal closures or catch size restrictions (Cunningham & Bostock, 2005). These may have had some localised benefits, but again the extent to which benefits have flowed to the poor is not well documented. Given the limited capacity for engagement in formal management systems, their limited voice to influence decisions and their inability to pay for the benefits that inclusion affords, they are unlikely to benefit from such schemes. In many cases, legislation to support such changes is likely to criminalise the livelihoods of the poor rather than to help them (Campbell *et al.*, 2006).

Compensating people for the loss of access to ecosystem services is growing in use and includes fisheries decommissioning schemes and payments for ecosystem services (PES). Whilst decommissioning schemes have the potential to reduce overall fishing pressure, they all too often fail for reasons linked to the political economy (Organisation for Economic Co-operation and Development, 2009). There are few documented cases where decommissioning schemes have been shown to work for the poor. As with many such measures, enforced compliance of re-entry restrictions by fishers is difficult to achieve. PES schemes have been used as a mechanism to incentivise good forest resource management (Chapter 14, this volume), but efforts specific to coastal areas are so far limited. To be effective they will, in the main, need to be used in association with other mechanisms (such as closed access systems or alternative livelihood schemes), otherwise they may encourage occupational migration into resource harvesting (Pagiola, 2008). Marine reserve use fees, fishing nursery protection charges and tourism levies offer opportunities for paying local communities for maintaining coastal resources in good condition, but the processes of linking buyers and sellers of ecosystem services often involve payments moving through formal and informal systems where political interference can redirect those funds. In a socially differentiated society where the poor have a limited say in decision making and few formal rights over resources, they will find it extremely difficult to benefit from such systems (Chapter 14, this volume).

Developing livelihood alternatives for poor people has often been seen as a solution to conserving coastal resources whilst achieving poverty reduction (WorldFish Center, 2008). The success of these efforts has been rather low (Campbell *et al.*, 2006). All too often, alternative livelihood initiatives are components of wider environmental programmes where they are used simplistically to address very complex change processes. Very few systematic approaches have been developed to support livelihood change for communities dependent on natural resources (Campbell, 2008).

The perceived degree of success of coastal resource management measures depends on who measures it, how it is measured and what is measured. Where the poor are excluded from that process, or success is measured against externally defined criteria, the significance of the results is likely to be questionable. Overall the process of coastal ecosystem management has not convincingly demonstrated its success in achieving both conservation and poverty reduction aims – especially at any scale on a sustained basis.

The factors influencing livelihood responses

Part of the difficulty in achieving sustainable benefit flows from ecosystem services to the poor in coastal environments is due to the complexity of this interface. This is characterised by complex feedback mechanisms and non-linear cause-and-effect relationships that confuse and confound responses: people learn, they react to knowledge and experience, they compete and they make choices in unpredictable ways, often based on perceptions of risk and discount rates which change daily.

This is further compounded by the interaction of the political economy with that interface. The lack of clarity over the rights and responsibilities of the multitude of users in this dynamic and changing environment creates opportunities for the poor to be excluded and marginalised. One of the anomalies of coastal development, which is seen in so many parts of the world, is that the poor are all too often unable to take up development opportunities that present themselves and are squeezed out in the process. The development of coastal tourism is an example where many of the job opportunities are taken up by outsiders leaving the local poor in a worse situation. This has been called the *development exclusion paradox* (Campbell *et al.*, 2006) where much of the poverty becomes interstitial poverty (Jazairy *et al.*, 1992) which is obscured from sight by the development processes which surround it, and is thus difficult to respond to. This paradox serves to highlight how, beyond the more direct linkages between the poor and coastal biodiversity and ecosystem services, the political economy surrounding these linkages plays a critical role in influencing how they develop and how they are likely to change over time. Understanding and addressing this political economy dimension are both key if attempts to maintain the flow of benefits from coastal biodiversity to the poor are to have any chance of success.

Even where the poor are visible and efforts are made to engage with them, their livelihoods and the factors that determine how they respond to their environment and to changes in their relationship to it are very complex (IMM *et al.*, 2005). The Sustainable Livelihoods Approach (SLA) was developed to begin to understand the complexity of poor people's livelihoods (Carney, 1998) and has proved to be very useful in this respect. The application of the SLA to coastal communities has demonstrated the large number of factors that determine and influence those livelihoods and livelihood choices (IMM *et al.*, 2005). These complexities need to be fully understood within the local context if effective management and development strategies are to be developed that help the poor to more sustainably use their coastal biodiversity and emerge from poverty in the face of a massive coastal change process.

Future directions

Understanding and responding to the complexity of the system that makes up the interface between biodiversity, ecosystem services, livelihoods and the political economy have comprised an ongoing challenge and one that is becoming more difficult as major external changes impact the coast. Attempts to break down these complex relationships into their component parts at a sectoral or sub-sectoral level fail to recognise the interconnectedness of the system, the interdependence of the variables that make it up, the feedback loops that influence outcomes and the emergent behaviour which influences the operation of the system. To positively influence the interaction between coastal biodiversity, coastal poverty and the ecosystem services that flow between them requires moving away from simplistic generalisations about the coastal poor and embracing the complexity which their interaction with coastal environments creates. In addition to addressing the complexity of interactions between biodiversity and poverty, there is a need to interact more directly with the political economy which influences distributional issues, decision making and prioritisation. New approaches must be adopted to understand and respond to coastal complexity. Understanding and using emergent behaviour, for example, might suggest that management measures focus more on developing participation, building trust, gaining commitment to change, internalising the political economy, enhancing communication and fostering common interdisciplinary discourses rather than just focussing on technical measures for conservation or social development.

Rather than focusing on even more narrow aspects of pro-poor conservation or separating conservation from poverty, the interrelationships between coastal biodiversity, poverty and ecosystem services need to be embedded in wider inclusive economic growth in the coast. These need to provide the opportunities for the poor to engage with viable livelihood alternatives that can make management measures work, and to increase the opportunity cost of their labour and other livelihood assets.

References

Allison, E. & Horemans, B. (2006) Putting the principles of the Sustainable Livelihoods Approach into fisheries development policy and practice. *Marine Policy*, 30, 757–66.

Blakie, P., Cannon, T., Davis, I. & Wisner, B. (1994) *At Risk: Natural Hazards, People's Vulnerability, and Disasters*. Routledge, London.

Brown, K., Daw, T., Rosendo, R., Bunce, M. & Cherrett, N. (2008) *Ecosystem Services for Poverty Alleviation: Marine & Coastal Situational Analysis*. Appraisal and Design Phase. UK Department for International Development (DFID) and UK Research Councils (ESRC/NERC), University of East Anglia Overseas Development Group, Norwich.

Burke, L., Kura, Y., Kassem, K., Revenga, C., Spalding, M. & McAllister, D. (2001) *Pilot Analysis of Global Ecosystems: Coastal Ecosystems*. World Resources Institute, Washington, DC.

Campbell, J. (2008) *Systematic Approaches to Livelihoods Enhancement and Diversification: A Review of Global Experiences*. IUCN, Gland, Switzerland; CORDIO, Kalmar, Sweden; and ICRAN, Cambridge.

Campbell, J. & Beardmore, J.A. (2001) Poverty and aquatic biodiversity. In *Living of Biodiversity: Exploring Livelihoods and Biodiversity Issues in Natural Resources Management*, ed. I. Koziell, & J. Saunders. International Institute for Environment and Development, London.

Campbell, J., Whittingham, E. & Townsley, P. (2006) Responding to coastal poverty: should we be doing things differently or doing different things? In *Environment and Livelihoods in Tropical Coastal Zones*, ed. C.T. Hoanh, T.T. Tuong, J.W. Gowing & B. Hardy. CAB International, London.

Cheung, W.W.L., Lam, V.W.Y., Sarmiento, J.L., Kearney, K., Watson, R. & Pauly, D. (2009) Projecting global marine biodiversity impacts under climate change scenarios. *Fish and Fisheries*, 10, 135–51.

Cunningham, S. & Bostock, T. (2005) *Successful Fisheries Management: Issues, Case Studies and Perspectives*. SIFAR and World Bank Study of Good Management Practice in Sustainable Fisheries, Eburon Academic Publishers, Delft.

Degnbol, P. & McCay, B. (2006) Unintended and perverse consequences of ignoring linkages in fisheries systems. *ICES Journal of Marine Science*, 64, 793–97.

Food and Agriculture Organization (FAO) (2008) *The State of World Fisheries and Aquaculture*. FAO Fisheries and Aquaculture Department, Rome.

Gell, F. & Roberts, C. (2002) *The Fishery Effects of Marine Reserves and Fishery Closures*. World Wildlife Fund, Washington, DC.

Innes, J. (1996) Developing the linkages: indigenous involvement in the management of the Great Barrier Reef World Heritage Area. *Reef Research*, 6, 1–4.

Integrated Marine Management (IMM) & Integrated Coastal Management (ICM) (2003) *Major Trends in the Utilisation of Fish in India: Poverty-Policy Consideration*. Output of the India Fish Utilisation and Poverty Project, RRNRRs, UK Department for International Development (DFID), London.

Integrated Marine Management (IMM), Community Fisheries Development Office (CFDO) & Community-Based Natural Resource Management Learning Institute (CBNRM LI) (2005) *Understanding the Factors that Support or Inhibit Livelihood Diversification in Coastal Cambodia*. Output from DFID-funded research in Cambodia, IMM Ltd, Exeter.

Integrated Marine Management (IMM), Integrated Coastal Management (ICM) & Forum for Integrated Rural Management (FIRM) (2003) *Policy, Institutional and Informational Factors Affecting Poverty in Three Coastal Communities in Andhra Pradesh*. Sustainable Coastal Livelihoods Project Report on Phase 2 Socio-economic Research. UK Department for International Development (DFID), London.

Intergovernmental Panel on Climate Change Secretariat (IPCC) (2007) *Fourth Assessment Report: Climate Change 2007*. IPCC, Geneva.

Jazairy, I., Alamgir, M. & Panuccio, T. (1992) *The State of Rural Poverty: An Inquiry into Its Causes and Consequences*. New York University Press and International Fund for Agricultural Development, New York.

Johannes, R.E. (1981) *Words of the Lagoon: Fishing and Marine Lore in the Palau District of Micronesia*. University of California Press, Berkeley.

Lokani, P. (1995) An oral account of overfishing and habitat destruction at Pororan Island, Papua New Guinea. *South Pacific Commission and Forum Fisheries Agency Workshop on the Management of South Pacific Inshore Fisheries*, 1, 251–63.

Konstantinidis, N. (2009) *The Political Economy of Resource Rent Distribution*. IBEI Working Papers 2009/19. CIDOB ediciones, Barcelona.

McAllister, D.E. (1991) What is the status of the world's coral reef fishes? *Sea Wind*, 5, 14–18.

McClanahan, T.R., Glaesel, H., Rubens, J. & Kiambo, R. (1998) The effects of traditional fisheries management on fisheries yield and the coral reef ecosystem of southern Kenya. *Oceanographic Literature Review*, 45, 561–2.

McGranahan, G., Balk, D. & Anderson, B. (2006) Low coastal zone settlements. *Tiempo*, 59, 23–26.

McGranahan, G., Balk, D. & Anderson, B. (2007) The rising tide: assessing the risks of climate change and human settlements in low elevation coastal zones. *Environment and Urbanization*, 19, 17–37.

Millennium Ecosystem Assessment (MA) (2005a) *Ecosystems and Human Well-Being: A Framework for Assessment*. Island Press, Washington, DC.

Millennium Ecosystem Assessment (MA) (2005b) *Ecosystems and Human Well-Being: Biodiversity Synthesis*. Island Press, Washington, DC.

Nicholls, R.J. (2002) Rising sea levels: potential impacts and responses. In *Global Environmental Change*, ed. R. Hester & R.M. Harrison, pp. 83–107. Issues in Environmental Science and Technology No. 1. Royal Society of Chemistry, Cambridge.

Organisation for Economic Co-operation and Development (OECD) (2009) *Reducing Fishing Capacity: Best Practices for Decommissioning Scheme*. OECD, Paris.

Ormond, R.F.G. & Roberts, C.M. (1997) Biodiversity of coral reef fish. In *Marine Biodiversity: Patterns and Processes*, ed. R.F.G. Ormond, J.D. Gage & M.V. Angel, pp. 216–57. Cambridge University Press, Cambridge.

Parry, M.L., Canziani, O.F., Palutikof, J.P., van der Linden, P.J. & Hanson, C.E. (eds.) (2007) *Climate Change 2007: Impacts, Adaptation and Vulnerability. Contribution of Working Group II to the Fourth Assessment Report of the Intergovernmental Panel on Climate Change*. Cambridge University Press, Cambridge.

Rengasamy, S., Devavaram, J., Prasad, R. & Arunodaya, E. (2003) A case study from the Gulf of Mannar. In *Poverty and Reefs: Volume 2 Case Studies*, ed. E. Whittingham, J. Campbell & P. Townsley, pp. 113–46. UNESCO, Paris.

Robinson, J.A., Torvik, R. & Verdier, T. (2006) Political foundations of the resource curse. *Journal of Development Economics*, 79, 447–68.

Royal Society (2005) *Ocean Acidification Due to Increasing Atmospheric Carbon Dioxide*. Royal Society, London.

Ruddle, K., Hviding, E. & Johannes, R.E. (1992) Marine resources management in the context of customary tenure. *Marine Resource Economics*, 7, 249–73.

Silvestri, S. & Kershaw, F. (eds.) (2010) *Framing the Flow: Innovative Approaches to Understand Protect and Value Ecosystem Services across Linked Habitats*. UNEP World Conservation Monitoring Centre, Cambridge.

Small, C. & Nicholls, R.J. (2003) A global analysis of human settlement in coastal zones. *Journal of Coastal Research*, 19, 584–99.

Sunde, J. & Isaacs, M. (2008) *Marine Conservation and Coastal Communities: Who Carries the Costs? A Study of Marine Protected Areas and Their Impact on Traditional Small-Scale Fishing Communities in South Africa*. Samudra Monograph. International Collective in Support of Fishworkers, Chennai.

United Nations Environment Programme (UNEP) (2006) *Marine and Coastal Ecosystems and Human Well-Being: Synthesis*. UNEP, Nairobi.

Whittingham, E., Campbell, J. & Townsley, P. (2003) *Poverty and Reefs: Volumes 1 and 2*. DFID-IMM-IOC/UNESCO, Paris.

Wilkinson, C. (2008) *Status of Coral Reefs of the World 2008*. Global Coral Reef Monitoring Network, Townsville, Australia.

WorldFish Center (2008). *Waves of Change: Lessons Learned in Rehabilitating Coastal Livelihoods and Communities after Disasters*. WorldFish Center, Penang.

Worsley, P. (1997) *Knowledges: What Different Peoples Make of the World*. Profile Books, London.

7

Linking Biodiversity and Poverty Alleviation in the Drylands – The Concept of 'Useful' Biodiversity

Michael Mortimore

Independent Consultant, Drylands Research, Somerset, UK

Introduction

It is more than seven decades since the Crown Agents for the (British) Colonies published *The Useful Plants of West Tropical Africa, an Appendix to the Flora of West Tropical Africa* (Dalziel, 1937). The magnitude, conscientiousness and accuracy of this compilation have astonished generations of its readers, notwithstanding some taxonomic revisions that have since taken place. Its author, John McEwen Dalziel, was formerly a member of the West Africa Medical Service, a background that clearly drove his interest in a vast number of indigenous medical treatments and nutritional benefits based on plants. Now succeeded by a database at Kew Gardens, the work contained thousands of names in local languages as well as information collected from field tours. Countless researchers have cause for gratitude to Dalziel for his achievement.

My reason for beginning with Dalziel is that – driven by an official policy to inventory the natural resources of every colony – the concept of 'usefulness' played a key role in the undertaking. The idea of the ecosystem received no explicit attention at a time when the science of ecology was dominated by Clementsian succession theory and sustainability had not yet entered the agenda. However, a species-by-species inventory couched in functional terms reflected accurately – as it still does – the bank of indigenous knowledge whose custodians were (and still are) the women and men who inhabit the thousands of villages in the region, and whose decisions determine

Biodiversity Conservation and Poverty Alleviation: Exploring the Evidence for a Link, First Edition.
Edited by Dilys Roe, Joanna Elliott, Chris Sandbrook and Matt Walpole.
© 2013 John Wiley & Sons, Ltd. Published 2013 by John Wiley & Sons, Ltd.

the conservation and exploitation of biodiversity. And, since value is placed on useful plants, these managers have good reason to conserve them. Sometimes value is measured in monetary terms, and sometimes in collection effort.

The Sahel sits at the heart of the so-called *desertification paradigm* in which biodiversity loss is believed to be associated closely with 'deforestation', 'overcultivation', 'overgrazing' and salinisation of irrigated soils – the four horsemen of the apocalypse in 'doomsday scenarios' of ecological degradation (Millennium Ecosystem Assessment (MA), 2005). But is biodiversity loss a necessary outcome of rural population growth, the impact of markets, the conversion of woodland to farmland, the intensification of agriculture and mobile pastoralism? Fieldwork in several Sahelian countries (Northern Nigeria, Niger and Senegal) convinces me that destructive exploitation of biodiversity is in no way intrinsic to small-scale farming, herding, hunting or collecting systems even under conditions of rapid population growth, and even in arid or semi-arid ecosystems, but is better understood as a distortion of processes of transition from wild to domesticated landscapes (Mortimore & Adams, 1999; Faye *et al.*, 2001; Mortimore *et al.*, 2001, 2005). The conservation of biodiversity should not, therefore, be seen as an external agenda necessarily imposed (for the sake of the planet) on profligate short-term resource users but should instead build on indigenous values and knowledge and seek to control such distortions. These may include, for example, barriers to secure access and benefits or perverse market incentives.

The time when 'fortress conservation' was a realistic option in West African dryland ecosystems has passed. Instead, the benefits of 'managed' biodiversity, within ecosystems that are already appropriated and used at varying degrees of intensity by human populations, deserve to be understood and supported (Adams, 1996). This is particularly relevant in the world's drylands, with their large human and livestock populations, environmental and climatic variability and economic risk. Such co-habitation of humans with nature is consistent with the idea of complex systems co-evolving and interacting over short and long time scales (Reynolds *et al.*, 2007). This chapter uses examples of local ecosystem service management in the Sahel to provide policy lessons from biodiversity management in ways that are consistent with human well-being – in both the short and long term.

Biodiversity in the Sahelian context

The semi-arid land of northern Nigeria and southern Niger is home to Hausa, Fulani (Peulh) and Manga communities practising farming with livestock, and visited by nomadic and specialist livestock herders. Suffering declining rainfall from the mid-1960s to the mid-1990s, and prone to unpredictable droughts and livelihood risk, this region is characterised by a high incidence of poverty, rapid population growth and rural population densities ranging from very low to more than 200/km² in some quite extensive areas.[1] Biodiversity is rooted in a diversity of ecosystems with different rainfall regimes and locally variable soils (Brookfield, 2001). The Sahel is no exception, within an all-important south-to-north aridity gradient, whereby average rainfall decreases from dry sub-humid through semi-arid to arid ecologies. In addition, rainfall in the Sahel is characterised by high variability, both in space and in time.

Over large areas, the coefficient of annual rainfall exceeds 30%. In combination with a variety of soil-forming processes and parent materials, and pronounced seasonality, this makes for a mosaic of micro-environments and natural plant communities, notwithstanding the diminution of biological activities during the long dry season (6–9 months). The studies on which this chapter is based (see Figure 7.1) follow the aridity gradient: two villages in Nigeria along the south-to-north rainfall gradient from 571 mm per year to 345 mm per year, and three in Niger following the gradient down to 250 mm per year (Mortimore & Adams, 1999; Mortimore *et al.*, 2006, 2008).

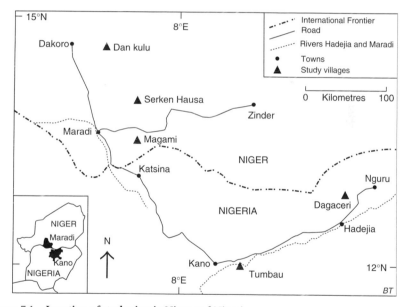

Figure 7.1   **Location of study sites in Niger and Nigeria.**

---

[1] The best known though by no means the only example is the Kano Close-Settled Zone of Nigeria, where a rural population of more than 10 million lives at an average density in excess of 140/km² (Mortimore, 1993).

Agro-diversity and 'wild' biodiversity

The case study sites exhibit high levels of agricultural biodiversity. At the Nigerian sites, 25 cultivars (cereals, beans, earth nuts, root crops, melons, vegetables, sugar cane, cotton and kenaf) were inventoried with the number of named and maintained landraces[2] ranging from 76 in the wetter site (including irrigation) to 23 in the drier (no irrigation). The greater part of this agro-diversity was found in the indigenous cereal grains: pearl millet (ranging from 12 down to 3 landraces) and sorghum (from 22 to 6). This may represent only a small proportion of the total number of genetically different strains – even though many landraces have the same names, and individual farmers manage them in different ways and select for different characteristics that suit their individual needs. Furthermore, cross-pollination with wild relatives sometimes occurs. Popular cereal landraces, their seed selected for their early-maturing, high-yielding, drought-resistant or other characteristics, occupy most of the cultivated land each year. But the stocks of 'old' and less popular landraces may be maintained by elderly people. By such selection, each farmer may be managing a separate 'gene pool' (Busso *et al.*, 2000). This may include descendants of improved varieties introduced from research stations.

Changes in landraces in southern Niger have been attributed to adaptation to declining rainfall. Between 1960 and 2000, at our four sites, farmers reported up to seven lost millet varieties and similar numbers for sorghum; but they also reported up to eight new varieties. The amount of change was greater in the more humid places, which had a wider gene pool. Similar adaptation occurred with bean and groundnut varieties. Thus the heritage of agro-diversity is dynamic, adaptive and specific to place and time, maintained by selection and controlled crossing in the face of an uncontrolled 'genetic anarchy' in the surrounding wild vegetation. Landrace erosion is not a new phenomenon (Brookfield, 2001), and neither is adoption or hybridisation in the interests of maintaining a 'useful' biodiversity.

The linkages between agro-diversity and poverty are ambiguous. On one hand, it is clear that manipulation of genetic diversity using indigenous knowledge plays a key role in adaptive agriculture based on the production of food staples such as pearl millet and sorghum. The capital costs of such genetic manipulation (in which intended and unintended effects are hard to distinguish) are low. Local farming practices can thus sustain poor families in a variable and challenging environment. On the other hand, an increase in farm incomes (development?) is often associated with a move into growing highly marketable and improved cultivars or varieties, together with an expansion of farm holdings, increased capital investment and additional hired labour. If such

---

[2] A *landrace* is a population that is named and maintained under local farming practices, and adapted to local agro-climatic conditions. It is variable between and within sites, is best known to farmers themselves and is not easily defined taxonomically (Zeven, cited in Busso *et al.*, 2000; Sauer, cited in Brookfield, 2001).

changes are accompanied by dependence on commercial seed supplies, it may herald redundancy for local agro-diversity – which may in turn have longer term implications for poverty, particularly in the context of climate change.

Beyond this agro-diversity the Sahel is also home to a rich 'wild' biodiversity. Even in the most intensively cultivated and most densely populated of the sites, 86 named types of grass and herbs were identified. Weeds, live hedges and pathway vegetation are assiduously collected during the rainy season for stall-fed ruminants. Nothing is wasted. Crop residues and tree foliage are similarly recycled through animals, whose dung – of course including the seeds – is redistributed to the fields. There are no baseline data to permit a quantitative analysis of change over time, but the abundance of named types suggests that impoverishment need not result from low external input intensification, and that many of the types identified have value.

An association between such wild, field and fallow biodiversity and poor people arises from the role it plays in drought-prone areas where household food security is periodically at risk. A wide range of plants in semi-arid ecosystems yield edible products (mostly leaves, but also roots, flowers or seed). Finding these during the 'hungry season' (when the grain stores are empty and the new crop is not yet harvested) was often women's work, and they conserved and passed on the knowledge between generations (Box 7.1).

Box 7.1 **Biodiversity and famine foods**

An inventory of 'famine foods' used during the Sahel Drought of 1972–4 was carried out in 125 households in five villages distributed widely in (then) Kano State (Mortimore, 1989) and yielded 47 species of herbs, grasses, trees and shrubs used commonly or less commonly as food sources. Other famine foods, known but not recorded in use at that time, extended the list to 68.

In a later study at one site (midway along the transect, with 330 mm rainfall per year, 1992–6), within an inventory of 121 useful plant species, 53 species yielded over 60 foods, 19 of them consumed only in times of scarcity (Harris & Mohammed, 2003). These foods are available during the pre-rainy season (5 foods), the rainy season (19), the harvest season (30) and the dry season (22). They cannot substitute for the cereal grains or cassava, but many supply vitamins or other substances of dietary significance, used to supplement the daily diet, generate income or substitute for meals.

In the Sahel, *poverty* is locally defined in terms of incapacity to cope with food emergencies. The role of 'wild' biodiversity as a 'safety net' is thus clear enough. Moreover, it may be threatened by land use change if the common woodland and grassland, or agricultural fallows on which many wild foods are found, are reduced. Some wild foods are also marketable, and with other non-timber forest products (NTFPs), efforts to sell them may be intensified during food emergencies. But very few opportunities exist with which to escape from poverty.

Biodiversity and rangeland degradation

A malign connection between degradation in grassland ecosystems and overgrazing by livestock is still perceived by many observers of events in drylands – and biodiversity loss is often assumed as an inevitable outcome. But rangeland ecosystems do not vary in accordance with an equilibrium model of biomass production and carrying capacity. Instead, driven by spatial and temporal variability in rainfall, both within and between growing seasons, they are highly variable and unpredictable both in growth and in species composition. The producers' strategy within such non-equilibrium systems is to constantly move livestock around – in response to the changing conditions in different locations – thus "exploiting environmental heterogeneity rather than attempting to manipulate the environment to maximise stability and uniformity" (Behnke *et al.*, 1993) (Box 7.2).

Box 7.2 **Mobility and biodiversity in the Sahel – the case of the WoDaaBe pastoralists**

The WoDaaBe nomadic pastoralists who live in the northern part of our study area have been intensively studied with regard to their grazing management and their breeding practices (Kratli, 2008a; International Institute for Environment and Development and SOS Sahel, 2010). The herders know where to take their cattle in order to take advantage of new growth at its optimal stage of development and how to stimulate feeding to get the most nutritious diet from the range. The cattle select from more than 40 species, including grasses, shrubs and trees, and prefer certain parts of the plant. Thus both mobility and biodiversity in the rangelands are crucial for feeding, and good feeding enables good reproductive performance. Learning – and not merely genes – is transmitted between generations of Bororo cattle as well as acquired through trial and error. The variability of the range is thus not viewed as a risk, constraint or 'environmental problem' but rather as a resource to be exploited.

Generalisations linking grazing (mis)management with loss of either biomass or biodiversity are simplistic. Biodiversity – as well as biomass – is a sought-after property in rangelands. Together with cattle-breeding skills, herd mobility and local knowledge, the exploitation of diversity in the range (including some trees) contributes to building herds and thereby incomes. Livestock production systems are labour and knowledge intensive; viable alternative systems do not exist. The loss of diversity would certainly increase poverty, and poverty reduction will continue to depend on incomes derived from marketing livestock products.

Trees, shrubs and 'deforestation'

Alongside land degradation, deforestation is also generally assumed to reduce biodiversity as well as woody biomass. Land use change (woodland clearance for farming) is the main agent of deforestation. The simplification 'More farmers, less forest' led the colonial forest departments in both Niger and Nigeria to assume the role of protectors (through 'gazetting' forest reserves and licensing the cutting of trees on privately owned farmland) and restorers (through plantations). A further confident prediction was that fuel wood cutting would replace both natural woodland and farmland trees, for example around the city of Kano (Trevallion, 1965). In response, the ownership of trees was alienated by the forest administration, and incentives for conservation were undermined (Cline-Cole *et al.*, 1990). The prediction that fuel wood demand would result in forest clearance failed to materialise, however (Box 7.3).

Box 7.3 **Biodiversity conservation on Kano farms**

During the major drought cycle of 1972–4, and again in 1983–5, tree mortality in the Kano region increased from moisture stress. The urban market for fuel wood had grown to over a million consumers, and selling branch wood was an important option for food-starved farming households. Yet air photography from March 1972 and September 1981 shows that in a study area of $190\,km^2$, using a random sample of 0.6% of the area, the density of tree canopies per hectare increased from 12.9 to 15.2 (an increase of 17.8%). Furthermore ground sampling in 1983–4, and controlling for season, showed that the air photographic interpretation *under*counted, because some trees growing in thickets were not always differentiated. In 1990, 75 tree species growing on farmlands and in forest reserves in the Kano Close-Settled Zone were inventoried, the great majority of them native to this or adjacent ecosystems (Nichol, in Cline-Cole *et al.*, 1990).

A transition can thus be observed over time in the management of farm trees. While woodland is abundant and natural biodiversity yields easily accessible NTFPs, woodland clearance for farming takes precedence and biodiversity conservation is not a priority. When the ratio of cultivated land to woodlands and fallows falls to a level at which valued products (food, fodder, medicine, craft and others) become marketable or costly to collect, the regeneration of useful trees is protected and adds value to farmland, which undergoes a gradual transition to farmed parkland. Key determinants are: consolidating individual tenure (of both land and trees); growing density of the farming population, and hence demand for land; fertilisation strategies for compensating the loss of fallows and market values of NTFPs. In time, the planting of exotic and indigenous species becomes attractive especially if nurseries are established. Common woodland becomes restricted to vestigial or infertile land.

A prior condition of this transition is the farmer's secure access to and benefits from biodiversity. There is a correlation between the emergence of private rights and the increasing value of land under conditions of more frequent cultivation and fixed boundaries. During the transition, the ownership of biodiversity tends to become privatised, though this process is far from complete even in the densely populated Kano region. Poverty reduction appears likely to correlate with the transition to protected biodiversity, but only for those who enjoy secure land rights (Box 7.4).

Box 7.4 **Biodiversity conservation in transition in Niger**

In southern Niger, cultivated land cleared from woodland by waves of incoming Hausa farmers (1920s–1970s) accounted for 59% of land area in 1975 and 73% in 1996 (Mahamane, 2001). Many indigenous species became scarce, while exotics promoted by development projects were planted around towns and villages (especially *Azadirachta indica*). By the 1990s, however, a conservation ethos was developing, focused on the practice of *défrichement amélioré* (protecting the regeneration of selected species on newly cleared farmland) (Jouet et al., 1996). This was promoted by development projects in the area, which confirmed the value of the local practice. The evolution of a *code rurale* governing rights to land (essentially based on the concept of 'rights to the tiller') motivated farming households once again to assume responsibility for trees on their farms. In 2005 new studies based on remotely sensed data announced extensive tree growth on up to 500,000 km^2 of farmland across southern Niger (Reij et al., 2006). This consisted mainly of spontaneous regeneration of the native species, *Faidherbia albida*, protected by the farmers.

Faunal diversity – livestock

As we have seen, nomadic pastoralism (as exemplified in the WoDaaBe of Niger) makes intensive use of biodiversity in rangeland and of structured diversity in livestock herds in order to exploit the heterogeneity found in the natural environment. As many animals also belong to the farmers (or agro-pastoralists), whose livelihoods and production systems extend across the full range of our environmental transect, diversity also characterises these animal production systems.

A standard typology is lacking in common usage; what is now known of the exploitation of diversity in breeding Bororo cattle of the WoDaaBe suggests that a traditional approach to characterising breeds may be beyond use, except for the very few breeds having broad regional distributions (e.g. Bororo, Azawak and White Fulani cattle; and Maradi or Sokoto Red goats). When asked about local perceptions of within-breed diversity, farmers emphasise social or economic values (Box 7.5). For example, in one Nigerian village cattle can be described in eight colour classes, in four classes based on shape and size of horn, or in only two based on body weight. In the same village, six types of sheep are recognised on the basis of various characters, but only two in terms of size. Six types of goat are recognised by their colours, though in two other villages only one type is recognised, the ubiquitous red goat. In the most arid of the villages, where livestock are the basis of the economy, a more coherent typology of two cattle, five sheep, five goat and five camel types was offered (Mortimore *et al.*, 2008).

Box 7.5 **Diversity and cattle breeding: the Bororo of Niger**

There is a two-way relationship between the animals' diet, as manipulated by the herders and as selected by the cattle, and a highly controlled breeding system based on matriarchal lines – which give each animal its identity – and the use of selected bulls (most male progeny are sold). The aim is to achieve reliability in good performance under the prevailing conditions, and success is attributed to the line of descent. Productivity, measured in quantitative terms, is not the prime goal. In this human–animal–rangeland circle, men, women and children live in intimate relationship with the cattle. The resilience of the system was borne out in the drought cycles of the 1970s and 1980s, when the Bororo Fulani survived conspicuously better than other pastoralists in the region (Kratli, 2008b).

Diversity also characterises the fodder plants used. In the village closest to the city of Kano (high population density and intensively farmed), where fattening cattle or sheep for the market is becoming so important that it challenges crop production as a source of income, fodder strategies are finely tuned. The preferred fodder is crop residues, following harvest, but at other times of the year, hedgerow plants, vegetation on fallow fields and weeds collected from farmers' fields are important. Together with the use of animal manure, the collection of fodder closes the nutrient cycle which is broken by only a relatively small quantity of marketed biomass (a fraction of cereal grain production and legumes) (Harris & Yusuf, 2001).

At lower population densities and under drier conditions, there is more residual shrub woodland but preference is still given to nutrient-rich crop residues when available. In all the villages, there is some mobility, sometimes involving short journeys to perennial river valleys, sometimes longer distance transhumance to more southerly locations in Nigeria. It is known that livestock ownership has shifted significantly since the Sahel Drought in favour of farmers (driving a trend from cattle to small ruminants at the macro scale), and the benefits of animal ownership are realised not only by the better off farmers but also by the poor, by women and even by children, for whom ownership of one or two animals is widely desired. As the farming systems intensify their use of labour, and convert woodland into farmland, the quality and the quantity of diverse fodder sources – partly a product of their biodiversity – increase in importance in poor peoples' livelihoods and offers a pathway, if not out of poverty, then towards its amelioration.

Biodiversity conservation with poverty reduction?

Given that biodiversity is intrinsic to Sahelian pastoral and agro-pastoral systems, one may say that, given the present state of knowledge, it is difficult to conceive an alternative, sustainable mode of exploitation. What kind of development can deliver an enabling environment for biodiversity conservation along with poverty reduction and enhanced well-being?

This case study in Nigeria and Niger provides multiple illustrations of biodiversity services that are valued by dryland peoples, though not necessarily in monetary terms. These values play a critical role: firstly, in underlining land use rationales; secondly, in justifying local knowledge and motivating innovation; thirdly, in adapting to changing environments and, fourthly, in gaining recognition for drylands in policy, for example by inclusion in national accounting (IUCN, 2009). Investing in dryland biodiversity is therefore essential to support the flow of 'Nature's benefits' to poor peoples' livelihoods (International Livestock Research Institute, 2007). Otherwise, their vulnerability to variable ecology, markets and employment opportunities will increase.

The biodiversity found in a given ecosystem is the result of not only ecology but also management. Poor people are themselves managers. Investment does not imply – as is often assumed – corporate (commercial) donor and public sector investment. Corporate investment – for example, in biofuel, industrial crops or food production – is necessarily large in scale, dependent on fossil fuel and agro-chemicals and based on mono-cropping. It has a high and negative impact on biodiversity. Indeed, perhaps the best way to minimise poverty is to secure biodiversity from further damage by economic interests often beyond the users' control.

Dryland peoples themselves have investment strategies, born of the necessity of passing on viable enterprises to the next generation. They are generally ignored because they are small in scale, intermittent, and private – the savings of poor people. They invest labour, skills and capital in such improvements as livestock, soil and water conservation and drainage, tree planting, fencing and reseeding. Such investments, made within the framework of existing land use systems, are by and large non-threatening to plant biodiversity (though farming intensification is unfriendly to wild animals).

Arguing from a biodiversity conservation perspective, therefore, a different business plan is needed for dryland investment. Instead of large-scale commercial operations, directly controlled by corporations (or the state), a model based on micro-investments by small farmers or livestock producers is more likely to protect biodiversity within a framework of agricultural intensification. There may still be a role for corporations, but as intermediaries in service provision to small and poor producers. The already noticeable drift towards dual-sector development, with many African countries granting privileged status to foreign investors, may be a help to government revenues but cannot be good news for biodiversity.

Where external actors really make a difference is by improving access by poor resource users to profitable markets. A radical change is needed from a view that conservation is best achieved by protecting ecosystems from the market to an acceptance that ecosystem services are more likely to be sustainably managed if they have value for market production – and biodiversity is such a service. There are documented cases from African drylands where market access has encouraged more sustainable agriculture (Tiffen *et al.*, 1994; Ariyo *et al.*, 2001). New or rapidly growing niche markets are emerging for natural products (e.g. in southern Africa), with new value chains linking small producers with national or global markets.

Given the complex interactions of co-evolving human and ecological systems, it seems essential that policies and interventions, whether those of governments or NGOs, should (i) recognise the managers, their priorities, capacities and constraints, and (ii) empower them with knowledge enhancement, opportunities, market links and incentives to invest, in the interest of creatively co-managing a 'useful biodiversity'. The negotiation of bylaws, local agreements or 'conventions,' governing the rights of all stakeholders to the use of contested natural resources, together with the decentralisation of governance, has positive implications for protecting useful biodiversity. It should not be assumed that local communities, in exercising their reclaimed autonomy, will value all species equally. But local institutions offer a better way forwards than draconian impositions of the state, for example indiscriminate bans on tree cutting, or allocations of land that disregard customary property rights.

Conclusion

Biodiversity (as a resource) is *intrinsic* to Sahelian agro-pastoral and pastoral systems, not merely as an omnipresent characterisation but also in terms of *functionality* in human-ecological systems. Biodiversity is not merely a property of a 'natural' ecosystem that exists (or did exist) independently from human systems, external to them and ripe for exploitation, harvesting or management. Rather, it is the product or outcome of co-evolving human and ecological systems over a very long time. Genetic management (crop breeding, animal breeding and wild plant protection and harvesting) is the key 'bridge' between human and ecological systems, as conceived in the Drylands Development Paradigm (Reynolds *et al.*, 2007).

The use of biodiversity in breeding, the management of biodiversity in grazing and the drawing of skills from local knowledge achieves a level of integration which goes beyond functionality – even to the aesthetic – and must be extremely ancient. The continuity of the system, as recent research shows (Kratli, 2008b), is a condition for the conservation of biodiversity, not a threat to its existence (as it has often been misrepresented).

Wild biodiversity, apart from its value for livestock, is a source of countless kinds of nutritional, medicinal and craft uses, and a safety net in times of food scarcity. The perceived value of these ecosystem services is reflected in a conservation ethic which can be discovered in Sahelian communities where an appropriate institutional framework for rights of access, realisation of benefits and collaborative action exists. The same is true of the useful trees of the Sahel. Here the temptation of governments to intervene 'to prevent deforestation' has been particularly strong notwithstanding an observed process, cycle or transition from woodland clearance, via shifting cultivation, permanent cultivation, protected natural regeneration and finally to tree planting, which logically reflects the increasing values of non-timber forest products. Scarcity is a relative rather than an absolute concept, and represents a construction put on a natural resource given a set of values, institutions and policies (Mehta, 2010). Contrary to much conventional wisdom, resource scarcity need not promote destructive exploitation. As the value of ecosystem services increases, there is good reason to expect sustainable management, provided that market and policy incentives are not distorted (Tiffen & Mortimore, 2002).

Given the complexity of dryland systems (both human and ecological), 'poverty reduction' when conceived merely in terms of financial incomes and assets lacks clarity – as Berkes describes in detail in Chapter 17. The primary determinant of poverty or well-being (as understood in the Millennium Ecosystem Assessment) is the state of health of the coupled human-ecological system. Thus interventions aiming to reduce poverty need a broader reference than monetary income. Sustainable use of biodiversity can help reduce poverty only in this wider system context.

The concept of *useful biodiversity* describes the value applied by local people to specific plant varieties or animal breeds. It falls short of the idea of conservation of the system as a whole. It is necessary now to move on from the simple concept of useful biodiversity in the Sahel to a broader concept of ecosystem conservation shared by local people, scientists and government.

References

Adams, W.M. (1996) *Future Nature: A Vision for Conservation*. Earthscan, London.

Ariyo, J.A., Voh, J.P. & Ahmed, B. (2001) *Long-Term Changei in Food Provisioning and Marketing in the Kano Region*. Drylands Research Working Paper 34. Drylands Research, Crewkerne, UK.

Behnke, R.H., Scoones, I. & Kerven, C. (eds.) (1993) *Range Ecology at Disequilibrium: New Models of Natural Variability and Pastoral Adaptation in African Savannas*. Overseas Development Institute, London.

Brookfield, H. (2001) *Exploring Agrodiversity: Issues, Cases and Methods in Biodiversity Conservation*. Columbia University Press, New York.

Busso, C.S., Devos, K.M., Ross, G., Mortimore, M., Adams, W.M., Ambrose, M.J., Alldrick, S. & Gale, M.D. (2000) Genetic diversity within and among landraces of pearl millet (*Pennisetum glaucum*) under farmer management in West Africa. *Genetic Resources and Crop Evolution*, 47, 561–8.

Cline-Cole, R.A., Main, H.A.C., Mortimore, M., Nichol, J.E. & O'Reilly, F.D. (1990) *Wood Fuel in Kano*. United Nations University Press, Tokyo.

Dalziel, J.M. (1937) *The Useful Plants of West Tropical Africa*. Crown Agents for the Colonies, London.

Faye, A., Fall, A., Mortimore, M., Tiffen, M. & Nelson, J. (2001) *Région de Diourbel: Synthesis*. Drylands Research Working Paper 23e. Drylands Research, Crewkerne, UK.

Harris, F. & Mohammed, S. (2003) Relying on nature: wild foods in northern Nigeria. *Ambio*, 32, 24–29.

Harris, F. & Yusuf, M.A. (2001) Manure management by smallholder farmers in the Kano Close-Settled Zone, Nigeria. *Experimental Agriculture*, 37, 319–32.

International Institute for Environment and Development (IIED) and SOS Sahel (2010) *Modern and Mobile: The Future of Livestock Production in Africa's Drylands*. IIED, London.

International Livestock Research Institute (2007). *Nature's Benefits in Kenya: An Atlas of Ecosystems and Human Well-Being*. World Resources Institute with ILRI, Washington, DC.

International Union for Conservation of Nature (IUCN) (2009) *Dryland Opportunities: A New Paradigm for People, Ecosystems and Environment*. IUCN, Gland, Switzerland.

Jouet, A., Jouve, P. & Banoin, M. (1996) Le défrichement amélioré au Sahel. Une pratique agroforestiére adoptée par les paysans. In *Gestion des terroirs et des ressources naturelles au Sahel*, ed. P. Jouve, pp. 34–42. CNEARC, Montpellier, France.

Kratli, S. (2008a) *Time to Outbreed Animal Science? A Cattle Breeding System Exploiting Structural Unpredictability: The WoDaaBe Herders in Niger.* STEPS Working Paper 7. Institute of Development Studies, Brighton, UK.

Kratli, S. (2008b) Cattle breeding, complexity and mobility in a structurally unpredictable environment: the WoDaaBe herders of Niger. *Nomadic Peoples*, 12, 11–41.

Mahamane, A. (2001) *Usages des terres et évolutions végétales dans le département de Maradi.* Drylands Research Working Paper No. 27. Drylands Research. Crewkerne, UK.

Mehta, L. (ed.) (2010) *The Limits to Scarcity: Contesting the Politics of Allocation.* Earthscan, London.

Mortimore, M. (1989) *Adapting to Drought: Farmers, Famines and Desertification in West Africa.* Cambridge University Press, Cambridge.

Mortimore, M. (1993) The intensification of peri-urban agriculture: the Kano Close-Settled Zone, 1964–86. In *Population Growth and Agricultural Change in Africa*, ed. B.L. Turner II, R.W. Kates, & H.L. Hyden, pp. 358–400. University Press of Florida, Gainesville.

Mortimore, M. & Adams, W. (1999) *Working the Sahel: Environment and Society in Northern Nigeria.* Routledge, London.

Mortimore, M., Ariyo, J., Bouzou, I., Mohammed, S. & Yamba, B. (2006) *Local Natural Resource Management in the Maradi-Kano Region of Niger and Nigeria: A Dryland Application of the Ecosystem Approach: A Study Carried out for the World Conservation Union (IUCN).* Draft Final Report. Drylands Research, Sherborne, UK.

Mortimore, M., Ariyo, J., Bouzou, I.B., Mohammed, S. & Yamba, B. (2008) Niger and Nigeria: the Maradi-Kano region. A case study of local natural resource management. In *The Ecosystem Approach: Learning from Experience*, ed. G. Shepherd, pp. 44–58. IUCN, Gland, Switzerland.

Mortimore, M., Ba, M., Mahamane, A., Rostom, R.S., Serra del Pozo, P. & Turner, B. (2005) Changing systems and changing landscape: measuring and interpreting land use transformations in African drylands. *Geografisk Tidsskrift.Danish Journal of Geography*, 105, 101–18.

Mortimore, M., Tiffen, M., Boubacar, Y. & Nelson, J. (2001) *Department of Maradi: Synthesis.* Drylands Research Working Paper 39e. Drylands Research, Crewkerne, UK.

Reij, C., Larwanou, M. & Abdoulaye, M. (2006) *Etude de la régénération naturelle dans la région de Zinder (Niger).* International Resources Group (IRG), Washington, DC.

Reynolds, J.F., Stafford Smith, D.M., Lambin, E.L., Turner, B.L.I., Mortimore, M., Batterbury, S.P.J., Downing, T.E., Dowlatabadi, H., Fernández, Herrick, J.E., Huber-Sannwald, E., Jiang, H., Leemans, R., Lynam, T., Maestre, F.T., Ayarza, M. & Walker, B. (2007) Global desertification: building a science for dryland development. *Science*, 316, 847–51.

Tiffen, M. & Mortimore, M. (2002) Questioning desertification in dryland Africa sub-Saharan Africa. *Natural Resources Forum*, 26, 218–33.

Tiffen, M., Mortimore, M., & Gichuki, F. (1994) *More People Less Erosion: Environmental Recovery in Kenya.* John Wiley & Sons Ltd, Chichester, UK.

Trevallion, B.M. (1965) *Metropolitan Kano: Report on the Twenty Year Development Plan 1963–1983.* Newman Neame for the Greater Kano Development Authority, London.

(8)

Biodiversity Isn't Just Wildlife – Conserving Agricultural Biodiversity as a Vital Contribution to Poverty Reduction

Willy Douma

Hivos, the Hague, the Netherlands[1]

Introduction

The majority of the chapters in this book have focussed on the role of 'wild' biodiversity in alleviating poverty and supporting local peoples' livelihoods. Yet agricultural biodiversity can be just as important – if not more so. This chapter explores the role of agricultural biodiversity in supporting the livelihoods of the poor – particularly in terms of improving the adaptive capacity of poor and vulnerable groups to maintain agricultural productivity and to cope with climate change. When the Green Revolution was introduced some 50 years ago, food production was improved through the development of high-yielding varieties of cereal grains, the expansion of irrigation infrastructure, the modernisation of management techniques and the distribution of hybridised seeds, synthetic fertilisers and pesticides to farmers (International Assessment of Agricultural Knowledge, Science and Technology for Development (IAASTD), 2008). Only recently has it been recognised that this high-input agriculture has had some drawbacks. Not only has it brought poverty to some, but also it has polluted groundwater, rivers and seas as far as the North Pole through the use of highly toxic chemicals (IAASTD, 2008). In addition, it led to the loss of many crop varieties developed by farmers over thousands of years. Biodiversity of major crop varieties

[1] The author would like to thank Mrs. Lena Katzmarski (Hivos) for reviewing and for her support in finalising the chapter.

such as rice and maize, livestock breeds and farm-related plants, trees and insects has been reduced (IAASTD, 2008). This is the result of neglect: biodiversity's important role in supporting the resilience of farmers and the agricultural sector at large has been neglected by decision makers and agri-business. Ensuring that biodiversity is understood as an important asset and brought back onto decision makers' agendas requires a change to the work and orientation of many stakeholders, from farmers and their organisations to extension services, research organisations, financial institutions, companies and governments.

This change is beginning to happen. International organisations such as the Food and Agriculture Organization (FAO), United Nations Environment Programme (UNEP) and United Nations Conference on Trade and Development (UNCTAD) have indicated that business as usual is no longer an option when addressing the role of agriculture in tackling hunger, poverty and inequality (FAO, 2011a). More profound changes in the way that governments deal with the agricultural sector are also needed. In this chapter, I argue that biodiversity needs (to continue) to play an important role in the agriculture of the future. Tools, processes and regulations must ensure that biodiversity is used to its full potential. There are a number of success stories, but to ensure that these islands of success become seas of change, we need to work hard. Mainstreaming and therefore scaling up the existing successes are urgently required.

This chapter builds on promising civil society initiatives that were financed by the Hivos-Oxfam Novib Biodiversity Fund.[2] After many years of work, several of these initiatives have shown what scaling up of successful approaches can offer to biodiversity conservation and poverty reduction.

Agro-biodiversity and poverty

The term *agricultural biodiversity* (also called *agro-biodiversity*) refers to the diversity of plants and animals used for food and agriculture production. Three levels of agricultural biodiversity can be identified – each of particular importance to farmers and rural people: genetic diversity within plants and animals, species diversity in and around fields, and ecosystem or landscape diversity (FAO, 1999, 2009a).

The world's genetic diversity has decreased tremendously over recent decades. There are over 7600 documented breeds of domestic livestock, the result of breeding activities by livestock keepers over thousands of years to breed animals well adapted to their environment. However, of these known breeds, 20% are now classified as being at risk of extinction and 9% are reported to be extinct (FAO, 2009b).

---

[2] The Hivos-Oxfam Novib Biodiversity Fund was established in 2000 by the Netherlands Ministry of Foreign Affairs to promote and strengthen the sustainable management of biodiversity in primary production processes that are accessible for and beneficial to small-scale producers and low-income groups. For a period of 10 years, the Fund supported internationally operating civil society organisations.

This equals a loss of one livestock breed every month. The effect of this loss is yet to be established to its full extent, but case studies illustrate how losses of specific breeds are linked to the disappearance of agricultural strategies adapted to climatic circumstances and the loss of a gene pool that cannot be replaced (IAASTD, 2008; FAO, 2009b). Policies and research still undervalue the potentials of existing diversity for adaptation to rapidly changing climatic conditions (International Fund for Agricultural Development (IFAD), 2004; IAASTD, 2008).

Crop diversity shows a similar but better documented decline. While the Green Revolution ensured higher productivity per acre, its effect on genetic diversity has been devastating. Many traditional rice varieties have been lost, replaced by high-yielding modern varieties. Official public gene banks have been only partially successful in conserving genetic diversity: some existing diversity has been maintained, but over the millennia, incremental developments introduced by farmers in a process of continuous adaptation to change have not been valued (Almekinders & Hardon, 2006). Furthermore, on-farm seed breeding has largely been replaced by institutionalised breeding by either agri-business or public research bodies and this has further reduced diversity and accessibility to these seeds (Almekinders & Hardon, 2006). Farmers often have difficulties accessing seed varieties in *ex situ* gene banks, as these are located far from villages. Seed banks in any case distribute only a small number of seeds, so they are unable to help farmers restock with traditional varieties. New breeds have boosted agricultural productivity, but simultaneously they displaced traditional cultivars, leading to the dramatic figures given here (IAASTD, 2008).

Current food security and productivity debates focus on only a handful of crops. More than half of the global human requirement for proteins and calories is met by just three crops: maize, wheat and rice. Only 150 crops are traded on a significant global scale. Yet surveys indicate that around the world more than 7000 plant species are cultivated or harvested from the wild for food (Consultative Group on International Agricultural Research, 2010).

The lives of small-scale producers are often closely interwoven with the biodiversity of their local habitats. Access to a broad spectrum of different wild (non-farmed) species, both terrestrial and aquatic, provides them with diverse food resources such as fruits and fish, medicines, building materials and natural pest control (Chapter 4, this volume). The neem tree, found in many villages in rural India, is a well-researched example. Its leaves are used to control pests in local seed storage systems. Easy access to and control over the availability of tree leaves provide important security for poorer farmers (Girish & Shankara Bhat, 2008). Growing a variety of crops (e.g. through crop rotation systems) instead of monocultures also increases productivity (Hine & Pretty, 2008), while attention to soil fertility decreases the risk of crop failure (UNCTAD-UNEP, 2008). Free access to a wide variety of species provides a safety net, allowing producers to spread risk through varied sources of income, reducing their vulnerability to shocks such as crop failures. Not surprisingly, smallholder farmers often have a longstanding tradition of maintaining and increasing on-farm diversity. As a result, farmers provide not only food, feed and fuel but also other products and services of global significance. They conserve and develop seeds and breeds with special characteristics, increase the water-holding capacity of soils and can reduce CO_2 emissions depending on their agricultural systems. However, these valuable services are difficult to maintain under the current dominance of intensive monoculture agriculture. Research conducted in Africa suggests that 65% of agricultural land is degraded, and farms have even collapsed due to soil degradation, an effect of farming methods that focus on monocultures rather than maintenance of soil fertility (IAASTD, 2008). Because of this ecosystem degradation, farmers have had to invest in alternatives to ecosystem services – including pesticides and fertilisers made with the help of fossil fuels. The effects, particularly on poorer people, are clear (Hine & Pretty, 2008).

Rural poverty on the increase

There are currently around 1.4 billion people living on less than US$1.25 per day – the internationally agreed poverty line (FAO, 2010b). Meanwhile, the number of hungry people has increased compared to the 1990s to almost 1 billion. Of these, 95% live in the global South, and at least 70% still live in rural areas. These facts are not likely to change in the near future (IFAD, 2011). Agriculture plays a vital role in most countries – over 80% of rural households farm to some extent, and the poorest households typically rely the most on farming and agricultural labour (IFAD, 2011).

Looking more closely at the agriculture sector, smallholder farms occupy 60% of the world's arable land. Out of 525 million farms worldwide, 85% are smaller than 2 hectares (IFAD, 2011). Their context is rapidly changing, as rural areas everywhere in the world are increasingly integrated into the global economy. Domestic food production has to compete with well-paying export products such as soy and other animal feed, biofuels including palm oil or maize, or food products for other nations. Commoditisation of food products has led to price spikes, resulting in social unrest and political upheavals (FAO, 2010b). In addition, climate change poses serious risks. Globally, 1.7 billion farmers are highly vulnerable to the effects of climate change, with crops lost to unexpected flooding or drought. Those who are already hungry are particularly vulnerable (Oxfam, 2009, 2011).

The global population is expected to grow to 9 billion in 2050, requiring an estimated 70% increase in food production (FAO, 2010b). Meanwhile there are a number of different dimensions to food security: sufficient production, access to food, nutritional quality of food and stability of access (FAO, 2011a). How to approach this challenge is a matter of debate, with recommendations varying from working with smallholder farmers as the potential producers of future food to scenarios in which farming becomes a large-scale industrial activity.

Although smallholder farmers currently play a vital role in food production, they are barely considered in future agricultural development plans. Including them, and ensuring that greater and more effective efforts are made to address the concerns of poor rural people as both food producers and consumers, are now being advocated by several organisations for different reasons. Whether from a poverty perspective (IFAD, 2011) or a food security perspective (IAASTD, 2008; FAO, 2010b), smallholder farmers are seen as the starting point for a transition to more resilient agriculture. Decision makers in government, research and banks need to determine how smallholders' potential can be tapped into so that the smallholders' position is improved and their role strengthened.

Focussing attention on sustainable agriculture

Interest in agriculture at the policy level is increasing again after a long period of neglect. Examples are the *World Development Report 2008* (World Bank, 2007), the plan by the African Union and African governments to spend 10% of their national budgets on agriculture (African Union, 2003) and the report on agro-ecology presented by the UN Special Rapporteur on the Right to Food before the Human Rights Council (de Schutter, 2011). Although important for agriculture at large, history has shown that political attention on agriculture does not easily translate into policies and practices favourable for smallholder farmers or poorer communities. It also remains unclear to what extent a role for biodiversity is envisaged in future agricultural scenarios even though its importance for improved resilience of agricultural systems to climate change is clear.

The International Assessment of Agricultural Science and Technology for Development, a comprehensive assessment of world agriculture, concluded that fundamental changes were needed in world agriculture, notably a shift towards sustainability. It looked not only at technical improvements but also at unfair trade rules and the need to revise the regulations on intellectual property rights (IAASTD, 2008). Meanwhile, the International Fund for Agricultural Development argues that there is an urgent need to invest more (and more wisely) in agriculture and rural areas based on a new approach to smallholder agriculture that is both market oriented and sustainable (IFAD, 2011).

While the interest in sustainable agriculture – an agriculture that values biodiversity – is increasing, we also need to understand better what precludes a more rapid shift. This is explored in the sections that follow.

Enhancing genetic diversity

Plant genetic diversity is probably more important for farming than any other environmental factor, simply because it is *the* factor that enables adaptation to changing environmental conditions such as plant diseases and climate change (FAO, 2010a, 2011b). However, support for maintaining such a valuable resource has rested predominantly with a limited number of research institutions and private companies with farmers largely excluded despite the knowledge they hold. Despite an international initiative to highlight farmers' roles in conserving plant genetic diversity (Box 8.1), farmers continue to be seen not as innovators but only as end users of research results.

Box 8.1 The Community Biodiversity Development and Conservation programme

In 1991 a group of researchers from the global North and South asked, in the Keystone International Dialogue on Plant Genetic Resources (PGR), why the world seemed to put all its eggs into the single basket of research institutes and breeding companies to provide for our genetic diversity. They wanted to show that farmers are mistakenly excluded from agricultural developments, and that they are needed to ensure that agriculture can continue to provide products to a growing world population under rapidly changing circumstances. As a result of this meeting, the Community Biodiversity Development and Conservation programme (CBDC) came into being. CBDC is an international initiative to understand and strengthen farmers' systems of plant genetic resource conservation and development in particular and biodiversity management in general (Andersen, 2008). Soon after its establishment an additional component was added: the selection and development of new varieties based on the continuously changing requirements of both the environment (e.g. droughts, flooding and temperature) and the markets.

The regional co-ordinator of the programme in Asia is Southeast Asia Regional Initiatives for Community Empowerment (SEARICE), an NGO based in the Philippines.[3] The programme's activities focus on Vietnam, Bhutan, Lao PDR, Thailand and the Philippines, and target primarily farmers, particularly smallholder farmers in marginal areas, who often fall outside the scope of national agricultural research institutes. In cooperation with national plant-breeding institutes and their gene banks, which provide farmers access to new germ plasma for breeding purposes, SEARICE supports farmers in developing new varieties. SEARICE's methodologies are technical and include Farmer Field Schools focusing on participatory plant breeding and participatory variety selection. In Lao PDR, the organisation's work has resulted in farmers and extension workers reporting a 10–20% increase in yields from the use of

---

[3] See www.searice.org.ph.

farmer-developed varieties. In one province in Lao, the programme was able to close the 3-month hunger gap for indigenous communities. It is important to note that the programme was effective even outside poorer regions with less favourable soils. In Vietnam, for instance, it also secured 90% of the community seed requirement in prime irrigated areas, where farmers formerly had to buy all their seeds. Access to high-quality seeds is important in stabilising and increasing production, the backbone of agriculture.

Current national seed regulations in many countries around the world act as an obstacle, making it difficult for farmers to use markets as motors for development. Regulations often require seeds to have a quality certificate that is difficult for farmers to obtain. In the European Union, for example, farmers are not allowed to use farm-saved seed from protected varieties on their own holdings, or else must pay a license fee to do so. While such a system is suitable for seed companies focusing on a limited number of varieties, it in fact blocks farmers' access to markets.

There have nevertheless been changes. Several governments have started to acknowledge the importance of farmer varieties. Bhutan's current 5-year plan and Vietnamese government approaches demonstrate a change in policies. Farmers' rights (Box 8.2) in national rules and regulations are now under development, in keeping with the International Treaty on Plant Genetic Resources for Food and Agriculture. The Treaty recognises the enormous contributions made by farmers worldwide in conserving and developing crop genetic resources, and it provides for measures to protect and promote their rights. Civil society organisations played an important role in raising the issue, and came up with innovative approaches. Multi-stakeholder processes in which farmers, researchers and government officers collaborated over a longer period played a crucial role. To make further progress and expand implementation, agribusinesses, agricultural research institutes and governments must invest in a more participatory approach to genetic diversity based on farmers' preferences. This requires an environment in which government regulations enable farmers to access both seeds and markets.

Box 8.2 **What are farmers rights?**

Farmers' Rights consist of the customary rights of farmers to save, use, exchange and sell farm-saved seed and propagating material; their rights to be recognised, rewarded and supported for their contribution to the global pool of genetic resources as well as to the development of commercial varieties of plants and their rights to participate in decision making on issues related to crop genetic resources (Andersen, 2006). The concept includes recognition of the fact that farmers have developed and continue to help develop genetic diversity. In many cases, farmers engage in conscious and creative practices as they 'select' and 'breed' their crops. Farmers' rights are essential to maintaining crop genetic diversity, which is the basis of all food and agriculture production in the world. Farmers' rights means allowing farmers to maintain and develop crop genetic resources as they have done since the dawn of agriculture, and recognising and rewarding them for this indispensable contribution to the global pool of genetic resources.

Greater species diversity increases productivity and reduces farmers' risks

Apart from genetic diversity, diversity of plants, animals and micro-organisms in and around the farm has been shown to reduce risks of food insecurity, which is important for all farmers but in particular for poorer farmers (Araya & Edwards, 2006; IFAD, 2011; Chapters 4 and 7, this volume).

Greater diversity on farms is influenced by the agricultural production system employed. Several different technical approaches, such as low external-input agriculture, organic agriculture and agro-forestry systems have shown positive effects on biodiversity (McNeely & Scherr, 2003; IAASTD, 2008; Crowder et al., 2010; Kalaba et al., 2010). These approaches make use of techniques such as composting, crop rotation, zero tillage, application of green manure and/or abstention from chemical pesticides and fertilisers. The broad term *sustainable agriculture* encompasses these technical approaches.

The main argument made against sustainable agriculture is its perceived low productivity. While that is often the case during the conversion period of 3–5 years – a serious obstacle that needs to be addressed – sustainable agricultural techniques have been shown to produce good results in the longer term: production levels are similar and have positive effects on soil and water quality (e.g. Rodale Institute, 2011). In Africa, a meta-analysis of 114 projects (comprising 2 million hectares and 1.9 million farmers) showed an average yield increase of 116% (Hine & Pretty, 2008), while Badgley et al. (2007) report a yield increase of 80% in 133 plant and livestock varieties in field and farm comparison studies in developing countries, and Gibbon and Bolwig (2007) note that organic conversion in tropical Africa is associated with yield increases, not yield reduction.

Hine and Pretty (2008) show that these productivity increases are mainly due to improvements in the supporting and regulating ecosystem services – in other words, an increase in soil fertility due to increased availability of nutrients and minerals highly related to active decomposition based on more soil life, greater biodiversity on the farm (e.g. pollinators as well as natural pest reduction in soils and crops) as well as improved water-holding capacities and erosion control due to better soil structure and vegetation cover. The improvements in yields through biodiverse agricultural practices potentially increase household incomes due to reduced costs and debts, more surplus produce to sell and, in specific cases, higher prices on domestic and export markets for quality products. Of particular importance for farming households is the reduction of risks as a consequence of the cultivation of a mix of crops and fewer fluctuations in yield under changing climatic conditions (Hine & Pretty, 2008).

Looking at the ecosystem level

Biodiverse agricultural systems not only increase soil fertility as referred to in this chapter's 'Greater species diversity' section but also, at the same time, reduce net greenhouse gas emissions through carbon fixation. Given that agriculture is currently responsible for 20–30% of global greenhouse gas emissions (counting direct and indirect agricultural emissions)(Intergovernmental Panel on Climate Change (IPCC), 2007), it is important to further explore and realise these potentials. It has been shown that composting and agro-forestry practices fix gaseous carbon: agro-forestry practices' mitigation potential is significantly different from that of conventional practices (Ajayi et al., 2011). The same has also been found to apply to organic agriculture. Organic agriculture significantly reduces greenhouse emissions (Müller, 2009; Niggli et al., 2009). Important aspects of organic practices include the use of organic fertilisers and crop rotations including legumes, leys and cover crops. CO_2 sequestration is also achieved through avoiding such things as open biomass burning, synthetic fertilisers and the related production emissions from fossil fuels.

This aspect of biodiverse agricultural practices has significance for the global community as a whole, but schemes to reward farmers adequately for this service to reduce greenhouse gas emissions are currently not in place. Recently, discussions around payments for ecosystem services (PES) (Chapter 14, this volume) have also started to consider how such payments could become an incentive for increased sustainability efforts in agriculture. Areas that require further exploration include permanency of soil carbon fixation, affordable data collection and identifying sectors willing to pay. Carbon fixation is just one aspect of improved ecosystem services with positive effects on poverty and the livelihoods of people. Soil biodiversity has the potential to improve soil structure and stability. This leads also to better water-holding capacities and reduced vulnerability to drought, extreme precipitation events, floods and water logging (Niggli et al., 2009). Landscape-level interdisciplinary research is required to establish further crucial determining factors and how these can be implemented in such a way that smallholder farmers can benefit.

Apart from mitigation, agricultural practices need to adapt to cope with the consequences of climate change (FAO, 2009c). Agricultural production could fall by about 30% in Sub-Saharan Africa alone (IPCC, 2007). Climate change will also affect biodiversity, putting more and more plant and animal species at risk of extinction (Chapter 18, this volume). Reduced genetic diversity means fewer opportunities for adaptation and innovation. Because smallholder farmers depend on biodiversity, their livelihoods are particularly threatened by climate change. Ensuring they retain access to and use of biodiversity (both domesticated and non-domesticated) will pay off both in terms of biodiversity conservation and more indirectly in terms of mitigating and adapting to climate change.

Do markets work for biodiversity and poverty?

Markets have been shown to play a potential role in encouraging biodiversity-conserving production processes. Research among consumers in the North showed that animal welfare and CO_2 neutrality have become as important as personal health reasons for buying products labelled as sustainable. Biodiversity is on the list of reasons, but ranks much lower (Willer & Kilcher, 2011). In the future, a more mature market for verified, biodiversity-friendly products might support a growth in biodiversity-conserving production and processing.

The challenges to making markets work for biodiversity are huge and diverse. First, with biodiversity conservation not being a target in agriculture, measuring the impact of sustainable practices on biodiversity has largely been ignored. The current standards – such as organic – do not target biodiversity, although this is a feature of forest and marine certification. Although these practices do have a positive effect, as shown by a number of studies (discussed in this chapter), this is not systematically pursued by the sustainable agriculture movement. Monitoring of effects on these important societal developments and issues is still at an early stage. There is no method to indicate and quantify what effect an increase in market share for sustainable products will have on biodiversity. This brings us to a second challenge, as markets for sustainable products need to be further developed. The volume and area under organic agriculture continue to grow, even during the heat of the financial crisis in 2009, showing consumer preference (Willer & Kilcher, 2011), but to expand current practices, changes in many stakeholders' policies and practices in the agricultural sector are urgently needed. Current subsidies and tax and tariff systems in general favour high fossil fuel–based inputs. Including environmental costs in prices would the improve competitiveness of sustainable products on the market.

There are examples of positive links between increasing sustainable production, access to markets and reducing poverty (Murphy, 2010). However, smallholder farmers tend to be excluded due to the costs and other difficulties involved in compliance, although the International Federation of Organic Agriculture Movements (IFOAM) is developing participatory guarantee systems (PGS) which are designed to reduce the costs of certification by peer review, helping farmers to access local and regional markets with their organic products (IFOAM, 2008).

Towards an integrated approach: next steps in enhancing the contribution of agricultural biodiversity to poverty reduction

Paths divert in policy debates around the world. Is smallholder farming – given its current importance – the starting point, and does it require investment to build capacity to produce sufficient food for the growing world population, or does society at large need to invest in larger scale farming practices? Whatever the results of these policy debates are, the current situation is one of predominantly smallholder farmers seeing little support from government, research and business. There are signs of change, however: IFAD urges governments to invest in smallholder farming, and companies such as Unilever and Mars seek collaboration with organised smallholder farmers (Hoeven, 2008). Within the near future we will see reports coming out telling us more about the actual effect of such approaches on poverty and biodiversity (Phalen et al., 2011).

Sufficient experience and evidence are already available that policy makers, research and business alike ought to work more seriously on agro-biodiversity-oriented agricultural systems that are not only green but also inclusive: taking the interests of smallholder farmers seriously and ensuring their inclusion in developments towards more sustainable systems. Assistance to small-scale producers should therefore address social, ecological and socio-economic issues and look at rights, skills and attitudes, preferably in an integrated approach. Policy interventions based on this approach need to involve civil society organisations – notably farmers, small-scale producers and women's organisations – acting as change agents on the issues of (i) empowerment and (ii) demand and markets for biodiversity-friendly produce.

It should be acknowledged that farmers are competent managers of biodiversity, that they need access to a variety of seeds and that they have the capacity to develop these seeds and their varieties, if we work with them properly and build on their knowledge. Despite their important role in biodiversity management, small-scale producers – particularly women – seldom have a voice in decision making on these issues, and their needs are often overshadowed by those of more powerful groups in national and international fora.

Policy interventions to support empowerment should create an enabling environment that results in providing smallholder farmers with more opportunities to engage successfully with a broad set of actors. Creating strategic alliances, networks and coalitions, which also involve private sector parties, governments and international bodies, has shown some promising results (IFAD, 2011).

At a broader policy level, the issue of empowerment embraces many different aspects. Here policy interventions should (i) concentrate on increasing political involvement of CSOs in national and international for a on the issues of food security and climate change, (ii) challenge the unsustainable practices of large-scale agriculture while meeting the technological needs of viable smallholder farmers and (iii) promote and expand systems that enable smallholder farmers, particularly women, to have sustainable access to large gene pools so they can select the crop varieties and/or livestock breeds best suited to their needs.

Hivos and Oxfam Novib in the Netherlands have published results of over 8 years of engagement with civil society organisations working on these issues (Hivos-Oxfam Novib, 2009). The road was an inspirational one, with very committed organisations working for change against the mainstream current. Hivos and Oxfam Novib are committed to continue working on these issues.[4]

The focus on empowerment and markets as drivers for change is likely to grow in the coming years, in both the national and international arenas. The lessons learned through the Biodiversity Fund over the past 8 years are likely to become increasingly relevant. The challenges for the stakeholders in these processes, from government policy makers to researchers, farmers' and women's organisations, include striking a balance between developing an enabling environment for a greener and more inclusive agricultural sector and developing incentives for those currently in power (e.g. the largest agri-businesses) to change their policies and practices.

It is time for development agencies, government and others to take smallholder farmers seriously as key stakeholders in the development of biodiverse food production systems. Civil society organisations have contributed significantly to evidence suggesting that conservation through use is a viable strategy for the management of biodiversity. Experience also shows that weaving biodiversity management into social change is a long-term process. It is good to know that there is a growing community involving different sectors intent on building from this experience.

References

African Union (2003) *Maputo Declaration on Agriculture and Food Security*. Assembly/ AU/Declaration 7(II). African Union, Addis Ababa, Ethiopia.

---

[4] See www.hivos.nl and www.oxfamnovib.nl.

Ajayi, O.C., Place, F., Akinnifesi, F.K. & Sileshi, G.W. (2011) Agricultural success from Africa: the case of fertilizer tree systems in southern Africa (Malawi, Tanzania, Mozambique, Zambia and Zimbabwe). *International Journal of Agricultural Sustainability*, 9, 129–36.

Almekinders, C. & Hardon, J. (eds.) (2006) *Bringing Farmers back into Breeding. Experiences with Participatory Plant Breeding and Challenges for Institutionalisation*. Agromisa Special 5. Agromisa, Wageningen.

Andersen, R. (2006) *Farmers' Rights and Agrobiodiversity*. Issue Papers: People Food and Biodiversity, GTZ Issue Paper series. GTZ, Eschborn.

Andersen, R. (2008) *Governing Agrobiodiversity: Plant Genetics and Developing Countries*. Aldershot, Ashgate.

Araya, H. & Edwards, S. (2006) *The Tigray Experience: A Success Story in Sustainable Agriculture*. Third World Network (TWN) Environment and Development Series 4. TWN, Penang.

Badgley, C., Moghtader, J., Quintero E., Zakem, E., Chappell, J., Avilés-Vázquez, K., Samulon, A. & Perfecto, I. (2007) Organic agriculture and the global food supply. *Renewable Agriculture and Food Systems*, 22, 86–108.

Community Biodiversity Development and Conservation programme (CBDC) (2006) *Pathways to Participatory Farmer Plant Breeding: Stories and Reflections of the Community Biodiversity Development and Conservation Programme*. Southeast Asia Regional Initiatives for Community Empowerment (SEARICE), Manila.

Consultative Group on International Agricultural Research (CGIAR) (2010) *COP10 Factsheets, 2010*. CGIAR, Rome.

Crowder, D.W., Northfield, T.D., Strand, M. & Snyder, W.E. (2010) Organic agriculture promotes evenness and natural pest control. *Nature*, 46, 109–12.

Food and Agriculture Organization (FAO) (1999) *Sustaining Agricultural Biodiversity and Agro-Ecosystem Functions, Opportunities, Incentives and Approaches for the Conservation and Sustainable Use of Agricultural Biodiversity in Agro-Ecosystems and Production Systems*. Workshop Report. FAO, Rome.

Food and Agriculture Organization (FAO) (2009a), *State of Food and Agriculture, Livestock in the Balance*. FAO, Rome.

Food and Agriculture Organization (FAO) (2009b) *Livestock Keepers – Guardians of Biodiversity*. Animal Production and Health Paper No. 167. FAO, Rome

Food and Agriculture Organization (FAO) (2009c) *Low Greenhouse Gas Agriculture, Mitigation and Adaptation. Potential of Sustainable Farming Systems*. FAO, Rome.

Food and Agriculture Organization (FAO) (2010a) *Second Report of the State of the World's Plant Genetic Resources for Food and Agriculture*. FAO, Rome.

Food and Agriculture Organization (FAO) (2010b) *State of Food Insecurity in the World*. FAO, Rome.

Food and Agriculture Organization (FAO) (2011a) *Greening the Economy with Agriculture: Taking Stock of Potential, Options and Prospective Challenges*. FAO@Rio+20 Concept Note. FAO, Rome

Food and Agriculture Organization (FAO) (2011b) Second *Global plan of Action for Plant Genetic Resources for Food and Agriculture*. FAO, Rome.

Gibbon, P. & Bolwig, S. (2007) *The Economics of Certified Organic Farming in Tropical Africa: A Preliminary Assessment*. DIIS Working Paper 2007. Danish Institute for International Studies, Copenhagen.

Girish, K. & Shankara Bhat, S. (2008) Neem – a green treasure. *Electronic Journal of Biology*, 4, 102–11.

Hine, R. & Pretty, J. (2008). *Organic Agriculture and Food Security in Africa*. United Nations Conference on Trade and Development (UNCTAD) and United Nations Environment Programme (UNEP), Geneva and New York.

Hivos-Oxfam Novib (2009) *Biodiversity, Livelihoods and Poverty. Lessons Learned from 8 Years of Development Aid*. Hivos and Oxfam Novib, the Hague.

Hoeven, H. (2008) *Sourcing and Procurement as Driving Forces for Sustainable Business*. Sustainable Trade Initiative (IDH), Utrecht.

Intergovernmental Panel on Climate Change (IPCC) (2007) *Fourth Assessment Report: Climate Change 2007*. Cambridge University Press, Cambridge.

International Assessment of Agricultural Knowledge, Science and Technology for Development (IAASTD) (2008) *Agriculture at a Crossroad*. Island Press, Washington, DC.

International Federation of Organic Agriculture Movements (IFOAM) (2008) *Participatory Guarantee Systems: Case Studies from Brazil, India, New Zealand, USA and France*. IFOAM, Bonn.

International Fund for Agricultural Development (IFAD) (2004) *Livestock Services and the Poor*. IFAD, DANIDA, World Bank, DAAS, University of Reading and national institutions in Bangladesh, Bolivia, Denmark, India and Kenya.

International Fund for Agricultural Development (IFAD) (2011) *Rural Poverty Report 2011*. IFAD, Rome.

Kalaba, K.F., Chirwa, P., Syampungani, S. & Ajayi, C.O. (2010) Contribution of agroforestry to biodiversity and livelihoods improvement in rural communities of Southern African regions: tropical rainforests and agroforests under global change. *Environmental Science and Engineering*, 3, 461–76.

McNeely, J. & Scherr, S. (2003) *Ecoagriculture: Strategies to Feed the World and Save Wild Biodiversity*. Island Press, Washington, DC.

Müller, A. (2009) *Benefits of Organic Agriculture as a Climate Change Adaptation and Mitigation Strategy in Developing Countries*. Discussion Paper Series. Environment for Development (EfD), Gothernburg.

Murphy, S. (2010) *Small-scale Farmers, Markets and Globalization*. Working Paper, Hivos-IIED Knowledge Programme on Small Producer Agency. IIED and Hivos, London and the Hague.

Niggli, U., Fließbach, A., Hepperly, P. & Scialabba, N. (2009) *Low Greenhouse Gas Agriculture: Mitigation and Adaptation Potential of Sustainable Farming Systems*. Food and Agriculture Organization of the United Nations, Rome.

Oxfam (2009) *People-Centred Resilience*. Oxfam Briefing Paper. Oxfam, Oxford.

Oxfam (2011) *Growing a Better Future: Food Justice in a Resource-Constrained World*. Oxfam International, Oxford.

Phalen, B., Onial, M., Balmford, A. & Green, R.E. (2011) Reconciling food production and biodiversity conservation: land sharing and land sparing compared. *Science*, 333, 1289–91.

Rodale Institute (2011) *The Farming Systems Trial: Celebrating 30 Years*. Rodale Institute, Kutztown, PA.

Schutter, O. de (2011) *Agro-ecology and the Right to Food*, Report presented at the 16th session of the United Nations Human Rights Council by the Special Rapporteur on the Right to Food, http://www.srfood.org/images/stories/pdf/officialreports/20110308_a-hrc-16-49_agroecology_en.pdf (accessed 25 April 2012).

United Nations Conference on Trade and Development and United Nations Environment Programme (UNCTAD-UNEP) (2008)*Organic Agriculture and Food Security in Africa*, http://unctad.org/en/docs/ditcted200715_en.pdf (accessed 25 April 2012).

Willer, H. & Kilcher, L. (eds.) (2011) *The World of Organic Agriculture: Statistics and Emerging Trends 2011*. IFOAM, Bonn, and FiBL, Frick.

World Bank (2007) *World Development Report 2008: Agriculture for Development*. International Bank for Reconstruction and Development and World Bank, Washington, DC.

Part III
Poverty Impacts of Different Conservation Interventions

(9)

Does Conserving Biodiversity Work to Reduce Poverty? A State of Knowledge Review

Craig Leisher[1], M. Sanjayan[1], Jill Blockhus[1], S. Neil Larsen[1] and Andreas Kontoleon[2]

[1]The Nature Conservancy, Monson, USA
[2]Department of Land Economy, University of Cambridge, UK

Introduction

In 1987, the former prime minister of Norway, Gro Harlem Brundtland, chaired a UN commission that produced the report entitled *Our Common Future*. The report stated that "poverty is a major cause and effect of global environmental problems." Though subsequent research and thinking have added complexity to this statement, increasing evidence of initiatives that create benefits for both poverty alleviation and conservation continues to add support to this premise.

The initial dominant view was that increasing living standards, healthcare and education would lead to decreased dependence on and degradation of natural resources (Wells & Brandon, 1992). However, this prediction has not been borne out in fact; China, Indonesia and Brazil, for example, have greatly reduced rural poverty while continuing to draw down their natural capital. Reductions in rural poverty alone are unlikely to break destructive resource use patterns, as natural resource consumption tends to increase with income, and better off people frequently gain the most from the exploitation of natural resources. However, the fact remains that, throughout the world, the rural poor are often highly dependent on the goods and services provided by biological diversity, and these goods and services are frequently taken for granted,

Biodiversity Conservation and Poverty Alleviation: Exploring the Evidence for a Link, First Edition.
Edited by Dilys Roe, Joanna Elliott, Chris Sandbrook and Matt Walpole.
© 2013 John Wiley & Sons, Ltd. Published 2013 by John Wiley & Sons, Ltd.

underpriced and overexploited. Reducing local poverty in conservation areas remains crucial both ethically, as conservation initiatives should help people and not hurt them, and practically, as conservation initiatives that do not provide tangible benefits to local people will not be socially sustainable.

The question then arises as to whether, and under what circumstances, conservation interventions can also benefit the poor, rural people affected by them. This is the subject of this chapter.

Approach

In order to assess the 'state of knowledge' on the extent to which conservation can contribute to poverty reduction, we reviewed over 400 documents collected from peer-reviewed publications, books and grey literature.[1] We covered documents from 1985 until early 2010 and gave greater weight to those documents that present multisite empirical evidence of a conservation intervention's poverty alleviation benefits. While we recognise the problems associated with different definitions of *poverty* (Chapters 1 and 4, this volume), to cast a wide net, any definition of poverty was sufficient for inclusion. We assessed 16 interventions and included 10 for which we found empirical evidence of poverty reduction benefits (Figure 9.1).

Several constraints became increasingly evident through the course of this review. The first is the limited number of studies generating hard evidence of poverty impacts. The second is that the generalisations made in this knowledge review miss the individual variations in costs and benefits from a conservation initiative. Thirdly, most of the evidence for poverty alleviation focuses on changes in income and thus does not fully reflect the multidimensional nature of poverty. Finally, there is the lack of uniformity of poverty indicators, so evidence that a mechanism was a route out of poverty depends greatly on how poverty impacts were measured.

Conservation interventions

Here, we describe the 10 conservation interventions shown to provide effective poverty alleviation benefits. We assess the way in which each intervention operates, the types of benefits it provides, the primary beneficiaries and the type of biodiversity responsible for the poverty reduction benefits. A summary of the results is provided in Table 9.1.

---

[1] The list of documents can be found at http://conserveonline.org/library/conservation-and-poverty-reduction-literature-0/view.html.

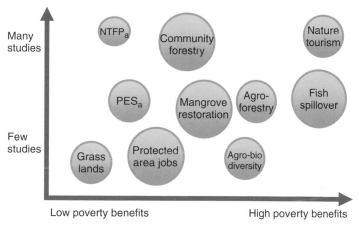

a. NTFP: Non-timber forest products;
and PES: payments for ecosystem services

Figure 9.1 **Evidence base, poverty reduction benefits and importance to biodiversity for specific conservation.**

Commercialisation of Non-Timber Forest Products (NTFPs)

Non-timber forest product (NTFP) projects support the sustainable gathering and sale of products from natural forests. These financial benefits give local people an incentive to protect forests, benefitting biodiversity. More recently, NTFPs have begun to be cultivated outside forests, reducing pressure on biodiversity. NTFPs that appear to generate poverty reduction benefits for the poor include honey, bamboo, fuel wood and mushrooms.

There is compelling evidence that the poorest of the poor are those who use NTFPs the most (Neumann & Hirsch, 2000), that NTFPs are often the employment of last resort (Angelsen & Wunder, 2003) and that NTFPs can prevent a decline deeper into poverty. Women are often the largest beneficiaries from the commercialisation of forest products (Shackleton & Shackleton, 2004).

However, NTFPs rarely sustainably reduce local poverty. NTFP projects have been criticised as at best a safety net (Wunder, 2001), and at worst a poverty trap through which poor people are kept poor (Dove, 1993). There are several reasons why this mechanism may fail to lift people out of poverty (see also Chapter 6). Firstly, NTFPs are often collected in open access areas, and thus overexploitation is common. Secondly, access to reliable markets tends to be poor for many forest dwellers. Thirdly, fluctuations in the quantity and quality of NTFPs cause an unpredictable income

Table 9.1 Summary of poverty reduction evidence for conservation mechanisms

Mechanism	Number of studies[a]	Poverty reduction benefits	Which groups benefit?	Other benefits?	Importance of biodiversity versus biomass for poverty reduction
Commercialisation of NTFPs	Many	Low	Very poor, better off	Nutritional, medicinal	Biomass
Community forestry	Many	Medium	Very poor, moderately poor, better off	Community organisation	Biomass
PES	Moderate	Low	Landowners	Property rights, capacity building, social cohesion	Biomass
Nature-based tourism	Moderate	High	Moderately poor, better off	Infrastructure, social services	Biodiversity
Locally managed marine areas	Moderate	High	Very poor, moderately poor, better off	Social cohesion	Biomass
Mangrove conservation	Moderate	Medium	Very poor, moderately poor	Reduced erosion, storm protection, more fish	Biomass
Protected area jobs	Few	Low	Moderately poor, better off	Multiplier effect of local jobs	Biodiversity
Agro-forestry	Moderate	Medium	Moderately poor, better off, landowners	Income stabilisation	Biomass
Grasslands	Few	Low	Not enough evidence	Social cohesion	Both
Agro-biodiversity	Few	Medium	Moderately poor, better off	Global benefits to agriculture	Biodiversity

[a]Many = More than 50 studies. Moderate = 10–50 studies. Few = Fewer than 10 studies.

stream. Finally, middlemen often take the bulk of the added value (Pandey, 2007). Some recommendations have been made for increasing the benefit of NTFPs for poor households, including creating secure tenure rights, ensuring harvest occurs sustainably, ensuring that products reach appropriate markets and establishing effective cooperation of producers with each other and outside contacts (Ros-Toten & Wiersum, 2003).

Benefits to biodiversity are also uncertain. NTFP programmes that have poverty alleviation benefits do so largely through increasing the quantity of NTFP or the value added. Biological diversity itself has little to do with the direct benefits except by underpinning the ecosystem that produces the NTFP.

Community-based forestry enterprises

In 2004, one-quarter of forests in developing countries were owned by communities, and this is likely to double by 2020 (Scherr et al., 2004) due to increasingly secure tenure of forest lands, improved governance, growing demand for forest products, market deregulation and new technologies that make small-scale harvesting and processing easier (Sunderlin et al., 2007). Community forestry projects have been seen to reduce poverty in rural communities and, by harvesting the timber at sustainable levels, communities also help conserve biodiversity.

The majority of poverty reduction benefits come through small-scale wood-processing programmes that supply fibre, pulp or construction timber to larger companies. Studies of community forestry enterprises in developing countries have found that forest income is a significant part of household income, and that such projects can be very profitable (Molnar et al., 2007). However, failure rates can be high; a study of small and medium-sized timber enterprises found that, on average, 75% of enterprises fail in the first 3 years (Mayers, 2006). Also, timber harvesting has rarely benefited the rural poor for several reasons. Firstly local and national elites capture much of the profit. Secondly, timber production is a long-term, high-risk investment, and the time horizons of rural poor households are necessarily shorter. Thirdly, timber production requires secure land tenure, which poor communities may not have (Sunderlin et al., 2005). Fourthly, dependence on forestry resources can be a poverty trap when access to markets is limited (Mahanty et al., 2006).

There is evidence in many cases that community forestry has led to increases in forest cover (e.g. Bray et al., 2003). However, as with NTFPs, benefits to the poor from timber come from the abundance of particular tree species rather than the diversity of tree species, and benefits frequently come from fast-growing and commercially valuable species.

Payments for environmental services (PES)

Payments for environmental services (PES) are voluntary transactions in which a landowner or manager receives payment for maintaining a specific environmental service and making it available to a consumer (Wunder, 2005; Chapter 14, this volume). PES can include bioprospecting, eco-labelling, conservation easements, watershed protection and carbon sequestration or storage (Richards & Jenkins, 2007). To date, the evidence of PES benefitting the poor lies in payments for watershed services and carbon sequestration and storage. In payments for watershed services, downstream water users pay upstream landowners to protect their water supply. In carbon sequestration and storage, CO_2 emitters pay landholders to conserve forest or reforest an area, thereby offsetting their CO_2 emissions.

It is the moderately poor smallholders and the better off landowners who generally benefit from a PES scheme, and most schemes do not explicitly aim at reducing poverty (Pagiola *et al.*, 2005). Although poverty alleviation benefits have been seen – for example, a payment for watershed services project in Ecuador provided significant income for poor sellers (Echavarría *et al.*, 2004) – a recent review of watershed programmes found that most payments were fairly insignificant, and that transaction costs were high (Porras *et al.*, 2008). There are also uncertainties about whether or not PES initiatives deliver the environmental services promised (Porras *et al.*, 2008).

In addition, the majority of PES initiatives are not dependent on biodiversity per se. There is evidence, however, that for some PES initiatives, biodiversity is important. Higher levels of plant biodiversity, for example, result in faster and higher concentrations of soil carbon storage (Tilman *et al.*, 2001). Therefore, biodiversity can enhance the delivery of some ecosystem services and hence make the initiative more attractive.

Nature-based tourism

Nature-based tourism offers a number of opportunities for reducing poverty. These include jobs in the tourism sector such as accommodation and guiding, and in new markets for local services and products including sales of crafts, cultural services, food and drinks. Tourism also brings infrastructure development, including roads and healthcare facilities that can benefit poor people.

Nature-based tourism has been seen to be an effective poverty reduction mechanism in many cases. A meta-review of 27 tourism case studies in Asia found income gains for all economic levels, with those already better off gaining most (Shah & Gupta, 2000). In Zambia, a World Bank study (2007) found that nature-based tourism had reduced poverty, though those with most assets benefitted more than the very poor by up to 50%. Successful tourism programmes also have indirect effects for poorer community members through downstream benefits for casual labourers, crafters and small businesses (Ashley & Roe, 2002). Many of these economic benefits flow to women, and women comprise a higher percentage of the tourism sector than the general workforce (Hemmati, 1999). Where tourism operators commit to hire and train local people, the poverty reduction benefits are much greater (Davis, 2005).

While nature-based tourism tends to rely on a very limited number of key species, it is also dependent on the overall attractiveness of the landscape. Biodiversity, and the maintenance of an intact ecosystem, is therefore a crucial component of successful operations.

Locally managed marine areas

This intervention involves protecting a key area of marine habitat from fishing in order to allow fish stocks to replenish. One of the key poverty reduction benefits of this approach is that as stocks increase, some fish will 'spill over' into adjacent areas where they can be caught by fishers. The greater fish catches generate income for fishers, while the no-take zone provides protected habitat for marine biodiversity. Gell and Roberts (2003) cite 14 marine protected areas where spill-over has been documented. A 2001 study of 45 sites in the Philippines found that the most likely predictors of poverty conservation benefits were a community with a small population, a perceived crisis in terms of reduced fish populations, successful alternative income projects, community participation in decision making and continuing advice from the implementing organisation (Pollnac et al., 2001).

The poorest of the poor can benefit from spill-over, and poverty reduction benefits can be significant. For example, spill-over from two community-managed marine areas in Fiji roughly doubled local incomes within 5 years of the establishment of the no-take zone, and women were the primary beneficiaries (Leisher et al., 2007). In addition, organizing a community to manage a no-take zone often strengthens social cohesion, which improves local security and empowers local decision making, two key elements of poverty reduction (World Bank, 2001).

A number of factors must be in place for such schemes to generate benefits. The size of the protected area matters, because if a no-take zone is too large, spill-over will not offset the losses to fishers from closing sections of the fishing grounds (Partnership for Interdisciplinary Studies of Coastal Oceans, 2007). High levels of connectivity between protected areas are also required for successful conservation of marine species (McLeod et al., 2009).

This mechanism depends more on biomass than on biodiversity. However, there is evidence that greater biological diversity creates greater fish biomass in marine ecosystems, and therefore biodiversity remains a factor (Worm et al., 2006).

Mangrove conservation and restoration

The restoration and conservation of mangroves can enhance the productivity of the mangrove and adjacent ecosystems and benefit local livelihoods (Walters *et al.*, 2008). In three mangrove areas in India, the most intact mangrove area yielded approximately four times as much income from shellfish and finfish per day as the less robust mangrove areas (Kathiresan & Rajendran, 2002). Mangroves also are a source of valuable wood products, provide protection against coastal erosion, capture sediments from rivers and serve as storm barriers (Sathirathai & Barbier, 2001). There is evidence that mangroves can be a route out of poverty through increased income from shellfish harvest (Soontornwong, 2006), and income gains have been seen specifically for women (Magalhaes *et al.*, 2007). More often, however, mangroves provide only modest direct poverty reduction benefits (Islam & Haque, 2004).

The benefits to the poor from mangroves are more dependent on the biomass than the biological diversity of the mangroves. However, though replanted mangrove stands may be less diverse, they provide much of the same structural functionality as native mangroves and have been shown to hold equivalent populations of commercial crab species (Walton *et al.*, 2007). One review concluded that fish and crustacean communities can have a "remarkable recovery of biodiversity in restored mangroves" (Bosire *et al.*, 2008).

Employment in protected areas

When a protected area is established, the management often hires local people to help operate the protected area. A study by WWF International (2004) surveyed more than 200 protected areas across 37 countries and found the average protected area has 40 permanent staff. Local jobs provided by a protected area can reduce poverty, and the protected areas can help conserve biodiversity. While the number of protected area jobs and the pay may be modest, the multiplier effect of the new jobs in the local rural economy can be significant (Fortin & Gagnon, 1999). Those hired by protected areas tend to be the moderately poor to the better off. The poorest of the poor rarely have the basic skills or the connections needed to secure protected area jobs.

However, while there is anecdotal evidence of the employment benefits from protected areas, there is little hard evidence showing that working for a protected area can be a route out of poverty. Many protected area jobs do not go to local people, and positions that require a knowledge of the local area, such as guides and guards, often pay too little to lift a local person out of poverty. There is also the uncertainty about the net benefits to local people from a protected area. Establishing a new protected area may generate new local jobs, but it can also negatively impact the livelihoods of other local people by restricting access to natural resources inside the protected area or involuntarily resettling people living inside the protected area (Brockington *et al.*, 2006).

Unlike several other mechanisms discussed in this chapter, this strategy depends on biodiversity, not biomass, to drive economic benefits. Therefore, if properly functioning, a protected area initiative should have direct benefits for biodiversity conservation.

Agro-forestry

Agro-forestry is the practice of integrating domesticated trees into agricultural landscapes. The poverty reduction benefits include improved soil fertility and crop productivity, diversification of food production for both commercial and subsistence needs, and tree products such as oils and medicines (Leakey *et al.*, 2005; Schreckenberg *et al.*, 2006).

Evidence from West Africa suggests that indigenous domesticated fruit trees can improve the livelihoods of poor households, especially for women (Schreckenberg *et al.*, 2006). In India, agro-forestry was found to augment income from rice monocultures by 2–3 times compared with non-agro-forestry households (Pandey, 2007).

However, uncertain governance of trees and tree products can undermine potential benefits from agro-forestry (Ashley *et al.*, 2006), and without access to markets, households are not able to capture many of those benefits (Leakey *et al.*, 2005). As the cultivation of trees requires a multi-year investment, those without secure property rights are less likely to participate in agro-forestry initiatives (Garrity, 2004).

Agro-forestry has benefits for biodiversity by providing structurally complex habitat for forest species, serving as biological corridors and reducing human pressure on natural forests (Bhagwat *et al.*, 2008). While agroforestry landscapes do not contain the same level of biodiversity as natural forests, they generally contain significantly more species than monoculture agriculture (Harvey *et al.*, 2006). However, by some estimates, up to 10% of tree species used in agroforestry systems are non-native, potentially negatively affecting local biodiversity (McNeely, 2004).

Grasslands management

There is evidence that many types of grasslands are ecologically dependent on grazing to maintain their biodiversity and productivity (Fratkin & Mearns, 2003). Livestock is also the primary form of wealth for a number of often poor, nomadic communities in grasslands, underpinning the importance of grassland management as a poverty conservation mechanism.

However, despite grasslands being used by poor people on nearly all continents, hard data for better grasslands management *lifting people out of poverty* (as opposed to acting as a safety net) are found in only one case, in which a rotational grazing scheme in the Mongolian grasslands was seen to reduce the number of poor households in the community by half (Schmidt, 2006). In Zimbabwe, Campbell *et al.* (2002) found that better off herders are generally the ones who have the assets to capitalise on the rapidly changing opportunities in dryland habitats. There is some scope, however, for animal husbandry in grasslands benefitting the poorest of the poor. This group often does not own livestock, but if they can acquire animals, it can be a pathway out of poverty (Peden *et al.*, 2003).

In many grasslands, there is evidence that grazing pressure, grass productivity and biodiversity are interdependent (Georgiadis *et al.*, 1989; Tilman *et al.*, 1996, 2001; Cardinale *et al.*, 2007; Guo, 2007). Increased productivity of the grasslands depends on the biological diversity of the grasses which in turn is dependent on the grazing of animals.

Agro-biodiversity conservation

Agro-biodiversity encompasses all the variety and variability of animals, plants and micro-organisms that are used directly or indirectly for food and agriculture (Food and Agriculture Organization, 1999). Initiatives for agro-biodiversity conservation generally help poor farmers diversify the species and varieties of native crops on their farms. There is a strong gender component to agro-biodiversity conservation, as it is often women who grow native species as companion crops to the household's cash crops (Momsen, 2007). For low-input agriculture in marginal lands, promoting agro-biodiversity has been shown to benefit poor farmers through better nutrition and more stable food production (Scherr & McNeely, 2007). Conserving traditional landraces can also help poor farmers address local challenges like droughts, floods and pests and thus help with food security.

There is, however, scant empirical evidence of an agro-biodiversity conservation project directly reducing local poverty (Kontoleon *et al.*, 2008), largely because these causal links are difficult to trace. The poorest of the poor rarely benefit from agro-biodiversity conservation initiatives because they rarely own land. Also, a primary goal of agro-biodiversity conservation is to conserve genetic diversity crucial to increasing global crop yields; benefits of conservation at the local scale may be much smaller.

Challenges and knowledge gaps

To augment the empirical evidence of conservation mechanisms with proven benefits, over 100 policy evaluation studies were examined to identify common challenges faced by conservation projects with poverty impacts. From these studies, four largely interrelated challenges were common. Firstly, better off households with higher physical and social capital are more likely to participate in a conservation initiative (Groom *et al.*, 2010). Secondly, elites often capture the benefits of an initiative; for example, a game management project in Zambia provided positive welfare impacts only to non-poor households (Bandyopadhyay & Tembo, 2009). Thirdly, these two factors lead to widening income disparities within rural communities (Vyamana, 2009). Finally, discrimination against women often prevents some of the poorest members of a community from benefitting (Jumbe & Angelsen, 2006).

Our state of knowledge review found the major constraint in assessing the link between biodiversity conservation and poverty reduction is the lack of hard evidence. Despite a wealth of case studies, the vast majority of the existing body of work does not use the analytical and empirical methods required to make reliable inferences about the actual impact of a conservation intervention on measurable poverty indicators (Weber *et al.*, 2010). Overviews and meta-analyses that have been carried out are largely inconclusive or unsubstantiated by rigorous analysis. This lack of evidence is exacerbated by limited political will among government and policy organisations, a multiplicity of objectives across donor and implementing agencies, a lack of knowledge of policy evaluation methods and a misconception that policy evaluation is expensive (Ferraro & Pattanayak, 2006).

Conclusions

It is clear from the review that there is considerable scope for increasing the poverty reduction benefits of conservation initiatives. One approach that has been mentioned prominently (e.g. Millennium Ecosystem Assessment, 2005) is to focus on meeting the basic needs of the poor via conservation initiatives. Despite good intentions, we believe such a focus may be counterproductive to conservation efforts and is unlikely to actually meet the needs of the poor. Rather, we believe conservation organisations should focus on creating equitable opportunities for socio-economic development within poor communities via activities that are intrinsically linked to the conservation of nature. It is the involvement and support of local people that make or break most conservation initiatives. Conservation needs to ensure that local people – many of whom are poor – benefit tangibly from biodiversity. For conservation to be relevant in the 21st century, this has to be a primary focus.

References

Angelsen, A. & Wunder, S. (2003) *Exploring the Poverty-Forestry Link: Key Concepts, Issues and Research Implications.* CIFOR Occasional Papers No. 40. CIFOR, Bogor, Indonesia.

Ashley, C. &. Roe, D. (2002) Making tourism work for the poor: strategies and challenges in southern Africa. *Development Southern Africa*, 19, 61–82.

Ashley, R., Russell, D. & Swallow, B. (2006) The policy terrain in protected area landscapes: challenges for agroforestry in integrated landscape conservation. *Biodiversity and Conservation*, 15, 663–89.

Bandyopadhyay, S. &. Tembo, G. (2009) *Household Welfare and Natural Resource Management around National Parks in Zambia*. Policy Research Working Paper No. 4932. World Bank Environment Department, Washington, DC.

Bhagwat, S.A., Willis, K.J., Birks, H.J.B. & Whittaker, R.J. (2008) Agroforestry: a refuge for tropical biodiversity? *Trends in Ecology and Evolution*, 23, 261–7.

Bosire, J., Dahdouh-Guebas, F., Walton, M. *et al.* (2008) Functionality of restored mangroves: a review. *Aquatic Botany*, 89, 251–9.

Bray, D.B., Merino-Pérez, L., Negreros-Castillo, P., Segura-Warnholtz, G., Torres-Rojo, J.M. & Vester, H.F.M. (2003) Mexico's community-managed forests as a global model for sustainable landscapes. *Conservation Biology*, 17, 672–7.

Brockington, D., Igoe, J. & Schmidt-Soltau, K. (2006) Conservation, human rights, and poverty reduction. *Conservation Biology*, 20, 250–2.

Brundtland, G.H. (ed.) (1987) *Our Common Future: The World Commission on Environment and Development*. Oxford University Press, Oxford.

Campbell, B., Jeffrey, S., Kozanayi, W., Luckert, M., Mutamba, M. & Zindi, C. (2002) *Household Livelihoods in Semi-Arid Regions: Options and Constraints*. CIFOR, Bogor, Indonesia.

Cardinale, B.J., Wright, J.P., Cadotte, M.W., Carrol, I.T., Hector. A. *et al.* (2007) Impacts of plant diversity on biomass production increase through time because of species complementarity. *Proceedings of the National Academy of Sciences of the USA*, 46, 18123–8.

Davis, T. (2005) *Local and Semi-Local Economic Impacts of Dive Tourism in Bunaken National Park, North Sulawesi, Indonesia*. Unpublished master's thesis, University of Washington, Seattle.

Dove, M.R. (1993) A revisionist view of tropical deforestation and development. *Environmental Conservation*, 20, 17–24.

Echavarría, M., Vogel, J., Alban, M. & Meneses, F. (2004) *The impacts of payments for watershed services in Ecuador*. International Institute for Environment and Development, London.

Ferraro, P.J. and Pattanayak, S.K. (2006) Money for nothing? A call for empirical evaluation of biodiversity conservation investments. *PLoS Biology*, 4, 482–8.

Food and Agriculture Organization (FAO) (1999) *Agricultural Biodiversity, Multifunctional Character of Agriculture and Land Conference*. Background Paper No. 1. FAO, Rome.

Fortin, M.J. & Gagnon, C. (1999) An assessment of social impacts of national parks on communities in Quebec, Canada. *Environmental Conservation*, 26, 200–11.

Fratkin, E. &. Mearns, R. (2003) Sustainability and pastoral livelihoods: lessons from east African Maasai and Mongolia. *Human Organization*, 62, 112–22.

Garrity, D.P. (2004) Agroforestry and the achievement of the Millennium Development Goals. *Agroforestry Systems*, 61, 5–17.

Gell, F.R. & Roberts, C.M. (2003) Benefits beyond boundaries: the fishery effects of marine reserves and fishery closures. *Trends in Ecology and Evolution*, 18, 448–55.

Georgiadis, N.J., Ruess, R.W., McNaughton, S.J. & Western, D. (1989) Ecological conditions that determine when grazing stimulates grass production. *Oecologia*, 81, 316–22.

Groom, B., Grosjean, P., Kontoleon, A., Swanson, T. & Zhang, S. (2010) Relaxing rural constraints: a 'win-win' policy for poverty and environment in China? *Oxford Economic Papers*, 62, 132–56.

Guo, Q. (2007) The diversity-biomass-productivity relationship in grasslands management and restoration. *Basic and Applied Ecology*, 8, 199–208.

Harvey, C.A., Gonzalez, J. & Somarriba, E. (2006) Dung beetle and terrestrial mammal diversity in forests, indigenous agroforestry systems and plantain monocultures in Talamanca, Costa Rica. *Biodiversity and Conservation*, 15, 555–85.

Hemmati, M. (ed.) (1999) *Gender and Tourism: Women's Employment and Participation in Tourism*. Summary of UNED UK's Project Report. UNED Forum, London

Islam, M.S. & Haque, M. (2004) The mangrove-based coastal and nearshore fisheries of Bangladesh: ecology, exploitation and management. *Reviews in Fish Biology and Fisheries*, 14, 153–80.

Jumbe, C. & Angelsen, A. (2006) Do the poor benefit from devolution policies? Evidence from Malawi's forest co-management program. *Land Economics*, 82, 562–81.

Kathiresan, K. & Rajendran, N. (2002) Fishery resources and economic gain in three mangrove areas on the south-east coast of India. *Fisheries Management and Ecology*, 9, 277–83.

Kontoleon, A., Pascual U. & Smale, M. (eds.) (2008) *Agrobiodiversity Conservation and Economic Development*. Routledge, Oxon, UK.

Leakey, R.R.B., Tchoundjeu, Z., Schreckenberg, K., Shackleton, S.E. & Shackleton, C.M. (2005) Agroforestry tree products (AFTPs): targeting poverty reduction and enhanced livelihoods. *International Journal of Agricultural Sustainability*, 3, 1–23.

Leisher, C., Van Beukering, P. & Scherl, M. (2007) *Nature's Investment Bank: How Marine Protected Areas Contribute to Poverty Reduction*. Nature Conservancy, Arlington, VA.

Magalhaes, A., Marinho da Costa, R., da Silva, R. & Pereira, L. (2007) The role of women in the mangrove crab production process in North Brazil. *Ecological Economics*, 61, 559–65.

Mahanty, S., Gronow, J., Nurse, M. & Malla, Y. (2006) Reducing poverty through community based forest management in Asia. *Journal of Forest and Livelihood*, 5, 78–89.

Mayers, J. (2006) *Poverty Reduction through Commercial Forestry: What Evidence? What Prospects?* Forests Dialogue Research Publication Series No. 2. Yale University School of Forestry and Environment Studies, New Haven, CT.

McLeod, E., Salm, R., Green, A. & Almany, J. (2009) Designing marine protected area networks to address the impacts of climate change. *Frontiers in Ecology and the Environment*, 7, 362–70.

McNeely, J.A. (2004) Nature vs. nurture: managing relationships between forests, agroforestry, and biodiversity. *Agroforestry Systems*, 61, 155–65.

Millennium Ecosystem Assessment (MA) (2005) *Ecosystems and Human Well-Being: General Synthesis*. WRI/Island Press, Washington, DC.

Molnar, A., Liddle, M., Bracer, C., Khare, A., White, A. & Bull, J. (2007) *Community-Based Forest Enterprises in Tropical Forest Countries: Status and Potential*. International Timber Trade Organization, Yokohama, Japan.

Momsen, J.H. (2007) Gender and agrobiodiversity: introduction to special issue. *Singapore Journal of Tropical Geography*, 28, 1–6.

Neumann, R.P. & Hirsch, E. (2000) *Commercialisation of Non-Timber Forest Products: Review and Analysis of Research*. CIFOR, Bogor, Indonesia.

Pagiola, S., Arcenas, A. & Platais, G. (2005) Can payments for environmental services help reduce poverty? An exploration of the issues and the evidence to date from Latin America. *World Development*, 33, 237–53.

Pandey, D.N. (2007) Multifunctional agroforestry systems in India: a review. *Current Science*, 92, 455–63.

Partnership for Interdisciplinary Studies of Coastal Oceans (PISCO) (2007) *Science of Marine Reserves*, 2nd edn. PISCO, Corvallis, OR.

Peden, D., Tadesse, G. & Mammo, M. (2003) *Improving the Water Productivity of Livestock: An Opportunity for Poverty Reduction*. International Livestock Research Institute (ILRI), Addis Ababa, Ethiopia.

Pollnac, R.B., Crawford, B.R. & Gorospe, M.L.G. (2001) Discovering factors that influence the success of community-based marine protected areas in the Visayas, Philippines. *Ocean & Coastal Management*, 44, 683–710.

Porras, I., Grieg-Gran, M. & Neveset, N. (2008) *All That Glitters: A Review of Payments for Watershed Services in Developing Countries*. Natural Resource Issues No. 11. International Institute for Environment and Development, London.

Richards, M. & Jenkins, M. (2007) *Potential and Challenges of Payments for Ecosystem Services from Tropical Forests*. Forestry Briefing 16, December. Overseas Development Institute, London.

Ros-Toten, M.A.F. & Wiersum, K.F. (2003) *The Importance of Non-Timber Forest Products for Forest-Based Rural Livelihoods: An Evolving Research Agenda*. AGIDS/UvA, Amsterdam, the Netherlands.

Sathirathai, S. & Barbier, E.B. (2001) Valuing mangrove conservation in southern Thailand. *Contemporary Economic Policy*, 19, 109–22.

Scherr, S. & McNeely, J. (eds.) (2007) *Farming with Nature: The Science and Practice of Ecoagriculture*. Island Press, Washington, DC.

Scherr, S., White, A. & Kaimowitz, D. (2004) *A New Agenda for Forest Conservation and Poverty Reduction: Making Markets Work for Low-Income Producers*. Forest Trends, Washington, DC.

Schmidt, S. (2006) *Conservation and Sustainable Management of Natural Resources in the Gobi: People Centred Conservation and Community-Driven Poverty Reduction*. Unpublished draft report, January. New Zealand Nature Institute Initiative for People Centred Conservation, Wellington.

Schreckenberg, K., Awono, A., Degrande, A., Mbosso, C., Ndoye, O. & Tchoundjeu, Z. (2006) Domesticating indigenous fruit trees as a contribution to poverty reduction. *Forests, Trees and Livelihoods*, 16, 35–51.

Shackleton, C. & Shackleton, S. (2004) The importance of non-timber forest products in rural livelihood security and as safety nets: a review of evidence from South Africa. *South African Journal of Science*, 100, 658–64.

Shah, K. & Gupta, V. (2000) *Tourism, the Poor and Other Stakeholders: Experience in Asia*. Fair Trade in Tourism Project. Overseas Development Institute, London.

Soontornwong, S. (2006) Improving rural livelihood through CBNRM: a case of self-organization in community mangrove management in Thailand. In *Hanging in the Balance: Equity in Community-Based Natural Resources Management in Asia*, ed. S. Mahanty, J. Fox, M. Nurse, P. Stephen & L. McLees, pp. 182–99. RECOFTC and East-West Centre, Bangkok.

Sunderlin, W.D., Angelsen, A., Belcher, B. *et al.* (2005) Livelihoods, forests, and conservation in developing countries: an overview. *World Development*, 33, 1383–402.

Sunderlin, W.D., Dewi, S.D. & Puntodewo, A. (2007) *Poverty and Forests: Multicountry Analysis of Spatial Association and Proposed Policy Solutions.* CIFOR Occasional Paper No. 47. CIFOR, Bogor, Indonesia.

Tilman, D., Reich, P.B., Knops, J., Wedin, D., Mielke, T. & Lehman, C. (2001) Diversity and productivity in a long-term grassland experiment. *Science*, 294, 843–5.

Tilman, D., Wedin, D. & Knops, J. (1996) Productivity and sustainability influenced by biodiversity in grassland ecosystems. *Nature*, 379, 718–20.

Vyamana, V.G. (2009) Participatory forest management in the Eastern Arc Mountains of Tanzania: who benefits? *International Forestry Review*, 11, 239–53.

Walters, B.B., Ronnback, P., Kovacs, J.M. *et al.* (2008) Ethnobiology, socio-economics and management of mangrove forests: a review. *Aquatic Botany*, 89, 220–36.

Walton, M.E.M., Le Vay, L., Lebata, J.H., Binas, J. & Primavera, J.H. (2007) Assessment of the effectiveness of mangrove rehabilitation using exploited and non-exploited indicator species. *Biological Conservation*, 138, 180–8.

Weber, J., Sills, E., Bauch, S. & Pattanayak, S. (2010) Do ICDPs work? An empirical evaluation of forest-based microenterprises in the Brazilian Amazon. Unpublished work.

Wells, M. & Brandon, K. (1992) *People and Parks: Linking Protected Area Management with Local Communities.* World Bank, Washington, DC.

World Bank (2001) *Attacking Poverty: World Development Report 2000/2001.* Oxford University Press, Oxford.

World Bank (2007) *Zambia Economic and Poverty Impact of Nature-Based Tourism.* Report No. 43373-ZM. World Bank, Washington, DC.

Worm, B., Barbier, E.B., Beaumont, N. *et al.* (2006) Impacts of biodiversity loss on ocean ecosystem services. *Science*, 314, 787–90.

Wunder, S. (2001) Poverty alleviation and tropical forests – what scope for synergies? *World Development*, 29, 1817–33.

Wunder, S. (2005) *Payments for Environmental Services: Some Nuts and Bolts.* CIFOR Occasional Paper No. 42. CIFOR, Bogor, Indonesia.

WWF International (2004) *Are Protected Areas Working? An Analysis of Forest Protected Area by WWF.* WWF International, Gland, Switzerland.

10

Protected Areas – What People Say about Well-Being

George Holmes[1] *and Dan Brockington*[2]

[1]School of Earth and Environment, University of Leeds, Leeds, UK
[2]School of Environment and Development, University of Manchester, Manchester, UK

Introduction

Since the early 1990s, a body of academic literature has emerged which examines the social impacts of protected areas – the consequences of conservation regulations on the well-being of people living in and around protected areas. With the exception of a few located in extreme environments, all protected areas have neighbouring, resident or formerly resident populations whom they effect. In a way, this literature addresses a historical lacuna – there were concerns expressed during the early decades of Yellowstone National Park in the late 19th century about the negative impacts on local populations (Jacoby, 2001), and resident peoples' well-being was at the heart of debates during the creation of Serengeti National Park and Ngorongoro Conservation Area in the 1950s (Neumann, 1998). However, in these early debates the issue had not previously attracted much scholarly attention. Since the turn of the 21st century, however, the literature has become more detailed, covering contemporary and historical examples from across the globe, and getting increasingly theoretically rich. It is a topic addressed in mainstream conservation journals, and a frequent subject in newer journals such as *Conservation and Society*. Since around 2003, the academic literature has been accompanied by, and interacted with, the rising concern and frequent debates on this issue during conservation conferences and within large conservation nongovernmental organisations (NGOs) (as reflected in Brosius, 2004;

Biodiversity Conservation and Poverty Alleviation: Exploring the Evidence for a Link, First Edition.
Edited by Dilys Roe, Joanna Elliott, Chris Sandbrook and Matt Walpole.

Terborgh, 2004; Wilkie *et al.*, 2006; Redford & Fearn, 2007; Mascia & Claus, 2009), as well as in more popular presses (Chapin, 2004; Dowie, 2008). This chapter aims to summarise current understandings of the social impacts of protected areas, with a particular emphasis on how they relate to conservation governance – the different techniques and approaches used to regulate protected areas.

Readers will note that there is a tendency in this chapter to report some of the more problematic aspects of protected areas. This is not because protected areas tend to have more negative aspects than positive ones. Far from it; indeed the abiding refrain is that they distribute both fortune and misfortune and the analyst's task is to understand both and their distribution. The reason for the balance of this chapter is primarily that it is a literature review. The scholarship which we are reviewing has tended to focus more on negative aspects than positive because these were the ones which were not, historically, well attended to. Conservation has tended to be seen as something which is generally "a good thing" and researchers have tended to try to redress this balance. The second reason is that this chapter has to be read in the context of the rest of this book which includes chapters that highlight the positive impacts which protected areas can afford (e.g. Chapters 9 and 18, this volume).

We begin by outlining where knowledge about the social impacts of protected areas can be found, before discussing briefly what these impacts are, how to understand them and how widespread they are. We then analyse the generalisations that can be drawn from these surveys, particularly what factors shape who is affected by social impacts, and how.

Who knows what about the social impacts of protected areas?

Knowledge about the social impacts of protected areas can be found in four general areas, the first of which is the academic literature. There are hundreds of case studies of individual protected areas, which use quantitative or qualitative methodologies, or a combination, and range from short journal papers to book-length ethnographic accounts. In addition, there are general surveys of these academic case studies (e.g. Brockington & Igoe, 2006; West and Brockington, 2006; Adams & Hutton, 2007; Holmes, 2007; Coad *et al.*, 2008; Galvin & Haller, 2008) whose comparisons can demonstrate the complexities and heterogeneity of impacts, as well as outline potential generalisations. Secondly, there are a number of large-scale surveys undertaken by international conservation NGOs to assess how well-protected areas are performing against a number of measures, including impacts on local communities (e.g. Dudley *et al.*, 2008; Leverington *et al.*, 2008). These collate information from appraisals of thousands of individual protected areas, gathered by NGOs or protected area authorities, and attempt to standardise them to make global- or continental-scale generalisations.

Their scale and ambition make them useful, although the methodologies of individual appraisals can be questionable. Thirdly, there are designed studies, which aim to systematically explore whether protected areas have had visible social impacts globally or regionally, mainly by comparing large-scale data sets on protected areas, poverty, population changes and other factors (e.g. de Sherbinin, 2008; Upton *et al.*, 2008; Wittemyer *et al.*, 2009). These large-scale studies are of limited use because they fail to capture the crucial causative factors operating at local scales, although there are honourable exceptions (Ribot *et al.*, 2004; Nelson & Agrawal, 2008; Andam *et al.*, 2010). Finally, there are numerous individual monitoring studies undertaken by conservation and development NGOs tracking changes across a broad range of variables in communities surrounding various protected areas (see Richardson, 2008, for a review). On-going efforts are trying to develop a standardised methodology for reviewing the social impacts of protected areas, which promise to produce good-quality data to be used for comparing the impacts of different protected areas and drawing conclusions about what combinations of processes lead to what social impacts (Schreckenberg *et al.*, 2010).

Overall, our knowledge is (i) fragmented (with important geographical and thematic areas missing) and (ii) overly large scale (neglecting important local dynamics and national political economies). The epistemic community studying social impacts is small and fragmented, and much data remain informal, yet to emerge in peer-reviewed form.

What are the social impacts of protected areas, and how widespread are they?

Before discussing what the studies noted in this chapter's 'Who Knows What about the Social Impacts of Protected Areas?' section show, it is useful to carefully consider definitions of key terms. Firstly, our understandings of what constitutes a protected area reflect the broad definition used in contemporary debates, extending beyond International Union for Conservation of Nature (IUCN) category I–VI areas (see Table 10.1) to include indigenous and community conserved areas (ICCAs), even though these are understudied and barely present in discussions of the social impacts of protected areas. Recognising the diversity of forms of protected areas is central to how we understand how they might interact with human well-being. Secondly, and without exploring its complexities in detail, we prefer to use the term *well-being* in lieu of *poverty*, to capture the idea that the real and important social impacts of protected areas extend beyond economic and livelihood issues into matters of culture, identity and community. Thirdly, care is needed when distinguishing between forced change or induced change and voluntary change in communities surrounding protected areas (Schmidt-Soltau & Brockington, 2007). There is a broad spectrum running

Table 10.1   **Protected area definitions**

IUCN category	Designation	Description
Ia	Strict nature reserve	"Strictly protected areas set aside to protect biodiversity and also possibly geological/geomorphological features, where human visitation, use and impacts are strictly controlled and limited to ensure protection of the conservation values."
Ib	Wilderness area	"Large unmodified or slightly modified areas, retaining their natural character and influence without permanent or significant human habitation, which are protected and managed so as to preserve their natural condition."
II	National park	"Large natural or near natural areas set aside to protect large-scale ecological processes, along with the complement of species and ecosystems characteristic of the area, which also provide a foundation for environmentally and culturally compatible, spiritual, scientific, educational, recreational, and visitor opportunities."
III	Natural monument or feature	Areas "set aside to protect a specific natural monument, which can be a landform, sea mount, submarine cavern, geological feature such as a cave or even a living feature such as an ancient grove. They are generally quite small protected areas and often have high visitor value."
IV	Habitat or species management area	Areas which "aim to protect particular species or habitats and management reflects this priority. Many Category IV protected areas will need regular, active interventions to address the requirements of particular species or to maintain habitats, but this is not a requirement of the category."
V	Protected landscape or seascape	"A protected area where the interaction of people and nature over time has produced an area of distinct character with significant, ecological, biological, cultural and scenic value: and where safeguarding the integrity of this interaction is vital to protecting and sustaining the area and its associated nature conservation and other values."
VI	Protected area with sustainable use	Areas which "conserve ecosystems and habitats together with associated cultural values and traditional natural resource management systems. They are generally large, with most of the area in a natural condition, where a proportion is under sustainable natural resource management and where low-level non-industrial use of natural resources compatible with nature conservation is seen as one of the main aims of the area."

Source: Adapted from Dudley (2008: 13–19).

from changes that are forced on local communities by protected area authorities, to changes which the communities themselves freely choose to make. For example, in discussions of evictions from protected areas, local communities are sometimes formally evicted directly to make way for protected areas (physical displacement), sometimes they unwillingly leave their homes because protected area legislation overly restricts their livelihoods and sometimes they choose to leave to pursue better opportunities elsewhere. There is often an overlap between these forms of change (Cernea, 2005; Brockington & Igoe, 2006). Black and white distinctions between what is forced on local communities and how they freely choose to respond to protected areas are unhelpful.

When considering social impacts, two key observations need to be considered. Firstly, protected areas as institutions do not preserve ecologies or communities in a state of suspended animation. Rather, they take dynamic landscapes, societies and economies, and alter them further, often radically. These dynamics are both direct, such as new regulations changing how people use natural resources, and indirect, such as the long-term economic and social consequences of ecotourism. They can be intentional or unforeseen, but either way protected areas are far from static. Secondly, the data so far indicate that the social impacts of protected areas tend to stay at a local level, in communities within or close to their boundaries. Upton *et al.* (2008) and de Sherbinin (2008) both found no noticeable national- or subnational-scale relationship between poverty and protected areas. Given that those who live near resources are those who are most likely to depend on them, it is expected that they will be the people most affected when resource use changes.

A number of key social impacts of protected areas are reflected in the literature. One of the most prominent is the removal of longstanding residents of protected areas and their surrounds, through either forced eviction or induced economic displacement. The long-term occupants and users of Yellowstone were evicted by the US Cavalry shortly after the park's creation, and the lost access to land and resources induced further migration from neighbouring communities (Jacoby, 2001). Eviction and economic displacement from protected areas have continued ever since. The former is more dramatic, but the latter (which entails restrictions to economic activities but not physical displacement) is more common (Brockington & Igoe, 2006). Many removals can be linked to certain ideas of wilderness and the authenticity and necessity of people-free conservation landscapes. The people affected may have their livelihoods disrupted, putting them in (or exacerbating their) poverty (Cernea & Schmidt-Soltau, 2006). They may become more vulnerable to economic or environmental shocks and pressures. Where compensation is offered to offset livelihood losses, this may be inadequate – McLean and Streade (2003) show how people evicted from Royal

Chitwan National Park in Nepal were resettled on land far inferior to that which they previously farmed. Evictions and displacement can disrupt social relations, breaking ties binding communities, as people move to disparate places. As well as having important livelihood impacts – these social relations can be the foundations for economic relations – they also have real and important psychological consequences. However, removals are not always bad: sometimes local people are in favour of relocation when the compensation schemes are sufficient to improve their standards of living and livelihood security, and when they are empowered to negotiate the terms of their movement (Karanth, 2007; Beazley, 2009).

The converse is also true: as well as their histories of exclusion, protected areas can also be a means for marginalised rural and indigenous people to secure access to land and resources, particularly through the less strict IUCN categories and ICCAs, and in countries whose political and legal structures favour indigenous claims (Galvin & Haller, 2008). This can still entail different and more complicated processes of exclusion, as making a site available to indigenous people may also require a process to determine who is indigenous and who is not. This can involve rather painful processes of demonstrating attachment to land and communities in courts (e.g. in Australia), or it can entail conflictual community-based processes of exclusion in which members of communities are admitted, and excluded, according to their ethnicity. The early negotiations over the creation of the Mkomazi Game Reserve in Tanganyika in 1951 involved such processes (Brockington, 2002).

Protected areas can limit and change local people's access to land and resources, reworking local livelihoods and society. Restrictions can induce or increase poverty, and it can increase vulnerability to economic and environmental shocks, by preventing people from accessing resources upon which they depend, or limiting the livelihood strategies upon which they can draw during either normal times or times of stress (McElwee, 2010), although the inverse can also be true – protected areas can increase resilience by ensuring that emergency resources are protected (McSweeney, 2005; Chapter 18, this volume). Gjertsen (2005) notes that marine protected areas can have a noticeable effect on nutrition levels in neighbouring communities, with both positive and (especially in cases where restrictions are stronger) negative consequences for well-being (see also Chapter 9, this volume). People may need to purchase resources, such as fodder and fuel, which they previously could obtain for free, and their livelihoods may become subject to violent or corrupt regulation by park guards (Allendorf, 2007). Although a wider issue in rural societies, human – wildlife conflict – the effects of predation and crop raiding by wild animals on lives and livelihoods – is particularly exacerbated around protected areas (see Peterson et al., 2010, for a global review). Protected areas and the programmes they introduce (e.g. ecotourism) often bring new processes and forces into local societies, such as market logics and waged labour, which can alienate and impoverish some locals (MacDonald, 2005). Protected areas can also affect psychological aspects of well-being, as rural livelihood processes are interlinked with social structures. Cash benefits from economic projects around protected areas may become captured by local elites, exacerbating inequality and increasing social tension (Buscher, 2010). Changes in social structure can consequently rework land tenure, altering access to resources and livelihoods (Coad *et al.*, 2008). Protected areas regulations can also restrict access to religious and culturally important resources and locations (Neumann, 1998; Norgrove & Hulme, 2006: Mombeshora & Le Bel, 2009). Finally, protected areas may often be involved in rewriting local histories, removing people from landscapes, reinventing them and otherwise transforming the cultural connections between people and places built over many years (Neumann, 1998; Jacoby, 2001; Brockington, 2002).

The literature is not very revealing on the frequency of these impacts across protected area networks. Brockington and Igoe's (2006) global review of historical and contemporary eviction found examples from only 184 protected areas, a tiny fraction of the more than 100,000 in existence. Even if we consider that there are only 5,000 category I–IV areas (the stricter forms which are most likely to involve removal) with a surface area of more than $100 \, km^2$ (thus big enough to have a significant widespread impact), this does not represent a widespread trend. However, detailed country-level studies (e.g. Lasgorceix & Kothari, 2009) indicate that the global survey significantly under-calculates incidences of removals. In other cases, available data are vigorously contested, for example in the debate between Schmidt-Soltau (2009) and Curran *et al.* (2009, 2010) about conservation in Central Africa.

There are some interesting meta-studies to draw upon. Dudley *et al.* (2008), in their collated global survey of more than 1000 protected area effectiveness evaluations, found that 75% of areas had an acceptable to positive impact on local communities, and only 5% had a strongly negative impact. This appears a strong indicator that negative social impacts are relatively rare, and the net effect of protected areas on local communities is positive. However, these evaluations suffer from some methodological flaws. They are based on quick assessments or surveys of park staff rather than detailed analyses from more independent observers. Some surveys are designed to primarily assess the benefits of protected areas, with less data gathered on the negative consequences (e.g. Stolton & Dudley, 2008). There is a tendency to aggregate impacts at community levels, rather than analysing at household levels where much of the differentiation in impacts can be observed. Mascia and Claus's (2009) meta-study of 96 marine protected areas found that roughly equal numbers of communities lost or gained access to fishing resources as a result of protected area regulations, and that only 16% experienced a loss in catches per unit effort. Overall, although the literature appears to show that severe impacts are rare, the lack of good-quality data makes it difficult to draw reliable conclusions.

What affects who experiences the impacts of protected areas, and why?

There are three factors which appear central to how protected areas affect the well-being of local communities: local existing micro-politics, protected area governance regimes and wider political economies.

Firstly, an important observation is that whilst the social impacts of protected areas are generally limited to communities nearby, they are also unevenly distributed within these communities. People benefitting greatly from the changes that protected areas bring may live next door to those who have suffered negative impacts. These different experiences are often based on existing social divisions, such as gender or ethnicity. For example, Dressler *et al.* (2006) shows that protected area restrictions in the Philippines disproportionately restrict the livelihoods of indigenous groups, but not those of immigrant groups. Impacts tend to reproduce social inequalities – the better off within a community accruing the benefits, and the costs falling on the poorest (Adams & Hutton, 2007). Richer and more powerful people within a community have the ability to capture the benefits that conservation can bring, whereas the most marginalised are often those whose livelihoods depend most directly on natural resources, and who can least afford to lose them (Chapters 9 and 15, this volume). Kabra's (2009) study illustrates the unequal impacts particularly well. She compares two case studies of evictions from protected areas from the same country (India) and shows that more socially marginalised *adivasi* tribal groups, who are more dependent on forest resources for livelihoods, suffer more from the after effects of evictions than other, less marginalised, socio-ethnic groups. Even efforts to minimise social impacts on local populations, such as community protected areas, have uneven consequences for local people, depending on whether they are considered a member of that community or not, and with what rights (Berkes, 2004). These important micro-scale inequalities are not recognised in much of the literature, especially in the surveys and analyses of NGOs, which tend to aggregate impacts at the community level.

Secondly, social impacts need to be considered in the context of different governance regimes for protected areas. Protected areas are heterogeneous; the IUCN recognises six broad management categories of protected area and four different governance regimes (government managed, cooperatively managed, private and community protected areas), each of which envisages a different relationship with local people. In addition, there are many customary and informal regulations over use of land and resources which may lack formal legal recognition by governments and lie outside of the IUCN system. All protected areas will have some social impact by their nature – they are there to govern resource use, and this will inevitably have consequences for local populations – but we can attempt to distinguish between the impacts of different types of protected area. Stricter protected areas, such as scientific and wilderness reserves (IUCN categories Ia and Ib) and national parks (IUCN category II), have greater potential for serious social impacts because their management objectives emphasise stronger restrictions on use of land and resources. These also tend to be the categories that emphasise centralised state or private sector control rather than devolution to local communities. Less strict categories, such as protected landscapes (category V) and managed resource areas (category VI), promise fewer negative social impacts and more benefits, emphasising community involvement and increased, if still restricted, access to resources. Yet the relationship is not so straightforward. Firstly, whilst IUCN categories aim to be international standards, there may be differences in how they are interpreted at the national level – policies for a category III area in one country may look very different from those in another. More importantly, the *de jure* regulations written in protected area law and management plans may substantially differ from the *de facto* regulations which are actually enforced, and which local people experience. For example, there is an important distinction between people illegally living inside protected areas in India, and therefore under threat of eviction, and those who have actually been evicted (Lasgorceix & Kothari, 2009).

Thirdly, social impacts need to be considered in their wider political and economic context, particularly the place of nature and indigenous and rural peoples in national economies and political discourses. The treatment of the San people in Botswana reflects wider intolerances of indigenous lifestyles within a modernising state (Bolaane, 2004). A historical parallel might be the treatment of Native Americans at the birth of protected areas in the United States (Jacoby, 2001). The economic importance of wildlife and hunting tourism in Tanzania, and the kinds of landscapes that tourists want to see, shapes how Tanzanian protected areas treat rural people (Neumann, 1998). Interestingly, Galvin and Haller (2008) suggest, based on a comparison of 13 case studies, that the social impacts of protected areas within one geographical region will share some general similarities in comparison with those of another geographical region because of a shared wider political and economic context. For example, they argue that because British and German imperialism left a legacy of ideas of wilderness in Africa, protected areas there tend to emphasise non-use, whereas in South America, where protected areas post-date colonialism and where indigenous people generally have clearly defined rights, protected areas tend to allow traditional indigenous uses. Social impacts not only are the product of protected area institutions themselves, but also are heavily influenced by wider trends, even if the impacts are manifested at the local level.

This wider political context is important when we consider the possibilities and potential for local people to react to conservation regulations. Many discussions of social impacts of protected areas consider them to be imposed on local people, whereas local people often have a voice and to varying degrees resist, contest and shape how protected area policies affect them (see Holmes, 2007, for a review). Local people can resist through formal means such as legal appeals and protest marches, but more commonly they lack the political resources to do this, and resist through more subtle, everyday ways, such as non-cooperation with park guards and sabotage. Whilst resistance can win some concessions for local people, they rarely succeed in mitigating significant social impacts. Protected areas are backed by powerful organisations such as the state and conservation NGOs, often with money, laws and legitimised violence to enforce regulations, whereas people living in and around protected areas are often amongst the weakest and most marginalised in society. These differences in power are crucial. There are cases (e.g. Karanth, 2007; Beazley, 2009) where resistance has significantly mitigated social impacts, but these are relative exceptions, the product of particular political and economic circumstances empowering rural people, such as the presence of campaigning NGOs and laws on forest rights. Similarly, although protest over the social impacts of protected areas can have an impact on the success of the area in protecting biodiversity (e.g. Western, 1994; Hulme, 1997) and others trace how discontent with protected areas drove local people to kill endangered wildlife as a way of sabotaging conservation efforts and protesting their consequences, local people rarely derail conservation through protest. Power imbalances between conservation authorities and local people mean that local people's consent is not always a pre-requisite for successful conservation – powerful states can generally impose regulations on dissenting but weak citizens (Brockington, 2004).

Conclusion

In recent decades, the increased attention paid to issues of the social impacts of protected areas has produced an increasing number of studies, coming from both academics and conservation practitioners, using a variety of methodologies across different scales. Whilst there is considerable heterogeneity within this, a number of generalisations can be drawn. Impacts tend to remain local, but how they are experienced is in large part a combination of pre-existing social divisions, protected area governance and wider political economies of conservation. Protected areas are transformative, taking dynamic societies and ecologies and altering them further, and their social impacts are constantly changing too, particularly as they are contested by local people.

Whilst our knowledge of these issues has expanded considerably in recent decades, there are significant areas of ignorance. Some parts of the world are under-represented in the international literature, although knowledge may be held by local experts in local languages. Information on protected areas outside of the control of governments, such as private lands and ICCAs, is largely absent. Potentially important data, such as GIS and remote-sensing analyses of changing land use patterns, have been underutilised in the debate. Finally, we lack a detailed understanding of the potential correlation and causal links between certain protected area policies and certain social impacts. This could be because as yet we do not have enough in-depth case studies to draw strong inferences. However, the complexity of what goes on in a protected area's relationship with local people and the importance of local social, cultural, political and economic contexts may mean that it is simply impossible to draw such detailed conclusions.

References

Adams, W.M. & Hutton, J. (2007) People, parks and poverty: political ecology and biodiversity conservation. *Conservation and Society*, 5, 147–83.

Allendorf, T. (2007) Residents attitudes toward three protected areas in southwestern Nepal. *Biodiversity and Conservation*, 16, 2087–102.

Andam, K.S., Ferraro, P.J., Sims, K.R.E., Healy, A. and Holland, M.B. (2010) Protected areas reduced poverty in Costa Rica and Thailand. *Proceedings of the National Academy of Sciences of the USA*, 107, 9996–10001.

Beazley, K. (2009) Interrogating notions of the powerless oustee. *Development and Change*, 40, 219–48.

Berkes, F. (2004) Rethinking community-based conservation. *Conservation Biology*, 18, 621–30.

Bolaane, M. (2004) The impact of game reserve policy on the river BaSarwa/Bushmen of Botswana. *Social Policy and Administration*, 38, 399–417.

Brockington, D. (2002) *Fortress Conservation: The Preservation of the Mkomazi Game Reserve, Tanzania*. James Currey, Oxford.

Brockington, D. (2004) Community conservation, inequality and injustice: myths of power in protected area management. *Conservation and Society*, 2, 411–32.

Brockington, D. & Igoe, J. (2006) Eviction for conservation: a global overview: conservation and society. *Conservation and Society*, 4, 424–70.

Brosius, J.P. (2004) Indigenous peoples and protected areas at the World Parks Congress. *Conservation Biology*, 18, 609–12.

Buscher, B. (2010) Anti-politics as political strategy: neoliberalism and transfrontier conservation in Southern Africa. *Development and Change*, 41, 29–51.

Cernea, M. (2005) Restriction of access is displacement: a broader concept and policy. *Forced Migration Review*, 23, 48–49.

Cernea, M.M. & Schmidt-Soltau, K. (2006) Poverty risks and national parks: policy issues in conservation and resettlement. *World Development*, 34, 1808–30.

Chapin, M. (2004) A challenge to conservationists. *World Watch Magazine*, November–December, 17–31.

Coad, L., Campbell, A., Miles, L. & Humphries, K. (2008) *The Costs and Benefits of Protected Areas for Local Livelihoods: A Review of the Current Literature*. Working paper. United Nations Environment Programme World Conservation Monitoring Centre, Cambridge.

Curran, B., Sunderland, T., Maisels, F., Asaha, S., Balinga, M., Defo, L., Dunn, A., von Loebenstein, K., Oates, J., Roth, P., Telfer, P. & Usongo, L. (2010). Response to 'Is the displacement of people from parks only "purported" or is it real?' *Conservation and Society*, 8, 99–102.

Curran, B., Sunderland, T.C.H., Maisels, S., Oates, J., Asaha, S., Balinga, M., Defo, L., Dunn, A., Telfer, P., Usongo, L., von Loebenstein, K. & Roth, P. (2009) Are Central Africa's protected areas displacing hundreds of thousands of rural poor? *Conservation and Society*, 7, 30−45.

de Sherbinin, A. (2008) Is poverty more acute near parks? An assessment of infant mortality rates around protected areas in developing countries. *Oryx*, 42, 26−35.

Dowie, M. (2008) *Conservation Refugees: The Hundred-Year Conflict between Global Conservation and Native Peoples*. MIT Press, Cambridge, MA.

Dressler, W., Kull, C. & Meredith, T. (2006) The politics of decentralizing national parks management in the Philippines. *Political Geography*, 25, 789−816.

Dudley, N. (ed.) (2008) *Guidelines for Applying Protected Area Management Categories*. IUCN, Gland, Switzerland.

Dudley, N., Mansourian, S., Stolton, S. & Suksuwan, S. (2008) *Safety Net: Protected Areas and Poverty Reduction*. A research report by WWF and Equilibrium. WWF International, Gland, Switzerland.

Galvin, M. & Haller, T. (eds.) (2008) *People, Protected Areas and Global Change: Participatory Conservation in Latin America, Africa, Asia and Europe*. Perspectives of the Swiss National Centre of Competence in Research (NCCR) North-South, University of Bern, Vol. 3. Geographica Bernensia, Bern, Switzerland.

Gjertsen, H. (2005) Can habitat protection lead to improvements in human well-being? Evidence from marine protected areas in the Philippines. *World Development*, 33, 199−217.

Holmes, G. (2007) Protection, politics and protest: understanding resistance to conservation. *Conservation and Society*, 5, 184−201.

Jacoby, K. (2001) *Crimes against Nature: Squatters, Poachers, Thieves and the Hidden History of American Conservation*. University of California Press, London.

Kabra, A. (2009) Conservation-induced displacement: a comparative study of two Indian protected areas. *Conservation and Society*, 7, 249−67.

Karanth, K. (2007) Making resettlement work: the case of Indias Bhadra Wildlife Sanctuary. *Biological Conservation*, 139, 315−24.

Lasgorceix, A. & Kothari, A. (2009) Displacement and relocation of protected areas: a synthesis and analysis of case studies. *Economic and Political Weekly*, 154, 39−47.

Leverington, F., Hockings, M. & Costa, K.L. (2008) *Management Effectiveness Evaluation in Protected Areas: A Global Study*. University of Queensland, International Union for Conservation of Nature and World Commission on Protected Areas, The Nature Conservancy and the World Wildlife Fund, Gatton, Australia.

MacDonald, K.I. (2005) Global hunting grounds: power, scale and ecology in the negotiation of conservation. *Cultural Geographies*, 12, 259−91.

Mascia, M.B. & Claus, C.A. (2009) A property rights approach to understanding human displacement from protected areas: the case of marine protected areas. *Conservation Biology*, 23, 16−23.

McElwee, P.D. (2010) Resource use among rural agricultural households near protected areas in Vietnam: the social costs of conservation and implications for enforcement. *Environmental Management*, 45, 113−31.

McLean, J. & Straede, S. (2003) Conservation, relocation, and the paradigms of park and people management: a case study of Padampur Villages and the Royal Chitwan National Park, Nepal. *Society and Natural Resources*, 16, 509−26.

McSweeney, K. (2005) Natural insurance, forest access, and compounded misfortune: forest resources in smallholder coping strategies before and after Hurricane Mitch, Northeastern Honduras. *World Development*, 33, 1453–71.

Mombeshora, S. & Le Bel, S. (2009) Parks-people conflicts: the case of Gonarezhou National Park and the Chitsa community in south-east Zimbabwe. *Biodiversity and Conservation*, 18, 2601–23.

Nelson, F. & Agrawal, A. (2008) Patronage or participation? Community-based natural resource management reform in Sub-Saharan Africa. *Development and Change*, 39, 557–85.

Neumann, R.P. (1998) *Imposing Wilderness: Struggles over Livelihood and Nature Preservation in Africa*. University of California Press, London

Norgrove, L. & Hulme, D. (2006) Confronting conservation at Mount Elgon. *Development and Change*, 37, 1093–116.

Peterson, M.N., Birckhead, J.L., Leong, K., Peterson, M.J. & Peterson, T.R. (2010) Rearticulating the myth of human-wildlife conflict. *Conservation Letters*, 3, 74–82.

Redford, K. & Fearn, E. (2007) *Protected Areas and Human Displacement: A Conservation Perspective*. Wildlife Conservation Society, New York.

Ribot, J. (2004) *Waiting for Democracy: The Politics of Choice in Natural Resource Decentralization*. World Resources Institute, Washington, DC.

Richardson, V. (2008) *Livelihood Impacts of Protected Areas: A Global Analysis of the Application of Assessments and Methodologies*. Unpublished MSc dissertation, Imperial College, London.

Schmidt-Soltau, K. (2009) Is the displacement of people from parks only purported, or is it real? *Conservation and Society*, 7, 46–55.

Schmidt-Soltau, K. & Brockington, D. (2007) Protected areas and resettlement: what scope for voluntary relocation? *World Development*, 35, 2182–202.

Schreckenberg, K., Camargo, I., Withnall, K., Corrigan, C., Franks, P., Roe, D., Scherl, L.M. & Richardson, V. (2010) *Social Assessment of Conservation Initiatives: A Review of Rapid Methodologies*. Natural Resources Issues No. 22. International Institute for Environment and Development (IIED), London.

Stolton, S. & Dudley, D. (2008) *The Protected Areas Benefits Assessment Tool: A Methodology*. WWF International, Gland, Switzerland.

Terborgh, J. (2004) Reflections of a scientist on the World Parks Congress. *Conservation Biology*, 18, 619–20.

Upton, C., Ladle, R., Hulme, D., Jiang, T., Brockington, D. & Adams, W.M. (2008) Are poverty and protected area establishment linked at a national scale? *Oryx*, 42, 19–25.

West, P. & Brockington, D. (2006) An anthropological perspective on some unexpected consequences of protected areas. *Conservation Biology*, 20, 609–16.

Western, D. (1994) Ecosystem conservation and rural development: the case of Amboseli. In *Natural Connections: Perspectives in Community-Based Conservation*, ed. D. Western & R.M. Wright, pp. 15–52. Island Press, Washington, DC.

Wilkie, D.S., Morelli, G.A., Demmer, J., Starkey, M., Telfer, P. & Steil, M. (2006) Parks and people: assessing the human welfare effects of establishing protected areas for biodiversity conservation. *Conservation Biology*, 20, 247–49.

Wittemyer, G., Elsen, P., Bean, W.T., Burton, A.C.O. & Brashares, J.S. (2008) Accelerated human population growth at protected area edges. *Science*, 32, 123–6.

11

Species Conservation and Poverty Alleviation – The Case of Great Apes in Africa

Chris Sandbrook[1] and Dilys Roe[2]

[1]United Nations Environment Programme World Conservation Monitoring Centre, Cambridge, UK
[2]International Institute for Environment and Development, London, UK

Introduction

Species have been at the heart of conservation practice since the beginnings of the formal conservation movement in the Western world in the 19th century. The last few years have seen considerable change in the focus of conservation activity, with more attention being paid to larger scale conservation of ecosystems and landscapes (Simberloff, 1998; Redford *et al.*, 2003). However, despite this trend, species conservation remains an important component of conservation practice in both terrestrial and marine environments (Entwistle & Dunstone, 2000). There are many conservation organisations dedicated to saving particular species, such as Save the Rhino, Save the Elephant and the International Gorilla Conservation Programme. There are also some species-specific donor funds, such as the great apes programmes of the Arcus Foundation and the US Fish and Wildlife Service. As we learn more about the importance of larger scale species assemblages and ecosystems, why does the conservation of individual species continue to play such a role?

There are various rationales for a focus on species in conservation. Firstly, and most importantly, some species are highly charismatic, and are able to generate considerable support for conservation from the general public, wealthy individuals and celebrities. Indeed, it is no accident that the species-specific conservation organisations and

Biodiversity Conservation and Poverty Alleviation: Exploring the Evidence for a Link, First Edition.
Edited by Dilys Roe, Joanna Elliott, Chris Sandbrook and Matt Walpole.
© 2013 John Wiley & Sons, Ltd. Published 2013 by John Wiley & Sons, Ltd.

donor funds mentioned here are all focussed on charismatic species (particularly large mammals), rather than other less charismatic but important species such as components of agro-biodiversity (Chapter 8, this volume). A classic example of the power of a species' charisma is the African elephant (*Loxodonta africana*), which has enthralled foreign tourists on African safaris for generations, and led to great international concern for its fate during the 1980s when poaching for ivory was at its peak. In some cases, the benefits of species charisma can be harnessed for conservation more broadly by using them as 'flagships' to raise awareness of, and generate resources for, a much wider set of species, an ecosystem or the work of an entire organisation (Walpole & Leader-Williams, 2002). An example of the latter is the Worldwide Fund for Nature (WWF), which has used the giant panda as part of its logo since the organisation was founded in Switzerland in 1961, even though WWF did not start working directly on panda conservation until much later.

Secondly, some species gain attention because they are critically endangered, and therefore seen as a top priority for conservation. For example, the giant ditch frog (or mountain chicken; *Leptodactylus fallax*) of Montserrat is hardly charismatic, but has received international conservation attention from the Durrell Wildlife Trust and Fauna and Flora International (FFI) in response to its threatened status. Finally, some species attract on-going conservation attention because public awareness about them has been raised significantly by charismatic researchers or conservationists who themselves have attracted public interest and become conservation 'celebrities' (Brockington, 2009). An example is the work of famous primatologists Dian Fossey and Jane Goodall, who have contributed enormously to the conservation of mountain gorillas and chimpanzees, respectively, and have both given their name to conservation organisations.[1]

This book is about the relationship between conservation and poverty. In this context, the question this chapter asks is 'What difference does a focus on species conservation make for poverty?' We tackle this question in the following sections. Firstly, we review in general terms what is known about the relationship between species conservation and poverty, both positive and negative. Secondly, we go into more detail using the case of great ape conservation in Africa as a case study of species conservation and its relationship with poverty. Finally, we conclude by identifying a set of factors which seem to shape this relationship, and provide recommendations for improving the poverty impact of species conservation work in the future.

The relationship between species conservation and poverty

It is interesting to note that traditional species conservation programmes have focussed on species that are of interest to conservationists rather than those that are

---

[1] It must be noted that the charisma of the gorillas and chimpanzees has undoubtedly contributed to the celebrity of the researchers who revealed them to the outside world.

of importance to poor people (Kaimowitz & Sheil, 2007). Their primary concern is with the conservation of wildlife species, and not with other issues such as the alleviation of human poverty. Indeed, species conservation programmes can in some cases be detrimental to people, and cause or exacerbate poverty. The clearest examples are associated with the establishment of protected areas, which are often set up to conserve a particular species found within them. The poverty impacts of protected areas, both positive and negative, are reviewed by Holmes and Brockington in Chapter 10 of this volume. Species conservation may also exacerbate poverty by preventing the hunting or trading of species of value to local people, or by protecting species that are detrimental to local livelihoods, such as crop-raiding elephants or cattle-rustling lions (Woodroffe et al., 2005). However, this does not mean that species conservation cannot include activities intended to reduce or alleviate poverty – and this is the focus of this chapter.

As with any other form of conservation, different species conservation programmes adopt different perspectives on the links between their activities and poverty alleviation – in line with the typology developed by Adams et al. (2004, described in Chapter 1). In practice, there are many circumstances under which poverty is considered to be a factor linked to threats to species of conservation concern, providing a rationale for many conservation initiatives to adopt the position that poverty is a critical constraint on conservation. For example, poor people may be hunting or collecting a threatened species for food or sale, converting its habitat or facilitating the spread of diseases to which the species is vulnerable. In response to such threats (real or perceived), species conservation programmes around the world have embarked on a range of activities intended to reduce conservation threats through reducing poverty. These include, among others, alternative or enhanced livelihoods projects such as promoting tourism, beekeeping, woodlots or piggeries; providing fuel-efficient stoves; providing financial benefits by sharing park revenues; setting up multiple-use zones within protected areas; providing payments to local people for conservation efforts (generally termed payments for environmental services (PES)) and developing the capacity of local institutions to promote good governance of natural resources (e.g. through community-based conservation) (Hutton & Leader-Williams, 2003; Walpole & Wilder, 2008; Gray & Rutagarama, 2011). The impacts of many of these interventions are discussed in more detail elsewhere in this volume (e.g. Chapters 9 and 13). The particular type of intervention selected will depend on the nature of the perceived relationship between poverty and the level of threat to a species, as well as on the feasibility of an intervention succeeding. For example, if poor people are hunting a species of conservation concern, it might make sense to set up a tourism venture based on that species to provide alternative revenue to hunting that incentivises conservation. This will be effective, however, only if there is sufficient demand from tourists, appropriate security and infrastructure and so on. Because many species conservation programmes target 'charismatic megafauna' of great interest to the citizens of rich countries, options like photographic tourism and safari hunting are relatively common in species conservation programmes.

Where species conservation activities have negative impacts for local people, conservation organisations may attempt compensatory interventions. This is based on a philosophical position commonly referred to as 'do no harm', and may not be expected in any way to improve conservation status. An example would be ensuring that people who have been evicted from a protected area are properly treated and compensated with land or other resources at their new location. This might not have any impact on the conservation of species in the protected area (although it could well do), but would be seen as the right thing to do to minimise any harm to the relocated population.

In the next section of this chapter, we explore some of these issues focussing specifically on the case of great ape conservation in Africa. We review the kinds of poverty reduction and alleviation initiatives that have been attempted by ape conservation organisations and their success – in terms of poverty impacts – to date.

African ape conservation and poverty alleviation

Background

Africa is home to four species of great ape. These are the bonobo or gracile chimpanzee (*Pan paniscus*), the chimpanzee (*Pan troglodytes*), the western gorilla (*Gorilla gorilla*)[2] and the eastern gorilla (*Gorilla beringei*).[3] Great ape ranges coincide with some of the poorest countries of the world – particularly in Sub-Saharan Africa. Great apes attract a great deal of conservation interest and funding, due to their close genetic relationship with humans and their status as global flagship species for conservation. Highly endangered great apes are often protected through strictly controlled and enforced conservation areas that can – intentionally or otherwise – have negative impacts on the livelihoods of the already poor local communities, through restrictions on resource access and so on. Great ape conservation organisations such as the International Gorilla Conservation Programme (IGCP) have been at the forefront of efforts to assess and monitor the socio-economic impacts of conservation, and to create meaningful conservation incentives for local people. The geographical overlap between great ape habitat and areas of high human poverty makes African great apes an ideal case study for exploring the relationship between species conservation and poverty reduction.

Great apes are distributed across 23 countries in Africa, although in two of these (Mali and Sudan) populations are very small and in a further two (Burkina Faso and Togo) populations are likely extinct (Table 11.1). Within the range states, certain

---

[2] Two subspecies exist – the western lowland gorilla and the Cross River gorilla.
[3] Two subspecies exist – the eastern lowland gorilla and the mountain gorilla.

Table 11.1   **The distribution of great ape subspecies in African range states**

Country	Number of ape subspecies	Ape subspecies	Population size	Forest area (km²) and % of total land forested
Angola	2	Central chimpanzee	200–500	
		Western lowland gorilla	Rare	
Burkina Faso	0 or 1	Western chimpanzee	Likely extinct	
Burundi	1	Eastern chimpanzee	300–400	940 (3.7%)
Cameroon	4	Nigeria-Cameroon chimpanzee	3380	238,580 (50.2%)
		Central chimpanzee	30,000	
		Cross River gorilla	150	
		Western lowland gorilla	15,000	
Central African Republic	3	Central chimpanzee	No census	229,000 (32%)
		Eastern chimpanzee	No census	
		Western lowland gorilla	No census	
Congo (Republic of)	2	Central chimpanzee	10,000	220,600 (64.6%)
		Western lowland gorilla	34,000–44,000, declining	
Cote d'Ivoire	1	Western chimpanzee	8000–12,000	71,170 (22.4%)
Democratic Republic of the Congo	6	Central chimpanzee	Low numbers	1,352,000 (59.6%)
		Eastern chimpanzee	70,000–100,000	
		Western lowland gorilla	Low numbers	
		Eastern lowland gorilla	Few thousand?	
		Mountain gorilla	183 in 2001	
		Bonobo	10,000–100,000	
Equatorial Guinea	2	Central chimpanzee	600–1500?	17,520 (62.5%)
		Western lowland gorilla	1000–2000	
Gabon	2	Central chimpanzee	64,000 (but declined since)	20,600 (80%)
		Western lowland gorilla	35,000?	
Ghana	1	Western chimpanzee	1500–2200	63,350 (26.5%)
Guinea	1	Western chimpanzee	8000–29,000	69,290 (28%)

(*continued overleaf*)

Table 11.1 (*Continued*)

Country	Number of ape subspecies	Ape subspecies	Population size	Forest area (km²) and % of total land forested
Guinea-Bissau	1	Western chimpanzee	600–1000	21,870 (60%)
Liberia	1	Western chimpanzee	1000–5000 (1970s)	34,810 (31.3%)
Mali	1	Western chimpanzee	Fewer than 4860	131,860 (10.8%)
Nigeria	2 or 3	Western chimpanzee	?	135,170 (14.6%)
		Nigeria-Cameroon chimpanzee	3050	
		Cross River gorilla	80–100	
Rwanda	2	Eastern chimpanzee	500	463 (1.8%)
		Mountain gorilla	~130	
Senegal	1	Western chimpanzee	200–400	62,050 (31.6%)
Sierra Leone	1	Western chimpanzee	Less than 2000	10,550 (14.7%)
Sudan	1	Eastern chimpanzee	200–400	
Tanzania	1	Eastern chimpanzee	1500–2500	388,110 (43.8%)
Togo	0 or 1	Western chimpanzee	?	
Uganda	2	Eastern chimpanzee	4950	41,900 (21%)
		Mountain gorilla	392	

[a]Source: Data drawn from sources cited in the *World Atlas of Great Apes* (Caldecott & Miles, 2005).

sites have become particularly famous for their great apes. This is often driven by the location of long-running field study (e.g. Gombe Stream National Park (NP) in Tanzania, Volcanoes NP in Rwanda, Tai Forest NP in Cote d'Ivoire and Bossou in Guinea) and/or tourism sites (the best known being the mountain gorilla parks of the Democratic Republic of Congo (DRC), Uganda and Rwanda). Many other sites with important ape populations are relatively unknown because they are inaccessible and therefore lack profile-raising tourism and/or research activities.

All great apes are listed by the International Union for the Conservation of Nature (IUCN) as endangered or critically endangered.[4] The main threats to their survival are "habitat loss, degradation and fragmentation due to logging and clearance for agriculture ... and hunting (particularly in West and Central Africa)" (Caldecott & Miles, 2005). These threats are exacerbated by the facts that apes reproduce slowly,

---

[4] Species are considered endangered if their population declines by 50–80% over 10 years or three generations (whichever is longer), and critically endangered if declines are more than 80% over a similar period; see http://www.iucnredlist.org/technical-documents/categories-and-criteria.

are susceptible to human diseases and are relatively easy to hunt. Despite their endangered status, formal protection of great apes in Africa is limited. Mountain gorillas and to some extent eastern lowland and Cross River gorillas are largely contained within protected areas, but these are the exception. Indeed, most great apes live outside existing or planned protected areas, often in logging concessions (Caldecott & Miles, 2005).

The range states of African great apes include some of the poorest in the world. Out of the 169 countries listed in the 2010 Human Development Index (HDI; UNDP, 2010), the highest ranking African great ape range state is Gabon (rank 93/169) and even in this case the rank reflects large oil revenues which boost GDP but have little impact on the poverty of the rural population living in close proximity to great apes. The majority of ape range states fall in the category of *low human development* (rank 128 and below) with DRC ranking the lowest at 168. It is also interesting to consider the population density and levels of urbanisation in the ape range states. Some countries with high urbanisation rates (Gabon, Congo and Liberia) also have low overall population densities, meaning that their rural areas (where great apes reside) have extremely small human populations. This is in stark contrast to countries such as Rwanda, Uganda and Burundi which have very high population densities and relatively very low urban populations, meaning rural areas are densely populated. Table 11.2 provides some statistics which provide some context for the socioeconomic conditions prevalent in African ape range states.

Table 11.2 **Socio-economic data for African ape range states**

Range state	Human Development Index (HDI) rank (1–169)[a]	Percentage population living on less than US$1.25 per day (PPP)[b]	Rural population as percentage of total[b]
Angola	146	54.8*	43.3
Burkina Faso	161	56.5	80.4
Burundi	166	81.3	89.6
Cameroon	131	32.8	43.2
Central African Republic	159	62.8	61.5
Congo (Republic of)	128	54.1	38.7
Cote d'Ivoire	149	23.7	51.2
Democratic Republic of the Congo	168	59.2	66
Equatorial Guinea	119	N/A	60.6

(*continued overleaf*)

Table 11.2   (*Continued*)

Range state	Human Development Index (HDI) rank (1–169)[a]	Percentage population living on less than US$1.25 per day (PPP)[b]	Rural population as percentage of total[b]
Gabon	93	4.8	15.2
Ghana	130	30	50
Guinea	156	48.8	70.2
Guinea-Bissau	164	70.1	65.6
Liberia	162	83.7	39.8
Mali	160	51.4	62.9
Nigeria	142	64.4	51.6
Rwanda	152	76.6*	81.7
Senegal	144	33.5	57.7
Sierra Leone	158	53.4	61.2
Sudan	154	N/A	56.6
Tanzania	148	88.5*	74.5
Togo	139	38.7	58
Uganda	143	51.5	87

Notes: Poverty data based on 2008 estimates except where marked '*', which are1998 estimates.

Sources: (a) United Nations Development Programme (2010); and (b) International Fund for Agricultural Development (2011).

Current experience in linking ape conservation with poverty reduction: a diversity of interventions

The UNEP-coordinated Great Ape Survival Partnership (GRASP) convened an inter-governmental meeting in 2005 to discuss the way forward for ape conservation – in the context of poverty. The resulting Kinshasa Declaration includes targets to:

- "Encourage the provision of long-term ecologically sustainable direct and indirect **economic benefits to local communities**, for example, through the introduction or extension of carefully regulated sustainable ecotourism enterprises in areas of great ape habitat, and the creation of long-term research projects operating in or near these areas" (Target 7); and
- "Developing ecologically sustainable local poverty-reduction strategies which recognise and **integrate the needs of local communities** sharing great ape habitats, while securing the lasting health of the environmental resources upon which they depend" (Target 10d).

Even prior to the Kinshasa Declaration, great ape conservation organisations have experimented with different ways to link conservation and poverty – for the various motivations discussed in this chapter. A review of experiences conducted in 2009 under the auspices of International Institute for Environment and Development's (IIED) Poverty and Conservation Learning Group uncovered a wide range of approaches taken by different organisations at different sites: from changing the behaviour and attitudes of communities towards conservation to changing the practice of conservation *vis-à-vis* communities; from finding alternatives to resources of conservation concern to generating benefits from resources of concern; and from enforcing conservation priorities to paying for them. Specific examples include (Sandbrook & Roe, 2010):

- Income generation:
  - As a means to incentivise investment in and tolerance of conservation: for example, employment and/or revenue shares in tourism enterprises (Box 11.1), revenue shares from park entrance fees and PES.
  - As a means to reduce pressure on natural resources though alternative livelihood strategies: for example, beekeeping, improved agriculture and piggeries and facilitating market access for community products.
- Providing for subsistence needs: for example, alternative sources of protein to bushmeat, energy alternatives to firewood, fuel-efficient stoves and multiple-use zones within protected areas.
- Providing social services: for example, human health and family planning initiatives, and support to schools, clinics and other community projects.
- Sustaining the natural resource base: for example, community involvement in protected area management, risk management and insurance and strategies to avoid or mitigate damage from wildlife (e.g. crop raiding or livestock predation).
- Capacity building: for example, enterprise training, book-keeping and agricultural extension.
- Governance and empowerment: for example, policy advocacy, and community involvement in protected area management.

Box 11.1 **Great ape tourism – a source of income and jobs for poor people**

In 2008, the Sabyinyo Silverback Lodge opened on the borders of Volcanoes National Park in Rwanda. It is a joint venture between Musiara Ltd, International Gorilla Conservation Programme (IGCP), African Wildlife Foundation (AWF), Rwanda Development Board (RDB) and Sabyinyo Community Livelihoods

Association (SACOLA). In its first year of operation it generated US$300,000 for SACOLA. The RDB also has a policy (since 2005) of investing 5% of the revenue from park entry fees into community projects (such as schools and clinics). This is worth over US$100,000 per year. Similarly in Uganda, parishes adjacent to Bwindi Impenetrable National Park received between $50,000 and $75,000 in total per year between 2005 and 2007, and spent the money on a range of projects including roads and health facilities.

Although the figures sound impressive, the actual impact on poverty levels may, however, be limited given the extremely high population density around some of these sites. The 5% revenue share from park fees in Rwanda, for example, works out at less than US$0.50 per year per person to the 300,000 park-adjacent people. Nevertheless, tourism can be one of the few opportunities available in remote rural areas and can have direct impacts on poverty through the creation of jobs and opportunities to sell goods and services, and indirect impacts through 'multiplier' re-spending of revenue in the host economy (Ashley *et al.*, 2001). In Bwindi, Uganda, for example, revenue accruing to one parish in the Bwindi tourism hub amounted to US$360,000 in 2004 – about four times the value of all other sources of revenue to the area combined. Similarly the Sabyinyo Silverback Lodge employs 45 local people, purchases local produce and supports local tourist service enterprises such as handicrafts, dancing, guiding and so on, while the Volcanoes NP employs nearly 200 people as guides, guards and trackers. Jobs such as these are rare commodities in remote rural areas and their importance should not be underestimated.

Other sites with ape tourism have fared less well in their ability to generate poverty benefits. A good example is the Dzanga Sangha site in Central African Republic which has been running since 1997 but has barely been able to cover running costs and has not generated poverty benefits beyond a small number of jobs. Similarly in Asia orangutan-based tourism has not taken off in the same way, despite the economic value of mainstream tourism in the region. The solitary nature of orangutans and their large home ranges make them extremely hard to habituate and thus guarantee sightings in the wild. For this reason, tourists are more likely to visit one of the many rehabilitation centres focussed on the care of orphan individuals with a view to them being someday being re-released into the wild.

Sources: Nantha and Tisdell (2009), Nielsen and Spenceley (2010), Sandbrook and Roe (2010) and Sandbrook (2010).

Great ape habitats in Africa can be divided broadly into those that are in relatively intact forests, with very low human population density, and those that are in forest fragments, with high human population densities between the fragments. The relationship between biodiversity and poverty is different in these different contexts, and this (alongside other contextual factors such as infrastructure, market development and governance) affects the choice of intervention employed – and the impact achieved. In the low forest–high population areas (such as Uganda, Rwanda and parts of Nigeria), the threats to apes and their habitat are taken to be poverty driven, represented by forest clearance by small farmers, subsistence hunting, redress for ape incursions onto farmsteads for crop raiding and so on. Here the conservation interventions employed tend to focus on poverty-relevant interventions such as support for social services (schools and hospitals, public health and family planning), income generation (enterprise development and tourism), reducing direct pressures on resources (fuel-efficient stoves and agricultural improvements) and problem animal management. In the high forest–low population areas (such as the Democratic Republic of Congo, Gabon and parts of Cameroon), the major threats come from commercial forestry and commercial hunting rather than direct localised poverty. Here interventions focus around corporate responsibility, such as certification schemes, community conservation and the provision of bushmeat alternatives.

From our review of experience, it was possible to identify a number of key trends including:

- There has been an increase in recent years in the number of initiatives that seek to hand over some form of control over natural resource management to local people, and with it access to resource use within protected areas.
- There continue to be a large number of *integrated conservation and development* (ICD)–type projects that seek to substitute another activity for natural resource use, on the assumption that such activities will replace rather than add to the existing resource-destructive activities.
- Ape tourism remains by far the most popular way of converting the presence of great apes into resources for local development activities, and it continues to be seen as a first option by many new projects, even where ape tourism seems unlikely to be viable and in areas where the potential direct benefits per capita are too low to have real impact on poverty levels.
- There are relatively few projects that work directly with forestry concessions, given that vast areas of ape habitat are within forestry concessions. However, rapid progress is being made in this area, particularly in the Congo basin countries covered by the Central Africa Regional Program for the Environment (CARPE) project (de Wasseige et al., 2009).

• Initiatives that seek to deliver general development benefits to local people, such as infrastructure like schools and hospitals, are far more common in areas with very high human population densities.

Lessons learned: What works and why?

Despite the wide range of interventions that have been employed to link conservation and poverty, there has been little analysis of poverty impacts – although there are some reasonable data on income generation – particularly from tourism (Box 11.1). It is therefore difficult to draw precise conclusions about what kinds of initiatives 'work' best. The experience of different conservation organisations has been highly variable and reflects the very diverse conditions and contexts where they find themselves working (Box 11.2).

Box 11.2 **Different approaches to linking ape conservation and poverty alleviation in Africa**

Cross River State, Nigeria: The Mbe Mountains provide critical habitat for Cross River gorillas, but the integration of this area into the Cross River National Park has been rejected by the nine surrounding communities. Instead these communities established their own conservation association (Conservation Association of the Mbe Mountains (CAMM)) with the aim of conserving the gorillas and simultaneously improving the livelihoods of the surrounding communities. The Wildlife Conservation Society (WCS) has been working with the communities to support this process. Three key interventions were selected to achieve these goals: micro-enterprise development, local institutions for sustainable resource management and local capacity building for sustainable production and effective marketing of agricultural produce. Key lessons learned from the experience to date have included the danger of raising high expectations of economic benefits from conservation in the short term and the challenges to financial sustainability that are presented by the scale of benefits needed for large associations.

Kahuzi-Biega National Park – DRC: Not all conservation approaches intend to reduce poverty and not all approaches to reduce poverty contribute to sustainable conservation – this dichotomy is recognised by Strong Roots in their work to support eastern lowland gorilla conservation through community involvement. Key to engaging communities is to recognise that different communities have different types of needs and different attitudes

---

towards conservation. This in turn determines the types of interventions that will work. Three types can be identified around Kahuzi-Biega NP:

- Those who recognise the importance of ape conservation but whose livelihoods are dependent on resource exploitation and so present a threat. Interventions that work: participatory planning, law enforcement and alternative livelihoods and livelihoods improvement projects.
- Those who assume that all levels of forest resource use are sustainable and that there is no need to be concerned about conservation. Interventions that work: participatory planning, environmental education and alternative livelihoods and livelihoods improvement projects.
- Those whose only interest is to reclaim land in the national park and have no interest in engaging with conservation initiatives. Interventions that work: law enforcement and environmental education.

Usually a combination of interventions is necessary to achieve success – law enforcement on its own is not enough, and neither is a sole focus on alternative livelihoods projects.

Volcanoes National Park, Rwanda: Dian Fossey Gorilla Fund International (DFGFI) has a number of initiatives to engage with communities. These include support to a local school and a commitment to local employment. In particular, however, DFGFI focuses on 'ecosystem health' (disease, nutrition, water and sanitation) and on working with marginalised and indigenous communities (land rights, income generation and agricultural improvement). A key challenge faced by DFGFI is the financial sustainability of these initiatives as well as their limited development skills. The main problem, however, is the rapidly increasing population in Rwanda in general and around the park. This places huge pressure on the park resources and also means that average farm sizes are declining – making household production levels harder and harder to sustain.

Source: Poverty and Conservation Learning Group (2010).

Although the experience of conservation programmes is diverse, a number of common challenges can be identified influencing poverty outcomes from species conservation work:

1. *The scale and causes of poverty*: It is very difficult to have a meaningful impact in areas with huge populations of poor people (e.g. in Rwanda where population density is growing and can reach over 800 people/km^2; Box 11.1) (Nielsen &

Spenceley, 2010) or in areas where the human population is reeling from recent or on-going conflicts or disasters (e.g. eastern DRC).

2. *Benefit sharing*: In common with all types of community conservation, a key challenge is that benefits tend to accrue to a subset of the target group, and can often end up in the hands of a small local elite.

3. *The availability of economic opportunities*: Where tourism is possible it can generate meaningful benefits, but in heavily forested or remote areas this is very difficult. Emerging financial mechanisms such as payments for Reducing Emissions from Deforestation and Forest Degradation (REDD) or access and benefit sharing (ABS) may have better future potential in these areas but are themselves full of challenges (Blom *et al.*, 2010).

4. *Local capacity*: Community organisations often lack the capacity to address power imbalances, claim rights and implement projects in line with regulations in order to benefit from land and biodiversity resources.

5. *Conservation organisation capacity*: Conservation projects tend to be very focussed on sites, and have limited engagement with national policy and governance processes that are critical for scaling up. Furthermore, conservation professionals often lack the development skills that are needed to build capacity at the local level and to engage in political processes.

6. *Impact on the poorest*: Development interventions often target the very poorest, but species conservation interventions with a poverty reduction component are more likely to target the people identified as key resource users, who may not be the poorest. This can reduce their poverty reduction impact. Alternatively, the poorest may be the key resource users but be unable to capture benefits of the intervention (Box 11.3).

Box 11.3 **Linking gorilla conservation and poverty alleviation in Uganda – two decades of experience**

Bwindi Impenetrable National Park – which protects a critical habitat for mountain gorillas – was created in 1991 despite huge resistance and resentment from local people. A series of ICD strategies were implemented in response. The poverty impacts of the different interventions highlight the problems experienced in reaching the poorest groups:

- **Community social infrastructure projects (schools, bridges and roads)**: all income quintiles benefitted.
- **Controlled access to park resources (multiple use) through sustainable harvesting**: For resources used for weaving, medicinal plants, beekeeping

and so on. The buyers benefited more than the poorer and more directly dependent harvesters due to commercialisation, elite capture and exclusion of primary producers.

- **Tourism**: This was the most effective strategy in generating local benefits but was highly location specific. It was also the strategy that showed the most marked failure in reaching the poorest.
- **Revenue-sharing 'gorilla levy'**: Deliberate efforts had to be made to target the poor.
- **Improved agriculture**: This improved food security and income, but the poorest again did not benefit since, having limited public visibility, they were difficult to reach through agricultural extension services.

The key lesson from the ICD experience was that targeting the poorest in poverty alleviation efforts is a major challenge due to their political and economic marginalisation making them hard to reach. Meanwhile these same people tend to be the most impacted by crop and livestock predation from wildlife due to the location of their land in marginal areas, and the least well equipped to deal with such problems.

Source: Blomley *et al.* (2010).

Conclusions and recommendations

Species conservation remains mainstream in conservation practice, despite recent trends in favour of larger scale approaches at the level of the landscape or ecosystem. Like any conservation intervention, species conservation projects can have major impacts on poverty, both positive and negative. The nature of the relationship between species conservation and poverty depends in large part on the perceived role of poverty in determining the status of, and threats to, the species in question. In some cases, species conservation involves the enforcement of rules which limit the livelihood opportunities of people and exacerbate poverty. In other cases, particularly where poverty is seen as a driver of conservation threat, species conservation programmes include interventions intended to deliver meaningful benefits for local people as an incentive for conservation, reducing poverty in the process. As the fundamental goal of most species conservation is the long-term survival of the target species, any links to poverty reduction tend to be driven by a perception that poverty is a direct or indirect threat to the species, and not by an underlying mission to reduce poverty.

The African great apes provide a useful case study to understand in more detail the relationship between species conservation programmes and poverty reduction. It is difficult to draw firm conclusions as to whether great ape conservation initiatives in Africa have successfully addressed poverty to date – largely due to a lack of monitoring data – but it seems clear that great ape conservation is most likely to contribute to poverty alleviation when:

- Human population densities are relatively low and demographic trends are stable.
- Poverty is an obvious driver of a threat to species conservation, so the conservation logic for working on livelihoods is clear.
- Poverty is on a manageable scale so that interventions can make a difference.
- There are opportunities to generate meaningful financial or non-financial benefits from conservation that are wanted by local people.
- The governance regime gives space for local interests.
- There is local capacity to fight for, manage and fairly distribute benefits between and within communities.
- Programmes are at an appropriate scale and not excessively short term.
- Appropriate development skills are available to the conservation organisation – ideally through partnership with a specialist development organisation.
- The relationship between poverty and biodiversity at the site is well understood.
- There is good stakeholder cooperation.
- Benefits target the poorest and are accessible to the poor.

However, these things rarely come together and often change over time. When these factors are absent there is a significantly higher risk of trade-offs rather than win–wins, and conflict can result.

Acknowledgements

Preparation of this chapter – and presentation of an early version of this material at the international symposium held at the Zoological Society of London in April 2010 – was supported by the Arcus Foundation which provides funding to the IIED Poverty and Conservation Learning Group.

References

Adams, W.M., Aveling, R., Brockington, D., Dickson, B., Elliott, J., Hutton, J., Roe, D., Vira, B. & Wolmer, W. (2004) Biodiversity conservation and the eradication of poverty. *Science*, 306, 1146–9.
Ashley, C., Roe, D. & Goodwin, H. (2001) *Pro-Poor Tourism Strategies: Making Tourism Work for the Poor: A Review of Experience*. Pro Poor Tourism Partnership No. 1. Overseas

Development Institute (ODI), the International Institute for Environment and Development (IIED) and the Centre for Responsible Tourism (CRT), University of Greenwich.

Blom, B., Sunderland, T. & Murdiyarso, D. (2010) Getting REDD to work locally: lessons learned from integrated conservation and development projects. *Environmental Science and Policy*, 13, 164–72.

Blomley, T., Namara, A., McNeilage, A., Franks, P., Rainer, H., Donaldson, A., Malpas, R., Olupot, W., Baker, J., Sandbrook, C., Bitariho, R. & Infield, M. (2010) *Development AND Gorillas? Assessing Fifteen Years of Integrated Conservation and Development in Southwestern Uganda*, Natural Resource Issues 23. International Institute for Environment and Development, London.

Brockington, D. (2009). *Celebrity and the Environment: Fame, Wealth and Power in Conservation*. Zed, London.

Caldecott, J. & Miles, L. (2005). *World Atlas of Great Apes and Their Conservation*. United Nations Environment Programme World Conservation Monitoring Centre, Cambridge.

de Wasseige, C., Devers, D., de Marcken, P., Eba'a Atyi, R., Nasi, R. & Mayaux, P. (eds.) (2009) *The Forests of the Congo Basin–State of the Forest 2008*. Publications Office of the European Union, Luxembourg. http://www.observatoire-comifac.net/edf2008.php?l=en (accessed 27 April 2012).

Entwistle, A. & Dunstone, N. (eds.) (2000) *Priorities for the Conservation of Mammalian Diversity: Has the Panda Had Its Day?* Cambridge University Press, Cambridge.

Gray, M. & Rutagarama, E. (2011) *20 Years of IGCP: Lessons Learned in Mountain Gorilla Conservation*. International Gorilla Conservation Programme, Kigali, Rwanda.

Hutton, J. & Leader-Williams, N. (2003) Sustainable use and incentive-driven conservation: realigning human and conservation interests. *Oryx*, 37, 215–26.

International Fund for Agricultural Development (IFAD) (2011) *Rural Poverty Report 2011*. IFAD, Rome.

Kaimowitz, D. & Sheil, D. (2007) Conserving what and for whom? Why conservation should help to meet basic human needs in the tropics. *Biotropica*, 39, 567–74.

Nantha, H.S. & Tisdell, C. (2009) The orangutan–oil palm conflict: economic constraints and opportunities for conservation. *Biodiversity and Conservation*, 18, 487–502.

Nielsen, H. & Spenceley, A. (2010) *The Success of Tourism in Rwanda – Gorillas and More*. World Bank and SNV (Netherlands Development Organisation), Washington, DC and The Hague.

Poverty and Conservation Learning Group (PCLG) (2010) *Linking Conservation and Poverty Alleviation: The Case of Great Apes*, http://pubs.iied.org/G02770.html?k=poverty (accessed 11 May 2012).

Redford, K.H., Coppolillo, P., Sanderson, E.W., Da Fonseca, G.A.B., Dinerstein, E., Groves, C., Mace, G., Maginnis, S., Mittermeier, R.A., Noss, R., Olson, D., Robinson, J.G., Vedder, A. & Wright, M. (2003) Mapping the conservation landscape. *Conservation Biology*, 17, 116–31.

Sandbrook, C.G. (2010) Local economic impact of different forms of nature-based tourism. *Conservation Letters*, 3, 21–8.

Sandbrook, C. & Roe, D. (2010) *Linking Biodiversity Conservation and Poverty Alleviation: The Case of Great Apes*. International Institute for Environment and Development, London.

Simberloff, D. (1998) Flagships, umbrellas, and keystones: is single-species management passé in the landscape era ? *Biological Conservation*, 83, 247–57.

United Nations Development Programme (UNDP) http://hdr.undp.org/en/data/profiles/ (accessed 11 May 2012).

Walpole, M. & Leader-Williams, N. (2002) Tourism and flagship species in conservation. *Biodiversity and Conservation*, 11, 543–7.

Walpole, M. & Wilder, L. (2008) Disentangling the links between conservation and poverty reduction in practice. *Oryx*, 42, 539–47.

Woodroffe, R., Thirgood, S. & Rabinowitz, A. (eds.) (2005) *People and Wildlife: Conflict or Coexistence?* Cambridge University Press, Cambridge.

12

Community-Based Natural Resource Management (CBNRM) and Reducing Poverty in Namibia

Brian T.R. Jones[1], Anna Davis[2], Lara Diez[3] and Richard W. Diggle[4]

[1]Environment and Development Consultant, Windhoek, Namibia
[2]Independent Consultant, Windhoek, Namibia
[3]Nyae Nyae Development Foundation of Namibia, Namibia
[4]World Wildlife Fund, Windhoek, Namibia

Introduction

While Namibia is ranked as a low- to middle-income country, it has a highly skewed distribution of income and an official unemployment figure of 51%. According to the Central Bureau of Statistics (CBS, 2008), 41.5% of Namibian households are poor (i.e. they have monthly expenditures of less than N$262.45 or approximately US$37 per adult equivalent) with the incidence of poverty in rural areas at 38.2%. The majority of the population lives in the rural areas and is dependent on natural resources for supporting day-to-day livelihoods.

Namibia gained independence from South Africa in 1990 and is still suffering from the legacy of South Africa's apartheid administration, including a dual land tenure system dividing the country into communal land and freehold land. Close to a million people live on communal land while a few thousand people own freehold land. Apartheid policies left the communal areas poor and underdeveloped.

In addition Namibia is the driest country south of the Sahara, with average rainfall varying from above 600 mm in the northeast to less than 25 mm in the

Biodiversity Conservation and Poverty Alleviation: Exploring the Evidence for a Link, First Edition.
Edited by Dilys Roe, Joanna Elliott, Chris Sandbrook and Matt Walpole.
© 2013 John Wiley & Sons, Ltd. Published 2013 by John Wiley & Sons, Ltd.

Namib Desert to the west. Rainfall is erratic both temporally and spatially. Drought is a regular occurrence and generally the soils are poor. Most of the country is unsuitable for rain-fed crop production. Generally, arable farming does not provide rural households in Namibia with significant cash income, although it is important for subsistence (National Planning Commission of Namibia (NPC), 2001). One of the most common livelihood strategies adopted by people living in drylands such as Namibia is diversification of economic activities (Anderson *et al.*, 2004).

Community-based natural resource management in Namibia

One such form of economic diversification is *community-based natural resource management* (CBNRM). In Namibia this is a conservation approach that provides local communities with incentives to manage natural resources, such as wildlife, sustainably. Through forming common property management institutions called *conservancies*, local communities receive rights from government over wildlife that enable them to gain income from a number of different use options, including different types of hunting and photographic tourism. The aim of the approach is to create sufficient economic and other benefits from the use of wildlife so that rural communities will view wildlife as an asset rather than a liability (Namibian Association of Community Based Natural Resource Management Support Organisations (NACSO), 2008). The premise is that if communities use their wildlife wisely, then the wildlife will provide income and other benefits now and in the future.

Legislation providing for communal area conservancies was passed by the Namibian National Assembly in 1996, and the first conservancies were registered by the Ministry of Environment and Tourism (MET) in 1998. By 2009 there were 59 registered conservancies in Namibia (see Figure 12.1) and several more communities were establishing conservancies. Formation of a conservancy is entirely voluntary and adds wildlife and tourism to existing community land uses on unfenced communal land.

According to NACSO (2010), conservancies and other CBNRM activities (e.g. harvesting plant products, producing crafts and thatching grass) generated around N\$42 million (close to US\$6 million) in 2009. As a result of this income-generating potential, CBNRM has been recognised by the Namibian government as an important strategy that can contribute to achieving national development goals including poverty reduction (Government of the Republic of Namibia, 2002; NPC, 2004, 2008). In addition, various international donors have supported CBNRM in Namibia partly as a mechanism for addressing poverty issues.

This chapter analyses the extent to which CBNRM is meeting these poverty reduction expectations in Namibia. Some early research found that most income generated by conservancies was being spent at the community (i.e. conservancy) level on operating

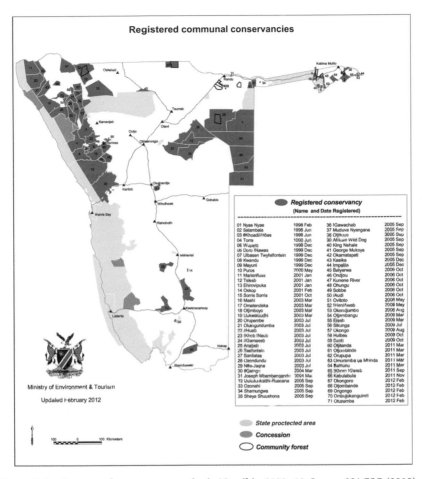

Figure 12.1   **Communal area conservancies in Namibia, 2009–10. Source: NACSO (2010).**

costs and on some social projects, with few direct benefits reaching households (Long, 2004). There were also concerns in government that benefits were being captured by local elites (Vaughan, 2006). This chapter analyses more recent data on conservancy benefits in order to assess the impact that these benefits have on reducing poverty.

In this analysis we recognise that the causes of poverty are multidimensional and include issues such as vulnerability, exposure to risk, voicelessness and powerlessness (e.g. World Bank, 2001). As a result, measures to reduce poverty should deal with these issues as well as the more common issues of income and access to health and education. We therefore try to assess some of the intangible impacts of conservancies on poverty as well as the tangible impacts such as income and jobs.

Benefits from conservancies

In 2009 the total cash income generated by conservancies was just under N$26 million or around US$3.7 million (NACSO, 2010). Not all of the 59 registered conservancies in 2009 were providing some form of benefit to members. This is partly because 28 of the 59 conservancies had been operating for only 4 years and had yet to build a significant income base. It is also true that several conservancies have low potential for generating large amounts of income. Those that generate the most income, and are able to provide the most benefits for members, are those in areas of high-tourism attractions and with high wildlife numbers and a diversity of wildlife. As a result, the highest earning conservancies are those in the north-west and north-east of the country, particularly in the Kunene and Caprivi Regions (see Figure 12.1). The following is an analysis of the various benefits that conservancies provide to their members.

Jobs

Jobs are clearly an important means for reducing poverty in Namibian rural areas. In 2009 conservancies employed 406 staff using their own funds and another 157 using donor funding. Tourism in conservancies provided another 789 full-time and 250 part-time jobs, and hunting operations generated 14 full-time and 53 part-time jobs. The value of conservancy-funded jobs was N$4.8 million in 2009 (NACSO, 2010) or about US$585,700.

In the Caprivi and Kunene Regions, wealth-ranking exercises showed that jobs were an important indicator of wealth and that one of the main indicators of poverty was a lack of a full-time job (Long, 2004). Murphy and Roe (2004) found that because there were very limited other formal employment opportunities, tourism jobs in rural areas were important. Tourism employment was seen as particularly beneficial because of its proximity to home, enabling employees to manage household activities as well (Murphy & Roe, 2004). Although some wages from tourism operations within conservancies could be low (particularly for campsite employees), they are "not insignificant for people living in remote, rural areas with few other means of accessing regular cash to diversify their livelihoods" (Murphy & Roe, 2004: 125). Research on community attitudes in five conservancies in 2008 led to the conclusion that of all the benefits provided by conservancies, "employment was the most significant, having the greatest impact at both household and individual levels" (Mosimane, 2009: 16).

Job creation is particularly important given Namibia's high rates of unemployment – 65.6% in Caprivi Region and 50.4% in Kunene, for example (Ministry of

Labour and Social Welfare, 2008). In Caprivi during 2009 nine conservancies generated 188 jobs with a total wage bill of N$2.4 million (approximately. US$343,000), while photographic and hunting tourism generated 256 jobs within the nine conservancies with a wage bill of just more than N$4 million or about US$571,000 (Diggle, 2010).

Cash

Although not a widespread practice, some conservancies choose to use profits from their wildlife and tourism income to provide cash either to villages or directly to members or households. Most cash distributions take place in Caprivi conservancies. Total cash distribution in Caprivi by eight of the nine registered conservancies in 2008 was almost N$499,000 or about US$71,000 (Diggle, 2010). Balyerwa Conservancy distributed N$99.20 (approximately US$14) to each conservancy member, and Kasika Conservancy distributed N$352 (approximately US$50) to each member. To put these figures into context, a 25 kg bag of maize meal in Caprivi currently costs between N$60 and N$75. Different households chose to use their cash income in different ways. Mosimane (2009) found evidence from five conservancies spread from north-east to north-west Namibia that some families used their cash to buy assets such as a gas freezer or goats, while others used the income to pay school hostel fees for their children. In a survey of 12 conservancies, it was found that most members favoured social projects instead of cash payments (Diez & Davis, 2007).

Where Caprivi conservancies provided cash to villages instead of directly to members, the villagers decide how to use the income. Some villages themselves decide to make payments to households, while others choose to use the income for social projects. The amounts going to villages in recent years have ranged from about N$2500 to N$20,000 (Diggle, 2010).

It is difficult to assess the impact of cash payments to members or households. However, according to official data, a household in Caprivi is likely to be poorer than a household living in any other region of the country (CBS, 2008) and unemployment rates are extremely high, so for some families even a little extra cash could help. Jones and Barnes (2008) estimated annual average household *cash* income to be around N$7560 (approx. US$1080).

Box 12.1 **Conservancies supporting some of Namibia's poorest**

The members of the Nyae Nyae Conservancy in north-eastern Namibia are almost all Ju|'hoansi San, one of the most poor and marginalised groups in

Namibia. According to government data, 63% of San people are poor and the San have the highest incidence of extreme poverty (NPC, 2008). One of the main ways that the conservancy supports members is through the provision of water and the protection of existing water installations from elephants. In addition the conservancy generated a total of 23 full-time jobs in 2009 with a wage bill of N$302,200 (approximately US$43,170). However, the conservancy also makes an annual cash payment to its members. From 2004 to 2009 the conservancy paid each member N$300, and in 2010 the amount was increased to N$400 per member (Nyae Nyae Development Foundation of Namibia, n.d.). Although these amounts are relatively small, they make a difference in a remote area where there are few jobs available except through the government and the conservancy.

In 2009 the conservancy spent 38% of its income of N$1,620,968 (approximately US$231,500) on direct cash benefits in terms of wages and payments to members. Mosimane *et al.* (2007) found that conservancy members valued the cash payments even though they did not lift them out of poverty. Most people used the income for food, a temporary benefit, but some members pooled their income and invested in livestock. Further, "Without this support, and with no other donors, some community members would not have had an income at all" (Mosimane *et al.*, 2007: 11). Generally the cash payments need to be viewed as 'dividends' from investment in wildlife which can provide some short-term relief to cash-strapped households.

There are other sources of income through CBNRM activities that bring cash to individuals. In 2009 documented income from craft sales was just more than N$1.2 million (approximately US$171,000), although the actual amount is thought to be higher (NACSO, 2010). Although the amounts to individuals are generally small, for poor people (mostly women) the amounts are "of great significance in alleviating poverty" (Murphy & Roe, 2004: 128), because they provide cash for much-needed household items or for supporting children at school.[1] Although this income is not directly generated by conservancies, craft sales are often linked to tourism within conservancies (NACSO, 2010). In addition, conservancies assist in the management of harvesting various indigenous plants that generate income for individuals. In 2009 sustainable harvesting of indigenous plants generated more than N$568,000 (approximately US$80,000) for 938 people in conservancies (NACSO, 2010).

---

[1] In most rural areas, if children go to school they need to stay in hostels for which a fee is charged.

Meat

Conservancies produce meat through trophy hunting and the hunting of game for personal consumption. In 2009 conservancies distributed around 330,000 kg of game meat, valued at about N$5 million or close to US$714,300 (NACSO, 2010). The amount of meat distributed differs between conservancies and is valued differently by members of different conservancies. Mosimane (2009) found that in #Khoadi//Hoas Conservancy the amount of meat distributed made little difference to residents, while in Torra Conservancy meat was the most significant benefit that community members received, with households receiving relatively large quantities of meat – sometimes a whole springbok carcass for household consumption. Mosimane found that where meat was distributed in sufficient quantities, it was an important factor in nutrition in the conservancies. In most cases meat distribution took place during the dry winter months, a period of high food insecurity in many households when there were limited options to supplement their diets.

Social projects and infrastructure

Most conservancies have elected to provide benefits to members through social projects rather than make payments to members or households. In recent years villages in Caprivi have spent their income from conservancies on a variety of activities including donations to the local school, housing for teachers, support to the Traditional Authority, developing fish ponds, road building, community transport (e.g. an engine-powered boat) and cultural festivals (Diggle, 2010).

In Kunene Region in the north-west, conservancies in 2009 used their income for activities including sports tournaments, support to schools and kindergartens, funding of medical treatment, women's projects, financial support to traditional authorities, financial support to the elderly, financial support for students from the conservancies, transport for the elderly and school children and funeral assistance (NACSO, 2010).

Marienfluss Conservancy represents the most remote community in Namibia. Its membership is made up of mostly semi-nomadic Himba pastoralists. The conservancy is located in the far north-west corner of Namibia on the margins of the Namib Desert and in rugged, inhospitable country (Figure 12.1). There are no schools or clinics in the conservancy which covers an area of 3034 km^2. The conservancy has helped fund the provision of facilities for a mobile school and a mobile clinic. It provides transport or cash to pay for transport to get sick people to hospital at the nearest clinic, in a town more than 200 km away. The conservancy has bought a house in the town so families have accommodation while waiting for family members to be treated at the clinic. Conservancy members rank this benefit as having an important impact on their well-being (Davis, 2007).

Wildlife stocks

Wildlife is increasing in many Namibian conservancies, particularly those in the north-west and the north-east (NACSO, 2010). Springbok, for example, in the north-west have increased from a few thousand in the early 1980s (a period of severe drought and heavy poaching) to around 160,000. Similar increases have taken place with Hartmann's mountain zebra and oryx. Elephant and black rhino have more than doubled in number, and black rhino are being re-introduced by government into some conservancies in Kunene Region. The increase in wildlife stocks represents a significant increase in natural capital which is an asset that communities can use to diversify livelihoods and provide meat and jobs. Increasing wildlife numbers also provide a solid foundation for the further development of tourism in conservancies. However, increasing numbers of elephants and predators also bring increasing conflict between humans and wildlife. The costs of living with wildlife for rural people are discussed in the 'The Costs of Living with Wildlife' section of this chapter.

Rights and institutions

Empowerment and the role of community institutions are increasingly seen as important components of poverty reduction. The World Bank (2000) suggests that local empowerment can be viewed as a form of poverty reduction in its own right, independent of its income effects. Conservancies are local institutions that are providing communities with increased opportunities to manage their own affairs. The successful conservancies earn their income independent of government, NGOs or donors, and conservancy members decide how to use that income to support wildlife and tourism as land uses and for community benefit. Most conservancies invest in wildlife management by employing their own game guards, taking part in game counts and monitoring wildlife trends and the incidence of human–wildlife conflict.

Conservancies are increasingly viewed by government as structures through which to channel services. Conservancies are also a platform for the development of *holistic rangeland management* and *conservation agriculture*, both of which help to improve food security and reduce human wildlife conflict (NACSO, 2010). Holistic rangeland management and conservation agriculture initiatives involve farmers (or groups of farmers) who are implementing measures such as planned grazing, herding, low-till agriculture and water conservation that provide improved returns to farmers while conserving soils and perennial grasses fundamental to subsistence agriculture.

The conservancy legislation provides legal rights over wildlife and tourism that are secure, empowering communities to take decisions about issues that affect them and providing an enabling environment for economic growth in remote rural areas.

Increasing numbers of rural women are playing leadership roles in conservancies, in cultural settings that have traditionally not accepted women in such roles. In 2009 four conservancies were chaired by women, 53% of conservancy treasurers or financial managers were female, 34.8% of conservancy committee members were female and 24.3% of conservancy staff members were female.

Limitations on conservancy benefits – elite capture

Some researchers have suggested that conservancies are prone to "elite capture" and are therefore less likely to benefit the poor (e.g. Vaughan, 2006; Hoole, 2008). It is not always clear whether they are referring to conservancy leaders benefitting themselves, or to control of decision making and benefits by the already wealthy and powerful. Household socio-economic survey data for seven Namibian conservancies were analysed by Bandyopadhyay *et al.* (2004) to investigate whether conservancy benefits were being controlled by elites. They found that "there is little evidence to suggest that better-educated or the asset-rich are gaining more from conservancies relative to their less-educated or poor counterparts. Thus we conclude that conservancies, if not pro-poor, are at least not being dominated by the elite" (Bandyopadhyay *et al.*, 2004: 20). Analysis of data from a later household survey conducted in eight conservancies came to the same conclusion (Bandyopadhyay *et al.,* 2008).

The costs of living with wildlife

As indicated in this chapter, increased numbers of species such as elephant and hippo and increased numbers of predators bring increased human–wildlife conflict (HWC), which has negative impacts on livelihoods. Available data suggest there has been a clear increase in HWC between 2003 and 2009 for all species in all conservancies. Although the increase is partly because of the increase in conservancies and improved data collection, it is clear that conflict is growing. In 2009 a total of 7659 incidents were reported (NACSO, 2010). In the north-west, incidents include attacks by predators on livestock and damage to water infrastructure by elephants. In the north-east the main problem is damage to crops. Jones and Barnes (2008) found that in Caprivi the total average annual value of losses due to wildlife (including both crops and livestock) was N$525 (approximately US$75) or 7% of total annual household cash income. They estimated the combined costs of HWC to communal farmers in northern Namibia to be around N$7 million (approximately US$1 million) annually. However, the levels of HWC suffered by households across this area differ widely with increased levels occurring closest to wildlife habitat or corridors used by wildlife.

In addition, there are differential impacts according to the status of house-holds – whether wealthy or poor. Elephant damage to the crops of a wealthier farmer with large croplands will have less impact as not all the crops are likely to be damaged. However, the same level of damage in terms of quantity of crops lost could have a much higher impact on a poor farmer with smaller areas of land. As most crops are consumed by the household and not sold, losses to wildlife can have significant negative impacts on household food security (Jones & Barnes, 2008). In the same way, the loss of three head of cattle will have a bigger impact on someone who owns only 10 compared to the same loss out of a herd of 50 or more. While the financial impact is the same for both, the overall negative impact on the poorer person is greater. At the same time the income generated by wildlife in conservancies does not always preferentially reach the households most affected by HWC.

If benefits are perceived to be high enough, conservancy residents appear willing to tolerate problem-causing species. In Nyae Nyae Conservancy, elephants damage infrastructure, compete with people for bush foods and are dangerous. However, "despite widespread fear, people said they wanted to live with elephants because they represented income and employment through tourism and trophy hunting. Most people said that, given the choice, they would prefer to live with elephants than without them" (Matson, 2005).

In order to address the problems caused by HWC, conservancies are taking a number of different measures. Many conservancies are developing their own HWC management plans which include a number of prevention and mitigation strategies such as zoning areas of land for different uses, protecting water installations, herding livestock, using chilli in different ways to keep elephants out of fields, protecting livestock kraals, moving people away from high-conflict areas and building protected access points where people can safely collect water or water their livestock on river banks (Jones & Barnes, 2008; NACSO, 2010).

With support from NGOs, several conservancies piloted a scheme to use some of their income to help offset crop or livestock losses incurred by households. Payments to offset losses (the amounts paid do not compensate for the full cost of the losses) depended on the owner of the livestock or crop field having taken reasonable steps to prevent damage from occurring. In 2009 seven conservancies spent a total of N$237,000 on payments as part of the mitigation scheme (NACSO, 2010).

In 2009, the Namibian government approved a 'National Policy on Human-Wildlife Conflict Management', which was developed by the MET. The policy document sets out the government's approach to managing HWC and includes the introduction of a national mitigation scheme similar to that piloted by the conservancies, the Human-Wildlife Conflict Self-Reliance Scheme (HWCSRS). In order to help conservancies establish their HWCSRS, in 2010 the MET provided each conservancy with N$60,000 as a start-up fund to which conservancies are expected to add funding of their own.

Beyond HWC, in some cases in southern Africa the introduction of CBNRM with associated restrictions on local hunting reduces access to game meat for some community members, particularly the poor, and this can lead to increased food insecurity or a reduction in income for the poor (Jones, 2009). In Namibia, Vaughan *et al.* (2004) found that wildlife provided a safety net resource for some residents in conservancies in Kunene and Caprivi Regions, providing both food and to a lesser extent income security; that wildlife users tended to be poorer or less secure households and that access to wildlife was being restricted by the activities of community game guards. However, the data suggesting that closer control and regulation by conservancy games guards were reducing the use of game meat by residents were obtained in only two conservancies in Kunene Region. In addition, there is no indication of the numbers of households that would be severely disadvantaged if they no longer had access to wildlife. This is an aspect that requires further research in order to understand the actual impact of restrictions on hunting in conservancies on the poor. At the same time, conservancies may legally institute their own quota and permit system to enable hunting by individuals. So far no conservancies have done this.

Conclusions

The most significant immediate and long-term contribution of CBNRM in Namibia to poverty *reduction* (i.e. lifting people out of poverty) is likely to come through the provision of jobs. Employment provides a steady income that can be used to build up household assets and a local cash economy. Also the CBNRM jobs are linked to considerable training and capacity building that develop new skills and in turn open up new employment opportunities. In addition, empowerment (i.e. devolving legal rights) and developing new civil society structures are important contributing factors to longer term reduction of poverty within conservancies. This is particularly significant given Namibia's apartheid legacy that left a generation of rural Namibia marginalised and poverty stricken. Furthermore, conservancies assist in decreasing vulnerability to livelihood shocks (particularly drought) through diversification of land uses and economic activity in remote rural areas. In addition the increased wildlife provides an important form of natural capital on which to build economic activity; without it, jobs would be far harder to secure.

Broadly viewed, CBNRM is contributing to Namibia's development goal of eradicating extreme poverty and hunger (see Table 12.1). In both international and Namibian terms, most rural Namibians are considered to be poor and the Namibian CBNRM programme is aimed at rural Namibians. However, with the exception of jobs, most benefits from CBNRM probably contribute more to poverty *alleviation* (i.e. short-term relief from the symptoms of poverty) than to poverty reduction. Cash payments to villages or households, for example, are small and received annually. Although only

Table 12.1   **The fit between CBNRM and reduction of poverty goals and strategies in Namibia's National Development Plan 3**

NDP 3 Goal: Eradication of extreme poverty and hunger		
Strategies	**Contribution of CBNRM**	**Status**
1. Strengthen and diversify the agricultural base of poor rural communities through measures that diversify and improve agricultural production to ensure food security and expanded livelihoods with attention to gender equity.	Increased attention to conservation farming and holistic range management as part of CBNRM activities in conservancies.	Holistic range management practised in 6 conservancies. Conservation farming increased crop yields in 3 conservancies and 1 community association in Bwabwata National Park.
2. Ensure poor communities, particularly those in rural areas, are able to broaden their income base by participating in non-farm activities while maintaining environmental sustainability.	CBNRM adds or expands wildlife and tourism as land uses and provides new business and income-generating opportunities.	19 community campsites, 5 official craft markets and 6 cultural villages in operation, as well as beekeeping, fish farming, horticulture and indigenous plant product harvesting as a result of CBNRM.
5. Increase access and improve quality of basic and general education in rural areas.	Conservancies support education through funds for classrooms, meat for hostel children, accommodation for teachers and support to mobile schools.	Apart from meat provision to schools, conservancies contributed more than N$89,000 to education through support to schools and bursaries.
7. Strengthen and sustain Namibia's safety nets for the temporarily and chronically vulnerable, including people with disability and those affected by HIV/AIDS.	Conservancies are developing their own HIV/AIDS policies and strategies, and some support orphans and vulnerable children.	23 conservancies implement their own HIV/AIDS policies, and 431 peer educators work in the 59 registered conservancies.

Table 12.1　(*Continued*)

NDP 3 Goal: Eradication of extreme poverty and hunger		
Strategies	Contribution of CBNRM	Status
8. Expand employment opportunities.	Conservancies create additional jobs themselves, and many new jobs are created through joint venture tourism and hunting enterprises. Many conservancy-linked jobs are in remote areas where few other jobs are available.	1195 full-time jobs and 250 part-time jobs provided by conservancies and hunting and tourism operations in conservancies.

providing short-term financial relief they are, nevertheless, still hugely important in securing communities' commitment towards setting aside conservation areas for wildlife and tourism. Further, there are costs to living with wildlife which for some households could outweigh the material benefits, and can increase food insecurity.

It is therefore unrealistic and misleading to frame CBNRM as a specific *poverty reduction* strategy. The benefits generated provide incentives for rural landholders to manage wildlife and other natural resources sustainably. As a result, CBNRM in Namibia is not specifically "pro-poor" in the sense that it deliberately targets the poor – rather, it addresses the needs of all landholders and resource users in providing incentives for them to adopt wildlife and tourism as land uses. CBNRM should, however, aim to ensure it does not lead to increased poverty through the costs associated with living with wildlife.

By lifting some people out of poverty, improving the livelihood opportunities of others and providing long-term institutional platforms that help drive economic growth, CBNRM is being recognised by the Namibian government as making an important contribution to rural development in Namibia. However, properly addressing poverty *reduction* on a large scale requires concerted action by government to generate jobs in towns and to address larger macro-economic problems.

References

Anderson, J., Bryceson, D., Campbell, B., Chitundu, D., Clarke, J., Drinkwater, M., Fakir, S., Frost, P.G.H., Gambiza, J., Grundy, I., Hagmann, J., Jones, B., Jones, G.W., Kowero, G.,

Luckert, M., Mortimore, M., Phiri, A.D.K., Potgieter, P., Shackleton, S. & Williams, T. (2004) *Chance, Change and Choice in Africa's Drylands: A New Perspective on Policy Priorities.* CIFOR, Bogor, Indonesia.

Bandyopadhyay, S., Guzman, C.G. & Lendelvo S. (2008) *Communal Conservancies and Household Welfare in Namibia.* World Bank, Washington, DC.

Bandyopadhyay, S., Shyamsundar, P., Wang, L. & Humavindu, M.N. (2004) *Do Households Gain from Community-Based Natural Resource Management? An Evaluation of Community Conservancies in Namibia.* DEA Research Discussion Paper No. 68. Ministry of Environment and Tourism, Windhoek, Namibia.

Central Bureau of Statistics (CBS) (2008) *A Review of Poverty and Inequality in Namibia.* CBS, National Planning Commission, Windhoek, Namibia.

Davis, A. (2007) *Benefit Distribution Research Findings – Marienfluss and Sesfontein Conservancies.* Unpublished report. Integrated Rural Development and Nature Conservation, Windhoek, Namibia.

Diez, L. & Davis, A. (2007) *Member Perceptions of Conservancy Benefit Provision.* Unpublished report. Nyae Nyae Development Foundation of Namibia, Windhoek.

Diggle, R. (2010) Unpublished data. Namibian Association of Community Based Natural Resource Management Support Organisations, Windhoek.

Government of the Republic of Namibia (GRN) (2002) *National Poverty Reduction Action Programme.* GRN National Planning Commission, Windhoek.

Hoole, A. (2008) *Place, Prospects and Power: Community-Based Conservation, Partnerships, and Ecotourism Enterprise in Namibia.* University of Manitoba, Winnipeg.

Jones, B.T.B. (2009) Community benefits from safari hunting and related activities in southern Africa. In *Recreational Hunting, Conservation and Rural Livelihoods*, ed. B. Dickson, J. Hutton & W.M. Adams, pp. 157–77. Conservation Science and Practice No. 4. Wiley-Blackwell & Zoological Society of London, Oxford.

Jones, B.T.B. & Barnes, J.I. (2008) Human-wildlife conflict: Namibia. In *Common Ground: Solutions for Reducing Human, Economic and Conservation Costs of Human Wildlife Conflict*, ed. W. Elliot & R. Kube, pp. 13–32. WWF Global Species Programme/Macroeconomics Program, Gland, Switzerland.

Long, S.A. (ed.) (2004) *Livelihoods and CBNRM in Namibia: The Findings of the WILD Project.* Final Technical Report of the Wildlife Integration for Livelihood Diversification Project (WILD), prepared for the Directorates of Environmental Affairs and Parks and Wildlife Management. Ministry of Environment and Tourism, Windhoek, Namibia.

Matson, T. (2005) *Human-Elephant Conflict Research Project: Nyae Nyae Conservancy and Khaudum National Park.* Project Update 30 October. Namibia Nature Foundation, Windhoek.

Ministry of Environment and Tourism (MET) (2009) *National Policy on Human-Wildlife Conflict Management.* MET, Windhoek, Namibia.

Ministry of Labour and Social Welfare (MLSW) (2008) *Namibia Labour Force Survey.* MLSW, Windhoek, Namibia.

Mosimane, A.W. (2009) *The Contribution of CBNRM Activities in Namibian Conservancies to Poverty Alleviation: A Synthesis Report.* WWF/MRCC, Windhoek, Namibia.

Mosimane, A.W., Hipandulwa, G. & Muranda, N. (2007) *Nyae Nyae Conservancy's Contribution to Poverty Alleviation.* WWF/MRCC, Windhoek, Namibia.

Murphy, C. & Roe, D. (2004) Livelihoods and tourism in communal area conservancies. In *Livelihoods and CBNRM in Namibia: The Findings of the WILD Project*, ed. S.A. Long. Final Technical Report of the Wildlife Integration for Livelihood Diversification Project (WILD), prepared for the Directorates of Environmental Affairs and Parks and Wildlife Management. Ministry of Environment and Tourism, Windhoek, Namibia.

Namibian Association of Community Based Natural Resource Management Support Organisations (NACSO) (2008) *Namibia's Communal Conservancies: A Review of Progress and Challenges in 2007*. NACSO, Windhoek, Namibia.

Namibian Association of Community Based Natural Resource Management Support Organisations (NACSO) (2010) *Namibia's Communal Conservancies: A Review of Progress and Challenges in 2009*. NACSO, Windhoek, Namibia.

National Planning Commission of Namibia (NPC) (2001) *Second National Development Plan (NDP2)*. National Planning Commission, Windhoek, Namibia.

National Planning Commission of Namibia (NPC) (2004) *Namibia Vision 2030: Policy Framework for Long-Term National Development Main Document*. Office of the President, Windhoek, Namibia.

National Planning Commission of Namibia (NPC) (2008) *Third National Development Plan (NDP3)*. National Planning Commission, Windhoek, Namibia.

Nyae Nyae Development Foundation of Namibia (NNDFN) (N.d.) Unpublished data. NNDFN, Windhoek.

Vaughan, C. (2006) Livelihoods, poverty and the Namibian community-based natural resources management (CBNRM) programme: what way forward? *Policy Matters: Poverty, Wealth and Conservation*, 14, 305–16.

Vaughan, C., Long, S.A., Katjiua, J., Mulonga, S. & Murphy, C. (2004) Wildlife use and livelihoods. In *Livelihoods and CBNRM in Namibia: the Findings of the WILD Project*, ed. S.A. Long. Final Technical Report of the Wildlife Integration for Livelihood Diversification Project (WILD), prepared for the Directorates of Environmental Affairs and Parks and Wildlife Management. Ministry of Environment and Tourism, Windhoek, Namibia.

World Bank (2000) *Community Driven Development in Africa*. World Bank, Washington, DC.

World Bank (2001) *World Development Report 2000/2001: Attacking Poverty*. Oxford University Press, New York.

(13)

Conservation Enterprise: What Works, Where and for Whom?

Joanna Elliott and Daudi Sumba

African Wildlife Foundation, Oxford, UK and Nairobi, Kenya

Why conservation enterprise?

Community-based natural resource management (CBNRM) starts from the belief that local communities are best placed to conserve natural resources, and that they will do so if the benefits of conserving them exceed the costs and if those natural resources can be directly linked to their quality of life. Benefits may be derived from commercial activities, such as tourism or hunting, or from subsistence activities. CBNRM strategies are associated with policy changes that give local communities control over their natural resources and allow them often exclusive rights to exploit them for profit. Conservation enterprises are one way to create benefit streams aimed at addressing conservation problems.

There has been criticism of the implementation of CBNRM, for example failures in decentralising rights to local communities, widespread conflicts reinforced by lack of political incentives, weak governance and the capture of benefits by more powerful groups (Roe *et al.*, 2009). Communities may be quick to see opportunities from CBNRM, but fail to realise them due to governance issues. Yet increasingly CBNRM has been shown to be a valid conservation and development approach and has been adopted widely in Sub-Saharan African countries as an important element of rural development strategies (Hulme & Murphree, 2001). CBNRM can deliver non-financial as well as financial benefits such as strengthened land and resources

Biodiversity Conservation and Poverty Alleviation: Exploring the Evidence for a Link, First Edition.
Edited by Dilys Roe, Joanna Elliott, Chris Sandbrook and Matt Walpole.
© 2013 John Wiley & Sons, Ltd. Published 2013 by John Wiley & Sons, Ltd.

rights, empowerment and greater security (see Chapter 12, this volume, for an example of the financial and non-financial benefits of CBNRM in Namibia).

The interest in conservation enterprise in Africa followed pioneering programmes such as the Community Areas Management Programme for Indigenous Resources (CAMPFIRE) in Zimbabwe and Administrative Management Design (ADMADE) in Zambia, which used direct economic benefits to reinforce conservation incentives. Increasingly CBNRM has meant supporting natural resource-based businesses at the community level (particularly tourism, but also forest and agriculture based), often with private sector partners.

Thus the evolution of conservation enterprise as a conservation strategy is tied to the growing understanding of the need to valorise biodiversity resources if they are to be adequately protected. Enterprise is a tool for enabling biodiversity to deliver commercial success in line with its sustainable use.

The African Wildlife Foundation's conservation enterprise programme

The African Wildlife Foundation (AWF) was founded 50 years ago in Kenya. Today it works across nine priority landscapes in 14 countries in Sub-Saharan Africa. It has supported conservation enterprise as a core conservation strategy since the late 1990s, defining it as "a commercial activity which generates economic benefits in a way that supports the attainment of a conservation objective". It may support single businesses or intervene in the whole value chain for a product. Sectors include tourism, agriculture and natural products, such as harvesting and processing honey. AWF enterprise teams act as brokers between the community and private sector partners, ensuring the commercial and conservation rigour of each enterprise while offering services such as due diligence and business planning, legal contracting, community mobilisation and raising capital.

In some cases AWF uses a revenue-based model to structure income for communities in single-enterprise (usually tourism) partnerships. The private sector manager pays fees to the community – including land leases or use rights fees, a percentage of bed-night fees and conservation fees – from the 'top line' of the accounts (revenue) rather than from net profit. Some fees are fixed, guaranteeing a minimum income for the community, whereas other fees vary based on visitor numbers.

In contrast, in value chain–linked enterprises, the benefits are structured in a way to strengthen producer added value, often through cutting out middle sections of the value chain, thus avoiding the need for complex 'benefit-sharing' contracts. An example would be enabling small-scale honey producers to sell their product directly to end markets.

Measuring conservation and poverty impacts

AWF has invested heavily in developing a sound impact assessment process to ensure experience is fed back into project and programme design. Conservation impact is measured through assessment of species and habitat indicators, often using GIS-referenced data. However, the development of socio-economic indicators is more difficult. AWF worked with the Overseas Development Institute to generate assessment methodologies (Ashley, 1998). The resulting AWF Program Impact and Assessment (PIMA) system incorporates many of these measures, including gender-specific measures of income and employment impacts. One of the nine sections in PIMA measures the livelihoods benefits to local communities (Box 13.1).

Box 13.1 **PIMA 'human livelihoods' impact indicators**

Number of business ventures: how many conservation enterprises or agreements benefitting both communities and conservation objectives have been identified, established and supported.

Amount of capital invested to develop enterprises: broken down by grant, debt and equity funds.

Commercial performance of enterprises: annual turnover, profit and return on investment as well as business-specific indicators such as occupancy or sales volume.

Local financial benefits from enterprises and related activities: employment, disaggregated by gender, and financial returns from dividends, profits, wages, fees and so on.

Local governance and empowerment impacts: the number of community institutions constituted or strengthened, women participating in conservation-based local institutions and enterprises and community organisation partners managing significant revenues.

Number of direct beneficiaries of AWF action: individuals in households, specific groups or organisations, disaggregated by gender.

Source: AWF (unpublished).

One recognised weakness of many impact analysis approaches is that they tend to gather data from the community, rather than at the household level where most resource use decisions are made. Where AWF has undertaken household-level impact assessment, focal group discussions have proved more useful than household-specific data. However, such assessments have been expensive and hard to replicate. AWF continues to refine its household-level assessment methodologies to ensure they are both meaningful and cost-effective. Another challenge is ensuring that assessments are fed back into project designs, with particular focus on making multidisciplinary approaches (social as well as biological sciences) accessible.

In partnership with the UK Department for International Development (DFID), the World Wildlife Fund (WWF) is trialling different methodologies to explore how people are affected. These include more participatory approaches such as Stories of Change (Wilder & Walpole, 2008), and community-based participatory impact assessments. While these mainly generate qualitative information, by applying methods such as participatory ranking or scoring, the feedback can be expressed numerically and transferred into quantitative changes.

What are the impacts of conservation enterprises?

Poverty reduction

By the end of 2009, AWF had enabled the investment of US$11 million throughout Africa from funding sources including grants, debt and private sector equity. This was spent on supporting 31 enterprises: 12 large tourism enterprises, 15 small spin-off enterprises (honey, crafts and resins), 2 agricultural enterprises and 2 livestock-based enterprises. These generate net community income of about US$1.9 million a year, employ about 255 community members full-time and benefit approximately 76,000 local people through associated capacity building, revenue sharing and community-designed social development projects.

Different types of AWF enterprises have different results:

- Tourism enterprises can generate substantial annual returns for communities: ranging from US$61,000 to US$378,300 per enterprise in 2009 or between US$4 and US$259 per head. These are lower than figures for some South American enterprises (Wunder, 2000; Stronza, 2007) probably because benefits are being shared across larger communities.
- Large value chain interventions show higher returns per head: a 'revolving debt' livestock enterprise in northern Kenya provides pastoralists with direct access to premium cattle markets (raising US$825 per capita), and the Kenya Heartlands Coffee partnership with Starbucks enables coffee farmers to sell certified conservation coffee (earning US$452 per capita).
- Small locally managed enterprises using local materials, such as the cultural *bomas* and handicrafts businesses run by women, can provide substantial benefits even where total revenues are smaller. These can empower disadvantaged groups through improved capacity, self-determination and higher individual benefits, and can be targeted to reduce poverty.

While communities appreciate financial benefits from enterprises, non-financial benefits often have higher livelihood value. In our experience, communities consistently decide to invest incomes from enterprises into communal benefits – such as education, security, water and health services – with huge multiplier effects. At Ololosokwan, the first tourism enterprise that AWF helped broker, a steady stream of community income for more than 10 years has resulted in improved education, healthcare and school infrastructure. Funds have been used by pastoralists to restock cattle after major droughts, increasing their resilience. Despite a lack of detailed household data, we have observed indicators of improvement such as housing quality, satellite dishes, more vehicles and an expanding town. Similar effects can be seen in other communities where enterprises have been running for more than 5 years.

One question about conservation enterprises is whether these would be viable without donor support when the full costs of developing them are factored in. In one case where AWF did track these costs – the Koija Starbeds Ecolodge – the analysis has shown an impressive return on investment that justifies spending donor money. AWF tracked the full costs of its support and found that an investment of US$70,000 (US$48,000 in grants; US$20,000 facilitation costs and a US$2000 community contribution) broke even after 5 years of operation and by 2006 had generated a 225% return on its costs (Sumba *et al.*, 2007).

AWF increasingly promotes the use of commercial debt as a financing mechanism for conservation enterprises, to prevent market distortion, ensure risks are assessed and met appropriately and stimulate entrepreneurship. The Sabyinyo Silverback Lodge in Rwanda pioneered this approach, with the community borrowing money to buy their equity in the partnership (see Chapter 11, this volume). They are currently repaying that loan. AWF sees enterprise funding evolving until debt is the primary funding source and grant funding is used to cover transaction costs such as community capacity building. AWF recognises that debt is riskier funding for communities than grant funding, and works with communities to ensure that the options and associated risks and benefits are well understood

Other organisations have also found that enterprises can deliver significant benefits (Box 13.2).

Box 13.2 **Poverty impact of SNV (Netherlands Development Organisation) conservation enterprises**

Khwai Development Trust, Botswana: Generated substantial income (US$510,843 between 2000 and 2002) from the auction of wildlife quotas to various hunting companies and individuals. In 2000 it accrued US$181,062

in revenue from community-based tourism enterprises and $488 per capita from joint-venture income.

Torra Conservancy, Namibia: Earned income from a wildlife tourism joint venture and trophy hunting. The money generated by tourism increased from US$77,375 in 1999 to US$188,307 in 2004. In 2003, each conservancy member was paid a dividend of US$74, equivalent to 14% of the average annual income in the region. Interestingly, the most common use of the income reported was to pay for school fees.

Nyae Nyae Conservancy, Namibia: Populated by one of the country's most poverty-stricken and marginalised communities. Since 1999, game numbers have increased, contributing significantly to members' livelihoods. In 2002–3 the conservancy provided 28% of all employment, as well as income from a hunting concession cash payment (US$99,953), handicraft sales (US$31,242) and game meat (US$14,708). The per capita benefit in 2002 was estimated at US$75.

Source: Spenceley (2008).

In 2010, WWF-UK analysed three enterprise projects aiming to benefit local communities from the sustainable harvesting, processing or production of natural resources (Studd, 2010). All three took place in areas of high poverty, with limited opportunities for employment and weak provision of government services; thus, despite being small scale, the relative importance of these enterprises is significant.

1. *Harvesting and trade of resin from oak-leafed commiphora* (Commiphora wildii*) for perfume in Namibian conservancies*: 275 participants generated income amounting to 25% of average annual household incomes for the region. Seventeen per cent of adult residents in five conservancies were involved in harvesting this valuable plant resin, with over 1006 beneficiaries in 2009 (Studd, 2010).
2. *Responsible Forest Management and Trade (RFMT) project in Peru, Nicaragua, Panama, Colombia and Bolivia*: Rates of out-migration declined due to the opportunities generated (Johnson, 2009). In Nicaragua, profits of US$116,000 between 2006 and 2008 were generated, 95% of it channelled through social investment programmes chosen by the communities. Employment numbers trebled from 100 to 300. The proportion of people living on less than US$1 per day fell significantly during the period (International Finance Corporation, 2009).
3. *Non-timber forest products and medicinal and aromatic plant harvesting in Nepal*: In the lowlands, 67% of people surveyed had improved food security, with 57% cultivating medicinal plants as their primary source of income. In the less populated

mountains, 95% of participants saw an increase in income as a result, improving food security by up to 6 months. Dependency on fuel wood also decreased.

A key finding is that while these enterprises are not enough on their own to take people *out* of poverty, they make an important contribution to improving socio-economic status. The most direct contribution to poverty reduction in each case was through the generation of income via employment or the sale of harvested products (Studd, 2010). Less tangible benefits came from the establishment or strengthening of community-based organisations. For example, an evaluation of community-based forest enterprises established under the RFMT found:

> *Without exception the communities visited reported that there had been a transformation in the community organisation and their sense of empowerment to act for themselves as owners of their own development process. The direct effects of support to the community organisational capacity ha[ve] also had collateral effects on awareness and exercise of citizens' rights as full members of democratic societies. (Johnson, 2009)*

These benefits are not necessarily down to the enterprise, but can also be linked to support agencies that often help local communities to organise themselves, understand their rights and negotiate with external agencies. Timing is also important – in the Namibia and Nepal projects, income is received in the 'lean' time of year, and provides a 'safety net'. "If we are hungry today, we can go and harvest [plant resin] and get money and tonight we can buy food", says Hepute Kapukire (aged 90, Marienfluss Conservancy, Namibia).

While non-governmental organisation (NGO) partners have published a significant volume of information about the benefits of conservation enterprises, it is surprisingly difficult to establish the *net* income earned by the community from each enterprise, and its share of net income flows. This is primarily due to the reluctance of private sector partners to make commercially sensitive information available, and is one reason why private sector–community joint ventures work best when benefits are defined from the enterprise's gross income (or revenue) line, rather than as a share in its profits. Most available research is case study based – with different methods of evaluation – precluding comparison and aggregation of data. There are few systematic assessments (see e.g. Dixey, 2008; Spenceley, 2008).

Impact on biodiversity

Nearly 200,000 acres of land have been brought under improved management through new community–private sector tourism enterprises supported by AWF. For example, 500 acres of critical corridor land have been secured through a conservation area by the

Koija Starbeds Ecolodge and are managed by community scouts paid by the lodge. An assessment after 4 years found that the health of the conservation area had improved, and 13 species of wildlife, including elephants, were now using the area frequently (Oguge, 2005). The AWF livestock project in northern Kenya is expected to improve the conservation management of an additional 3 million acres of pastoralist land. In the Democratic Republic of Congo, AWF has helped improve both livelihoods and forest conservation in the Lopori Maringa Wamba landscape. By assisting communities to resettle and restart agricultural activities through providing market access for products using the Congo River as the main marketing channel, improved forest management has been established, and bushmeat hunting and slash-and-burn agriculture have been reduced, through 'quid pro quo' agreements. These agreements directly link improve agricultural returns with improved forest conservation actions at the community level.

In order for a conservation enterprise to have a positive impact on biodiversity, it must be designed to do so, and then implemented and monitored accordingly. The benefits delivered by the enterprise must be clearly linked to the needs identified by the community and the intended conservation gains. Precise contracts or 'conservation covenants' (which spell out the conservation goals and outcomes as well as the benefit flows) and active enforcement are keys to ensuring that the enterprise delivers as planned. This principle reduces the risk of negative conservation impacts that may result from successful enterprises. For example, without proper conservation requirements the more profitable livestock industry in northern Kenya could result in higher concentration of livestock in the project area, thereby compromising land conservation and long-term sustainability goals. Similarly if conservation enterprise enables farmers to increase productivity and generate higher benefits from a unit of land from high-value cash crops, without appropriate safeguards to enforce the principle of sustainable resource management, this could encourage households to expand their cultivated land at the expense of the natural ecosystem.

Reaching the poorest

How benefits are shared is critical to giving people incentives to conserve biodiversity: benefits need to be felt by all members of the community, particularly those – usually the poorest – making decisions about resources or forgoing resource use. Some types of enterprise can target specific individuals, for example training women for a handicraft enterprise or improving value chains to benefit poor farmers. However, tourism joint ventures, which still account for the largest share of AWF conservation enterprises, are not targeted at the poorest members of a community. Tourism enterprises tend to employ the elite within a community (i.e. those with the required education and skills). Generally it is the private sector partner who determines who is employed, though the community may be able to nominate beneficiaries. Employment opportunities as game scouts and guards and in construction projects tend to be more equally spread.

AWF-supported tourism enterprises have created between 1 and 55 new jobs each. While each job is important, clearly this is not sufficient to reduce local poverty levels significantly, a point made in earlier assessments of community-based tourism (Kiss, 2004). This confirms that it is difficult to produce appreciable wealth for large numbers of people in poor rural areas through individual tourism enterprises (Young, 2006).

In the past, AWF has largely left local leaders to decide how the income is spent and the extent to which different individuals and households within the community benefit. However, the growing concern that benefits should be tied more explicitly to conservation goals is leading AWF to try to address these issues with community partners during the design process, and to document and monitor the agreed benefit allocation approach. In most AWF-supported enterprises, no household-level dividend is paid out. Instead, money is invested in social services such as education, health, water and rural transport. This approach spreads the financial benefits among community members, which can make it more difficult to establish direct links between the sustainable resource management practices required and the benefits received. There are also non-financial benefits: in northern Kenya enterprises have helped secure community conservancies with increased security and reduced cattle rustling, thereby improving the livelihoods of all community members. Other non-financial benefits valued highly by communities include capacity building for community members, empowerment and the right to participate in community institutions.

Sometimes, even with transparent mechanisms such as regular and open sharing of management accounts and more formal scrutiny by compliance boards (with representatives of both private sector and community partners) for managing benefits, local elites still benefit disproportionately. In some isolated cases, people in positions of power, like the village chief, have used their traditional authority to capture benefits or to sell or lease land in contravention of other agreements (although see Chapter 12, this volume, for a different perspective on elite capture in Namibia).

Other organisations such as WWF and Oxfam (Box 13.3) have found that successful community-led enterprise depends on harnessing skills, resources and entrepreneur-ship, rather than targeting activities to the poorest members of the community (Studd, 2010). Reaching the poorest requires the successful delivery of complementary gov-ernment services, sound community benefit-sharing practices and the trickle-down effect of successful community businesses. Studd (2010) concludes that the context and the design of the intervention determine who benefits, and highlights how the poorest can be targeted in some cases. For example, in a project in the mountain areas of Nepal, the poorest people (primarily women) were given priority when issuing permits for collecting juniper leaves for processing into essential oils. Similarly in Namibia, those conservancies with less potential for developing high-profit tourism or hunting enterprises were targeted for the trial of *Commiphora* harvesting. Again, women were targeted initially but when men realised it was profitable the number of men participating rose substantially (Studd, 2010)!

Box 13.3 **Using enterprise to reach the poorest: Oxfam's perspective**

Oxfam recognises that enterprise development might not suit the poorest of the poor and people in very vulnerable conditions, but enterprises run by or involving poor people do create opportunities for poorer people. In Palestine, one of the honey producer co-operatives includes very poor people who have few assets but can still benefit from the co-operative's marketing channels and technical assistance. In other cases, smallholders create job opportunities for landless people or people with few assets as temporary workers.

Oxfam also takes into account a wider web of institutions and structures which need to be dealt with to create an enabling environment for sustainable enterprises. For example, appropriate access to land and basic infrastructure (e.g. water and roads) is essential. Another element is business services, such as credit and market information, which usually do not reach the poor. Women and other groups face specific barriers, such as cultural barriers (e.g. 'Women should not be involved in marketing') and lower literacy levels.

Source: Nishant Pandey, Oxfam, personal communication, 2009.

Evidence from Tanzania and Kenya shows high levels of elite capture from conservation tourism enterprises. Household surveys around parks in Kenya found that few families were benefitting significantly from wildlife conservation, apart from in the Maasai Mara, and that the situation is worse in Tanzania (Homewood *et al.*, 2009; Chapter 15, this volume). Even in the Maasai Mara, Thompson (2009) found that those with livestock wealth and land-allocating authority captured 60–70% of all income from wildlife.

Evidence from Namibia and South Africa suggests that enterprises can contribute to poverty reduction (although Chapter 12, this volume, describes how the most common contribution is to improved livelihoods rather than poverty reduction per se). Evaluation of nature tourism in Zululand in 2002 found that it provided better opportunities for impoverished people than other industries, with more unskilled and semi-skilled jobs and higher returns on capital than the economy as a whole. For nature tourism 26% of expenditure was spent in small, micro-sized and medium-sized enterprises, and 14% was spent in local communities, versus 15% and 11% for the economy as a whole (Spenceley, 2009).

What types of conservation enterprises work best?

AWF has found six characteristics of conservation enterprises that combine livelihood benefits with conservation gains:

1. *Clear conservation logic*: Too often enterprises assume that delivering livelihood gains to communities will lead directly to improved conservation practices. However, experience indicates that specific conservation gains are most likely to be realised and sustained where they have been negotiated, agreed and contracted (e.g. through a conservation easement, a wildlife corridor or a reduction in use of a specific resource).
2. *Commercial success*: Evaluations of NGO-supported enterprises have pinpointed lack of commercial logic as a frequent weakness. AWF enterprises are designed and managed by pro-conservation venture capitalists and experienced business managers, with communities supported by AWF community conservation officers and processes. Multiple revenue streams and benefits must be secured contractually and delivered transparently, and preferably not be spread too thinly among a large number of people. The enterprise deal must be well structured, including legally binding agreements that protect the interests of all parties. Contracting terms must be prudent, with the contract long enough for the private sector to recoup capital costs and turn profitable, but not so long that a community becomes 'trapped' in a deal. Enterprises that provide opportunities for 'spin-off' enterprises – such as those supplying inputs to the business (e.g. food for a tourist lodge) or complementary products (e.g. cultural tourism experiences for tourists at a lodge) – can help communities tap into value chains and strengthen benefit flows. Each enterprise is taken through a thorough 'due diligence' process which includes business planning, market research, risk analysis, competitor analysis, deal negotiating, deal structuring and financing, training and capacity building, ensuring market access and supporting marketing and promotion.
3. *Right private sector partner*: Many private sector businesses, particularly those operating in poor rural parts of Africa, see themselves as having a socially beneficial as well as profit-making (and environmentally neutral) role to play. AWF chooses companies with a track record in social responsibility, including experience of working with communities as partners, not just as suppliers of labour and other inputs.
4. *Sound community partner with appropriate governance in place*: Enterprises are more likely to succeed where the community partner has a well-articulated and functioning management structure and remains engaged throughout. The ideal partner is a strong, representative and inclusive community institution with strong leadership that is accountable to and able to negotiate on behalf of a

clearly defined community. As this condition is rarely met, a key step is to give community institutions support with strategic planning and by drawing up procedures covering governance (calling meetings, decision-making protocols, appointment and retirement of trustees and directors), financial management (controls and approvals) and accountability (communicating the nature and extent of benefits created and distributed, and preparing, auditing and communicating accounts). The legal form of the community partner can bolster good governance practice – for example, in Kenyan group ranches the community interest can be in the form of a corporation in which every household holds shares and receives dividends, circumventing the risk of elite capture.

5. *Contractual community ownership and enforcement of benefit streams*: Community members must feel that the benefits from the enterprise are enough to justify sacrifices made. Benefits can flow from equity, leases, rents, other payments and employment, and should be monitored by a multi-stakeholder enforcement committee or other formal mechanism.

6. *Transparent intra-community benefit-sharing arrangements*: Conflicts over resource management issues – both between the operator and the community and among community members – can arise as an enterprise is developed. Rapid increases in benefits from a successful enterprise can create social impacts that lead to conflict. Equitable benefit-sharing systems must be agreed and executed transparently by community leaders, and negotiated as part of the planning process.

Other organisations have similar findings (Box 13.4). Fauna and Flora International (FFI) notes these elements of a successful enterprise:

access to markets, start up capital, continued access to financial services, production skills and business skills, good quality products, good packaging and presentation, strong institutions, competitiveness, profitability, security of tenure over resources, and a stable and supportive legal and political environment. (FFI, 2010)

Box 13.4 **Successful conservation enterprises: lessons from WWF**

Enterprise characteristics

1. A strong link between the enterprise and the conservation of natural resources so as to reinforce people's role as stewards of natural resources.
2. No negative environmental impact from the enterprise.
3. A high-value 'product' which is easy to harvest, grow or use, plus ideally the ability to add value locally.
4. Linked into existing community structures and/or help from local champions or entrepreneurs.

5. Part of a wider strategy for livelihood diversification.
6. Appropriately matched to local capacity, other livelihood strategies, traditional knowledge and practices; adaptable to local conditions and flexible in the application of management regimes. Tools and approaches need to be adapted to meet local needs and not just applied off the shelf.

Conditions for success

1. The right policy and legislative environment.
2. Local ownership and support by the community.
3. Sufficient investment in community management and technical capacity before starting. It may be advisable to work with a partner with enterprise development skills.
4. Effective links between the producers and the marketplace.
5. Mechanisms to enable the poorest members to benefit immediately after harvest, rather than waiting until products are sold. This allows people to see benefits immediately.
6. Appropriate research and development into potential environmental impacts or limits to harvesting beforehand.
7. Appropriate regulated processes (e.g. accountability and financial record keeping).

Source: Studd (2010).

The role of the NGO as independently funded 'broker' to help build trust between the partners and provide technical support and other services to the enterprise can be an important determinant of success.

The right legislative, policy and macro-economic environment is also key. In Tanzania, the Wildlife Management Area (WMA) directives and laws passed by the Director of Wildlife in November 2007 made it illegal for WMAs to negotiate local game-viewing deals and required the channelling of revenue centrally. This presented a challenge to successful enterprise development. Improved government capacity and supporting policy frameworks can also help. In Namibia, the new Concessions Unit in the Ministry of Environment and Tourism has helped improve the quality of concession contracts. In South Africa, the SANParks commercialisation strategy has greatly improved the tendering of ecotourism public-private partnerships in national parks (Spenceley, 2009).

Key challenges for successful enterprise development

Careful assessment of potential weaknesses can turn them into strengths, or prevent unviable enterprises from being initiated at all. Challenges include:

- *Poor choice of private sector partner*: Unscrupulous partners may exploit weaknesses in community institutions or be unwilling to meet environmental sustainability goals. Avoid a 'build it and they will come' mentality; deals must be vetted against current and forecasted market conditions.
- *Community partner problems*: High expectations may mean that community members are unwilling to wait for annual dividends. Weak institutions, and poorly defined and/or fragmented communities, may need long-term support – but NGOs can find this hard to sustain. Fragmented communities may experience high levels of conflict over benefit-sharing processes. Communities may lack funds to invest, and even when they are provided with grants they may be uncomfortable with the concept of 'equity' shares in businesses. Weak benefit-sharing mechanisms and high risk of elite capture of benefits can mean that the anticipated livelihoods impacts and conservation gains remain unrealised.
- *Grant funding undermines entrepreneurship*: Grant funding for enterprises can take away the entrepreneurial element and weaken the overall commercial proposition, as well as its sustainability.
- *External factors*: Fluctuations in the tourism industry can make tourism enterprise a risky business for communities, as it is driven by international events and crises, seasonality and foreign policy such as travel advisories or sanctions. Market fluctuations can be challenging, particularly when you want to set a fair price. This means that the financial planning of the operations has to be carefully done, especially when the margins are slim (Studd, 2010). Markets may change, for example the emergence of cheaper products from Asia. Weaknesses in government policy can inhibit whole enterprise sectors. The policy environment may need to be influenced.
- *The need to work within environmental limits*: This is particularly challenging for large-scale projects across whole-product value chains (e.g. livestock, or ecotourism in vulnerable areas), or where projects are ready to scale up. Project design must avoid over-extraction and be managed within the limits of ecological sustainability (Studd, 2010).
- *Choosing the right conservation intervention*: Enterprise development is one of several possible ways to address conservation priorities. AWF uses participatory threat analysis and zoning at the landscape level to identify the types of intervention needed. WWF experience suggests that influencing the policy environment, or

engaging in value chain and supply chain interventions rather than individual enterprises, can generate conservation gains faster, with lower opportunity costs and at a more significant scale. However, the actual cost depends on the need for capacity building at household and community levels.

Conclusions

At times it appears that there is no such thing as a win–win in development practice: rarely are all stakeholders satisfied with their rights and benefits compared with those of their neighbours or neighbouring communities or competitors, rarely are sustainable flows of benefits established through time that demonstrably and unequivocally lead to sustained improvements in conservation outcomes and rarely is there no 'leakage' where resource consumption forgone in one area does not lead to pressure on resources in another area. Yet where we are concerned with renewable resources, with areas where the opportunity costs of conservation are low (e.g. marginal rainfall areas and remote areas) and with areas of low human population, experience suggests that conservation enterprise can deliver 'win–wins' for specific groups of people and resources, in defined geographic areas and for the duration of the life of the enterprise.

Acknowledgements

We would particularly like to thank Kate Studd (WWF), Anna Spenceley (SNV), Giuseppe Daconto (CARE), Helen Schneider (FFI) and Nishant Pandey (Oxfam) for their personal contributions of background papers, ideas and examples for this chapter, with WWF in particular contributing substantial analysis of several case studies. However, responsibility for the contents of this chapter rests solely with the AWF research team.

References

African Wildlife Foundation (AWF) (Unpublished) PIMA 'human livelihoods' impact indicators. AWF, Washington, DC.

Ashley, C. (1998) *Handbook for Assessing the Economic and Livelihood Impacts of Wildlife Enterprise*. Washington, DC and London,African Wildlife Foundation and Overseas Development Institute.

Dixey, L. (2008) The unsustainability of community tourism donor projects: lessons from Zambia. In *Responsible Tourism: Critical Issues for Conservation and Development*, ed. A. Spenceley, pp. 323–41. Earthscan, London.

Fauna and Flora International (FFI) (2010) *Livelihoods and Conservation in Partnership 5: The Role of Enterprise Development in Conservation*. FFI, Cambridge.

Homewood, K., Kristjanson, P. & Chenevix Trench, P. (eds.) (2009) *Staying Maasai? Livelihoods, Conservation and Development in East African Rangelands*. Springer, New York.

Hulme, D. & Murphree, M. (eds.) (2001) *African Wildlife and Livelihoods: The Promise and Performance of Community Conservation*. James Currey, Oxford.

International Finance Corporation (IFC) (2009) *Final Evaluation Report of the Project 'Strengthening Sustainable Supply Chain Lumber in Nicaragua'*. IFC, World Bank Group, Washington, DC.

Johnson, J. (2009) *Lessons Learned from the Responsible Forest Management and Trade Programme in Latin America*. WWF, Santa Cruz de la Sierra, Bolivia.

Kiss, A. (2004) Is community based ecotourism good use of biodiversity conservation funds? *Trends in Ecology and Evolution*, 19, 232–7.

Oguge, N. (2005) *Monitoring and Evaluation of Community-Based Natural Resource Management Programmes: Biological Databases and Range Conditions in Koija, Tiemamut and Kijabe Group Ranches of Laikipia District*. Final report. African Wildlife Foundation, Nairobi.

Roe, D., Nelson, F. & Sandbrook, C. (2009) *Community Management of Natural Resources in Africa: Impacts, Experiences and Future Directions*. International Institute for Environment and Development, London.

Spenceley, A. (2008) Local impacts of community-based tourism in southern Africa. In *Responsible Tourism: Critical Issues for Conservation and Development*, ed. A. Spenceley, pp. 285–303. Earthscan, London.

Spenceley, A. (2009) Conservation Enterprises – What Works, Where and for Whom? Contributions from SNV. Unpublished paper.

Stronza, A. (2007) The economic promise of ecotourism for conservation. *Journal of Ecotourism*, 6, 210–30.

Studd, K. (2010) Summary of Lessons Learned from WWF about Conservation Enterprises. Unpublished paper.

Sumba, D., Warinwa, F., Lenaiyasa, P. & Muruthi, P. (2007) *The Koija Starbeds Ecolodge: A Case Study of a Conservation Enterprise in Kenya*. AWF Working Papers. African Wildlife Foundation, Nairobi, Kenya.

Wilder, L. & Walpole, M. (2008) Measuring social impacts in conversation: experience using Most Significant Change method. *Oryx*, 42, 529–38.

Wunder, S. (2000) Ecotourism and economic incentives: an empirical approach. *Ecological Economics*, 32, 465–79.

Young, T. (2006) Declining rural populations and the future of biodiversity: missing the forest for trees. *Journal of International Wildlife Law and Policy*, 9, 319–34.

Part IV
Distributional and Institutional Issues

Payments for Environmental Services: Conservation with Pro-Poor Benefits[1]

Sven Wunder[1] and Jan Börner[2]

[1]Center for International Forestry Research (CIFOR),
Rio de Janeiro, Brazil
[2]Center for Development Research, University of Bonn, Germany

Introduction

Environmental degradation and biodiversity loss in developing countries often spatially coincide with rural poverty. Innovative environmental management tools like payments for environmental services (PES) are thus naturally also scrutinised for their potential to bring benefits to the poor. Building on a conceptual framework for the analysis of poor people's participation in and benefits from PES schemes, this chapter reviews the recent literature on how payment schemes affect poor people's livelihoods.

A widely used definition describes PES as (1) a *voluntary* transaction where (2) a *well-defined* environmental service (ES) or corresponding land use is (3) being 'bought' by a (minimum of one) ES *buyer* (4) from a (minimum of one) ES *provider* (5) if and only if ES provision is secured (*conditionality*) (Wunder, 2005). This is a generalised model for PES, but in practice many incentive schemes do not necessarily comply with all five of these criteria. Poverty-wise, public sector schemes tend to have different access filters and less payment differentiation mechanisms than ones with private sector buyers. People can also be paid either for conserving pre-existing environmental services (*use-restricting* schemes) or for their restoration (*asset-building* schemes). Whether PES is for actively doing something, or for *not* doing something, has likely implications for local economic activity, employment and thus also poverty.

---

[1] This chapter provides summary arguments from Wunder (2008), updated with more recent case studies.

Biodiversity Conservation and Poverty Alleviation: Exploring the Evidence for a Link, First Edition.
Edited by Dilys Roe, Joanna Elliott, Chris Sandbrook and Matt Walpole.
© 2013 John Wiley & Sons, Ltd. Published 2013 by John Wiley & Sons, Ltd.

The links between PES and poverty can be summarised in four sequential questions (Wunder, 2008):

1. *Participation filters*: To what extent do poor people participate in PES schemes as buyers and sellers of environmental services?
2. *Effects on sellers*: If the poor become service sellers, does this make them better off?
3. *Effects on users*: Do poor service buyers (and non-paying poor service users) become better off from PES?
4. *Derived effects*: How are other, non-participant poor affected by PES outcomes?

In the PES literature, most discussion has been limited to questions 1 and 2, both of which are exclusively concerned with the fate of service providers. It is important to broaden the view to the other two questions and groups for a more complete view of PES effects on poverty, which is the focus of this chapter.

Do poor people sell environmental services?

Poor people face explicit PES access rules and underlying structural constraints, but will also consider whether they desire to participate or not. Figure 14.1 graphically illustrates the selection process leading to participation decisions.[2]

Eligibility

To be eligible for PES, service providers must, firstly, dispose of legitimate rights and means to exclude others from modifying ES provision on land under their control. Secondly, their land must be vital for service delivery (i.e. bear significant potential for ES restoration or conservation). These land-based eligibility criteria already exclude billions of poor people and direct them to the group of 'non-participant poor': the urban poor, the rural landless and those who have land but lack access control or 'strategic ES value' (i.e. those for whom non-action makes little difference for service provision).

Requiring formal land titles is probably the most common anti-poor enrolment criterion. Costa Rica abandoned a previous formal land title requirement for PES to eliminate this bias (Pagiola, 2008). Enrolment without formal title is often possible, as long as landholders have locally recognised tenure, and a proven ability to exclude potential intruders. In contrast, all national-level public PES-like programmes in

---

[2] The first three levels (eligibility, desire and ability) draw partially on Pagiola *et al.* (2005).

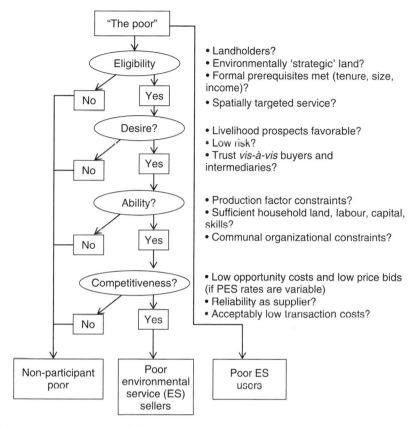

Figure 14.1   **Participation of the poor in payments for ecosystem services (PES) schemes.**

developing countries (Costa Rica, China, Ecuador, Mexico, South Africa and Vietnam) employ pro-poor targeting measures; for example, in the South African Work for Water programme, radical pro-poor screening makes only the poor and unemployed eligible (Turpie *et al.*, 2008).

Whereas spatial targeting filters out a large absolute number of poor, it can increase the poor's relative participation, because environmentally sensitive lands often spatially coincide with remote, sparsely populated, poverty-struck areas. Comparative poverty mapping in seven tropical developing countries showed a marked spatial correlation between closed forests and high poverty rates – although seldom high population *numbers*, and thus not high poverty *densities* (Sunderlin*et al.*, 2007; Milder *et al.*, 2010).

Desire and ability

Not only need the poor be eligible, but also they need to have the motivation and desire to participate in PES. Do they expect net gains from participation, considering both opportunity and potential transaction costs? Both income and wider livelihood factors need to be scrutinised. For example, it might not be desirable to enrol in a PES set-aside scheme if land retirement involves a high risk of third-party invasion. Similarly, landholders are sometimes suspicious of outsiders offering contracts that involve land use caps, mistaking these for the first step towards land expropriation. Initial trust building through PES intermediaries or ES buyers may then be necessary to convince landowners to participate. The implications of non-voluntary participation in PES schemes are discussed in this chapter.

Despite their desire, poor people may still be unable to enrol. This would apply in particular to *asset-building PES* requiring active restoration investments. Poor households may lack the necessary capital, skills or labour, as well as access to credit or technical assistance, to implement the changes required by the PES scheme. Studies on biodiversity and carbon asset-building schemes in Nicaragua and Colombia, however, find little evidence for ability constraints to the participation of poor service providers (Pagiola *et al.*, 2008; Rios & Pagiola, 2010). Instead, Southgate *et al.* (2007) found food security concerns to limit smallholders' willingness to enrol land in conservation set-aside PES schemes in Ecuador and Guatemala.

Other ability constraints may be at the village level, external to household decision making. For instance, under communal semi-open land access regimes, neighbours may exert strong pressures against strict set-aside conservation schemes (Sommerville *et al.*, 2010).

Competitiveness

Providers with low opportunity costs are usually more competitive. Poor people tend to engage in low-input, often subsistence-oriented production systems, and are thus generally more likely to have lower opportunity costs than large, commercially oriented landholders. Conversely, smallholders could also work their plots more intensively, which would draw in the opposite direction.

Probably the most important competitive factor is the high PES transaction costs that service buyers face in working with a large number of smallholders, compared to just a few large service providers. Transaction costs can be a 'killer assumption' for intended pro-poor PES schemes, as experienced in the field of carbon services – the Clean Development Mechanism in particular (Smith & Scherr, 2002; Rios & Pagiola, 2010). Ecuador's PROFAFOR carbon sequestration programme introduced a minimum plot size of 50 ha to reduce excessive transaction costs. This excluded some poor smallholders, yet the scheme still included many collective contracts with marginalised highland communities (Wunder & Albán, 2008).

Do poor people gain from selling environmental services?

The welfare effects on ES sellers will generally be determined by 'the rules of the PES game' (payment rates and modes, conditionality and monitoring), which can be either pre-set by the buyers (as in most public, nation-wide schemes) or negotiated upfront among buyers and sellers (see Figure 14.2). PES contracts are voluntary, so rational individual service providers can be made outright worse off only if they were *de facto* forced into participation (thus violating PES criterion 1), were cheated or were just surprised by the ex-post adverse livelihood impacts (e.g. underestimating opportunity costs).

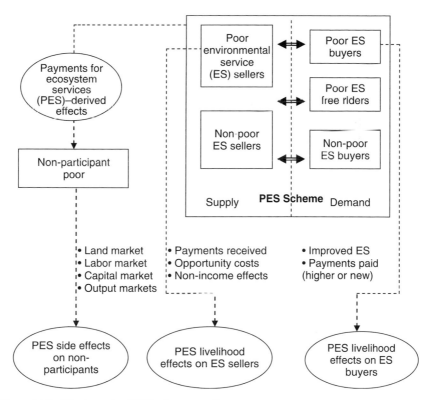

Figure 14.2 **The impacts of PES schemes on the poor.**

Voluntary, rational participants

In almost all cases, PES providers decide voluntarily to participate – and, *de facto*, exit contracts when they perceive that disadvantages have come to dominate. One partial counterexample has been the Chinese Sloping Land Conversion Programme, where farmers in some regions seem to have had no option to deny participation, and may have experienced net welfare losses (Bennett, 2008). Likewise, non-voluntary public forest protection contracts in Vietnam (under Government Programme 327 and its successor, Government Programme 661) typically made up only 1–2% of total household income, which was unlikely to cover provision costs (Wunder *et al.*, 2005). German *et al.* (2010) found for the Trees for Global Benefits Programme in Uganda that some farmers receiving carbon payments experienced higher than expected losses in farm productivity.

Deliberately forcing ES providers into participation would clearly violate the first criterion of the PES definition given in this chapter, and should be a *sine qua non*: a scheme with forced service providers should not count as PES. It could well result in net losses for forced service providers. Normally, service providers have a good approximate sense of their activity returns, and thus of their opportunity costs. However, providers may sometimes also choose voluntarily to participate even when they expect net losses. This could be when PES constitutes 'the lesser of two evils' (e.g. *vis-à-vis* credible regulatory threats or the compensation of an already illegal but weakly sanctioned activity). Another scenario could be where providers expect PES to be an effective entry point to other benefits, for example increasing collaboration with a municipality or establishing relations with a foreign cooperation agency.

However, convincing case study evidence of poor farmers being generally 'PES-trapped' into unfavourable long-term contracts is notably missing (Landell-Mills & Porras, 2002). Not only does PES offer a new income source in often cash-poor areas but also the cash flow is potentially more stable than that from common alternative sources, such as cash crops with heavily fluctuating output prices (Wunder, 2006).

Income gains

Even if poor PES providers are likely to become better off, questions remain as to 'how much' they gain from participation. ES buyers will often be in a better negotiating position – on account of being fewer in numbers, better informed and more initiative seeking than ES providers. Increasing organisation and information levels among ES providers might improve their bargaining position. However, there are also reverse cases where relatively few sellers hold unique assets that many buyers are willing to compete for. For instance, in the Zimbabwean Community Areas Management Programme for Indigenous Resources (CAMPFIRE) programme, which can be classified as a PES-like programme, communities controlling land areas with attractive wildlife and landscape beauty values have auctioned off land access rights for safaris and trophy hunting to the highest bidding tour operators (Frost & Bond, 2008).

Several studies report PES transfers to represent significant contributions to household income, but few provide net income change estimates. In Costa Rica, PES payments accounted for more than 10% of household income in more than one quarter of participants (Pagiola et al., 2005); in the Virilla watershed, PES payments averaged 16% of cash household incomes (Miranda et al., 2003). In the Osa Peninsula, the PES scheme lifted half of recipients above the poverty line, and became the primary cash income source for 44% of households (Muñoz, 2004). In Pimampiro, Ecuador, watershed protection payments to poor upland settlers made up 30% of recipient households' spending on food, medicine and schooling (Echavarría et al., 2004). PROFAFOR carbon projects in low-income, high-altitude areas of Ecuador, and the Huetar Norte project in a disadvantaged region of Costa Rica, both created employment and a valuable plantation asset for future incomes (Milne, 2000; Miranda et al., 2003; Albán & Argüello, 2004), with expected internal rates of return of 12–27% over a 30-year horizon (Grieg-Gran et al., 2005).

In Bolivia, four of five PES-like schemes jointly targeting biodiversity conservation and landscape beauty through communal ecotourism produced local annual income gains of between US$77 and US$640 per household. For the Noel Kempff project, estimates are in the range of US$100–250 (Milne et al., 2001; Asquith et al., 2002). Bolivia's Los Negros watershed PES produced in-kind benefits from beehive transfers with the low monetary equivalent of US$3.50–7.00 per hectare per year, but with very low opportunity costs, net gains were likely positive (Robertson & Wunder, 2005). At least for poor people living in remote, disadvantaged regions, the relative size of income contributions from PES or PES-like schemes tends to be significant.

Of course, PES transfers need to be evaluated against the opportunity costs of compliance with PES conditions, which can also be substantial if schemes target areas with high-value agricultural potential. Estimating future income streams represents a challenge not only for service buyers; some providers may find participation less attractive as expected returns from alternative land uses increase over time (e.g. with rising commodity prices or the construction of a new road). In such cases, long-term income effects will depend on participants' ability to either flexibly renegotiate or exit PES contracts at little cost. De facto, both individuals and communities usually find it easy to either formally exit or default on PES contracts, including because sanctions in most cases are restricted to a cancellation of current and future payments, rather than inflicting fines or repaying past transfers.

Non-income gains

Frequently PES recipients gain more than just income from participation; non-monetary side benefits can be equally important (Rosa *et al.*, 2003). Two factors stand out: land tenure consolidation, and increases in human and social capital.

Regarding tenure, one global review had feared that PES could induce more powerful groups to crowd out smallholders from their land whenever insecure property rights existed (Landell-Mills & Porras, 2002). While in theory this could have happened, PES participation usually *increases* smallholder land tenure security *vis-à-vis* neighbours or squatters. Moreover, the benefits of participation in most current PES schemes are not large enough to really attract the interest of the powerful (Rosa *et al.*, 2003; Robertson & Wunder, 2005). Future large-scale carbon payments, for example, under an international mechanism for Reduced Emissions from Deforestation and Forest Degradation (REDD) might, nonetheless, become more susceptible to rights abuses in the presence of unequal local power relationships.

Secondly, experiences show that PES participants tend to increase their human and social capital by improving internal organisation, for example through collective bargaining and action *vis-à-vis* service buyers (Rosa *et al.*, 2003; Grieg-Gran *et al.*, 2005). Some benefits are provided in advance (training courses, help in starting an association etc.); others accrue through PES 'learning by doing' (e.g. through negotiating with ES buyers or intermediaries). The Chalalán Ecolodge showed both effects: substantial upfront training was provided by Conservation International to the community that now fully controls the operation, and significant internal coherence and entrepreneurial spirit have been added (Robertson & Wunder, 2005). This empowerment is to the advantage of local people in any other business dealings with the outside world.

Inequality and social costs

Despite the predominance of positive socio-economic PES outcomes, some negative social side effects can also occur. Paying some and not others (e.g. enrolling only 'strategically located' landowners) could create jealousies and raise inequality. This is a feature that is probably shared with most development initiatives: not every individual is equally equipped and motivated to take advantage of new income-generating options. In addition, in some hybrid cases between PES and traditional projects, benefit flows can create paternalistic expectations of even greater flows, and tensions when these hyper-expectations are not met (Robertson & Wunder, 2005).

Nevertheless, collective PES payments benefitting all households may create equality but not equity: households that bear disproportionate opportunity costs are under-compensated and made worse off, yet cannot individually reject the PES deal. This can create social tensions, as in the case of the Zimbabwean CAMPFIRE programme where landowners and users directly adjacent to wildlife priority areas often lost out (Frost & Bond, 2008), and in the Noel Kempff project where one community particularly dependent on logging was disproportionably hit (Asquith *et al.*, 2002). Fine-tuning compensations in collective deals to individual opportunity costs may prevent such social tensions.

Do poor environmental service users benefit from PES?

It is frequently, and often correctly, assumed that ES providers and sellers are much poorer than ES users and buyers. If, however, watershed downstream areas are populous, the number of poor ES buyers represented by one municipal water plant may often exceed that of upstream poor ES sellers by orders of magnitude, which also has poverty implications. In a watershed PES scheme in Chaina, Colombia, ~5600 water users, many of them poor farmers, depend for their water quality on the action of 12 upstream settler households currently increasing sedimentation through deforestation of a 440 ha sub-watershed (Borda Almanza *et al.*, 2009). At a larger scale, the aqueduct of Táchira in western Venezuela serves ~90,000 people, but is similarly being threatened by sedimentation from ~90 upstream households' farming, and a small scale PES scheme has recently been piloted to reverse the trend. Hence, maximising sellers' PES rates alone would here have little poverty alleviation effect: the major welfare benefits from PES accrue through reduced drinking water fluctuations and pollution, improving the numerous poor water users' health. Helping the poor is thus here best achieved by making sure the PES scheme is efficient in delivering the service it promises.

How do poor non-participants fare?

For most poor non-participants, PES effects will be neutral, but some may be significantly affected through impacts on land, labour, capital and output markets, which will depend on the PES scheme's scale. PES schemes will usually directly affect the market for (or the availability of) land. To the extent that PES schemes are achieving effective conservation (e.g. reducing open access lands and stabilising agricultural frontiers) they could particularly hurt the landless who are looking for opportunities to invade land for homesteading – the scenario prevalent, for instance, in the Los Negros watershed discussed in this chapter. While most landowners there were poor, the landless were arguably even poorer, thus pointing to important conflicts of interest between different strata of poor (Robertson & Wunder, 2005).

Regarding labour–market effects, subgroups of poor may self-engage (or be employed) in environmentally degrading activities, such as logging company workers, firewood cutters and charcoal makers, extractors overharvesting non-timber forest products (NTFP) or farm hands hired for clearing land and cultivating converted soils. These poor people often depend on natural resources in open access lands for their livelihoods, in ways that come to degrade the environment. To the extent that the PES scheme is *use restricting* (i.e. it caps total economic activity levels, as in the Los Negros conservation set-aside PES), the poor are likely to lose out in terms of employment or informal sector income. For instance, PES restrictions were found to likely hurt traditional herder and NTFP harvester groups in India (Kerr, 2002). Conversely, if ES provision is *asset building*, such as by establishing silvopastoral systems in treeless pastures (Pagiola *et al.*, 2008) or planting trees in degraded landscapes with few productive alternatives (Albán & Argüello, 2004), one can expect a net expansion in rural jobs that benefits unskilled rural labour, thus alleviating poverty.

How important are these PES-derived effects from production-factor and output markets for the poor? Few empirical studies on these linkages exist. One economy-wide study on the Costa Rican PES programme suggests small to negligible effects (Ross *et al.*, 2006). Effects in the timber sector could nonetheless be a prominent exception. For instance, laid-off logging and sawmill workers constituted the most weighty local welfare loss, and the main reason for community compensations being implemented, in Noel Kempff (Asquith *et al.*, 2002). In other words, the non-participating poor should not be forgotten in PES livelihoods impact assessments.

Conclusions

The empirical evidence on welfare impacts of PES in developing countries remains sketchy, both because many schemes are still young, and because little systematic 'with and without PES' welfare data have been gathered. Yet, as suggested in this chapter, some patterns are emerging.

Poor providers' participation in PES schemes

Access to PES schemes depends upon eligibility, desire, ability and competitiveness. Among these filters, informal and/or insecure land tenure and the high buyer transaction costs of working with numerous smallholders may introduce a partial anti-poor bias. One might address this by skipping or modifying inappropriate access restrictions (such as formal land title requirements), or by developing collective smallholder-bundling schemes that reduce transaction costs. If ample participation of poor providers is a prime concern, explicit poverty targeting (e.g. maximum income thresholds) and subsidies (e.g. pro-poor carbon premiums) may be considered, though targeting tends to come at the cost of environmental efficiency losses.[3] Conversely, other pre-existing filtering conditions are clearly pro-poor, such as formal caps on land size enrolled in most public PES schemes, and in particular the spatial overlap between marginal production areas and environmentally sensitive lands. To the extent that these pro-poor filters dominate in the first place, we may not need to worry much about the poor's access to PES schemes.

Welfare effects of PES on poor providers

Welfare gains from participation in PES scheme generally require participation to be voluntary. PES schemes are then very likely to benefit poor providers. Large singular cash transfers often could have detrimental local development impacts, but there is growing evidence from other fields (e.g. education, demobilisation and emergency aid) that well-designed conditional cash transfers can efficiently achieve targeted sectoral goals while stimulating recipient welfare.[4] Both income and non-income gains from PES can be substantial, but per capita income gains are seldom impressively large: typically ES buyers are in a better negotiating position to appropriate the 'gains from trade'. Pro-poor premiums can be one pathway to increase provider incomes.

Poor service users' benefits from PES

Poor ES users – in some cases ES buyers, and in others free riders – can benefit significantly, so PES poverty assessments should not focus exclusively on ES sellers, as sometimes the former may outnumber the latter, especially in watershed protection. If so, making sure that the promised service is delivered is certainly the best way to help the poor. Donors and NGOs could help organise multiple poor service users as ES buyers – say, poor fishermen paying for mangrove restoration. Although 'the poor paying the non-poor' could also be a highly remunerative strategy for the former, this constellation is usually very challenging to implement.

---

[3] One example is the Mexican Payment for Hydrological Environmental Services (PSAH) program, which since 2004 has progressively achieved higher poverty-targeting levels, but correspondingly has become less focused on zones truly threatened by deforestation, which, all else being equal, lowers the program's additionality (Muñoz-Piña et al., 2007).

[4] These cash transfer programs include conditional payments for school enrolment in Mexico, Brazil and Argentina (e.g. Heinrich, 2007). In two Mozambican programs, flood victims and demobilized soldiers receiving cash have generally used their money wisely, administrative costs were at 5–10% low and the overall poverty alleviation impact was impressive (Hanlon, 2004). A cash transfer program in Malawi has had similar positive experiences (F. Ellis, personal communication, 2007).

PES effects on the non-participant poor

Derived effects are usually mixed, and minor in size. However, specific effects are that landless people can suffer if open access to land is closed, and people working in environmentally degrading activities can lose their job. This feature is not exclusive to PES; it is shared with any efficient conservation initiative. If implementers on fairness or political feasibility accounts are highly concerned with these derived effects, they could take compensatory measures that minimise third-party losses.

How important is PES on aggregate for poverty alleviation? In most cases, PES has positive but quantitatively small poverty reduction effects. Those claiming that PES will become a poverty trap are clearly barking up the wrong tree, while those believing in its major poverty alleviation potential may underestimate the limitations posed by narrow spatial targeting. Arguably, the main bottleneck impeding PES from reducing poverty more is its hitherto limited implementation scale. Analogous to the growth versus distribution development debate, the key trick is probably not how to redistribute the PES cake towards a greater share for the poor, but how to increase the size of the entire cake. The debate on REDD, nonetheless, suggests that with larger scale schemes also comes a need for specific safeguards to avoid elite capture and violations of rights of both poor participants and non-participants under weak local institutional conditions.

Conversely, to the extent that multiple side objectives (e.g. concerns for poverty, human rights, gender equity and indigenous people) are increasingly squeezed into the PES equation, PES will lose efficiency in achieving its main target: to maintain or improve environmental service provision. Conversely, if schemes become overregulated at the cost of efficiency, private sector interest would be lost. A reduced PES scale would limit not only environmental impacts but also, eventually, pro-poor effects. Policy and decision makers should thus not 'put the (poverty) carriage before the (PES) horse'. Poverty reduction is an important PES side objective, and safeguards can be taken to address it properly – but it should never become the primary goal of PES.

References

Albán, M. & Argüello, M. (2004) *Un análisis de los impactos sociales y económicos de los proyectos de fijación de carbono en el Ecuador. El caso de PROFAFOR – FACE.* Report No. 1 84369 506 5. International Institute for Environment and Development, London.

Asquith, N.M., Vargas-Ríos, M.T. & Smith, J. (2002) Can forest-protection carbon projects improve rural livelihoods? Analysis of the Noel Kempff Mercado Climate Action Project, Bolivia. *Mitigation and Adaptation Strategies for Global Change,* 7, 323–37.

Bennett, M.T. (2008) China's sloping land conversion program: institutional innovation or business as usual ? *Ecological Economics,* 65, 699–711.

Borda Almanza, C.A., Moreno-Sánchez, R.d.P. *et al.* (2009) *Pagos por Servicios Ambientales en Marcha: La Experiencia en la Microcuenca de Chaina.* Departamento de Boyacá, Colombia, Centro para la investigación forestal internacional, CIFOR.

Echavarría, M., Vogel, J., Albán, M. & Meneses, F. (2004) *The Impacts of Payments for Watershed Services in Ecuador.* Report No. 1 84369 484 0. International Institute for Environment and Development, London.

Frost, P.G.H. & Bond, I. (2008) The CAMPFIRE programme in Zimbabwe: payments for wildlife services. *Ecological Economics*, 65, 776–87.

German, L.A., Ruhweza, A., Mwesigwa, R. & Kalanzi, C. (2010) Social and environmental footprints of carbon payments: a case study from Uganda. In *Payments for Environmental Services, Forest Conservation and Climate Change: Livelihoods in the REDD?* ed. L. Tacconi, S. Mahanty & H. Suich, pp. 160–84. Edward Elgar Publishing, Cheltenham.

Grieg-Gran, M., Porras, I.T. & Wunder, S. (2005) How can market mechanisms for forest environmental services help the poor? Preliminary lessons from Latin America. *World Development*, 33, 1511–27.

Hanlon, J. (2004) It is possible to just give money to the poor. *Development and Change*, 35, 375–83.

Heinrich, C.J. (2007) Demand and supply-side determinants of conditional cash transfer program effectiveness. *World Development*, 35, 121–43.

Kerr, J. (2002) Watershed development, environmental services, and poverty alleviation in India. *World Development*, 30, 1387–400.

Landell-Mills, N. & Porras, I.T. (2002) *Silver Bullet or Fool's Gold? A Global Review of Markets for Forest Environmental Services and Their Impact on the Poor.* International Institute for Environment and Development, London.

Milder, J.C., Scherr, S.J. & Bracer, C. (2010) Trends and future potential of payment for ecosystem services to alleviate rural poverty in developing countries. *Ecology and Society*, 15, 6.

Milne, M. (2000) *Forest Carbon, Livelihoods and Biodiversity. A Report to the European Commission.* CIFOR, Bogor, Indonesia.

Milne, M., Arroyo, P. & Peacock, H. (2001) *Assessing the Livelihood Benefits to Local Communities from Forest Carbon Projects: Case Study Analysis Noel Kempff Mercado Climate Action Project.* Unpublished report. CIFOR, Bogor, Indonesia.

Miranda, M., Porras, I. & Moreno, M. (2003) *The Social Impacts of Payments for Environmental Services in Costa Rica.* Report No. 1 84369 453 0. International Institute for Environment and Development, London.

Muñoz, R. (2004) *Efectos del programa de servicios ambientales en las condiciones de vida de los campesinos de la Península de Osa.* Universidad de Costa Rica, San José.

Pagiola, S. (2008) Payments for environmental services in Costa Rica. *Ecological Economics*, 65, 712–24.

Pagiola, S., Arcenas, A. & Platais, G. (2005) Can payments for environmental services help reduce poverty? An exploration of the issues and the evidence to date. *World Development*, 33, 237–53.

Pagiola, S., Rios, A.R. & Arcenas, A. (2008) Can the poor participate in payments for environmental services? Lessons from the Silvopastoral Project in Nicaragua. *Environment and Development Economics*, 13, 299–325.

Rios, A.R. & Pagiola, S. (2010) Poor household participation in payments for environmental services in Nicaragua and Colombia. In *Payments for Environmental Services, Forest Conservation and Climate Change: Livelihoods in the REDD?* ed. L. Tacconi, S. Mahanty & H. Suich, pp. 212–43. Edward Elgar Publishing, Cheltenham.

Robertson, N. & Wunder, S. (2005) *Fresh Tracks in the Forest: Assessing Incipient Payments for Environmental Services Initiatives in Bolivia.* CIFOR, Bogor, Indonesia.

Rosa, H., Kandel, S. & Dimas, L. (2003) *Compensation for Environmental Services and Rural Communities: Lessons from the Americas and Key Issues for Strengthening Community Strategies.* PRISMA, San Salvador, El Salvador.

Ross, M., Depro, B. & Pattanayak, S. (2006) *Assessing the Economy-Wide Effects of the PSA Program.* Workshop on Costa Rica's Experience with Payments for Environmental Services, San José, Costa Rica, September.

Smith, J. & Scherr, S. (2002) *Forest Carbon and Local Livelihoods: Assessment of Opportunities and Policy Recommendations.* CIFOR, Bogor, Indonesia.

Sommerville, M., Jones, J.P.G., Rahajaharison, M. & Milner-Gulland, E.J. (2010) The role of fairness and benefit distribution in community-based payment for environmental services interventions: a case study from Menabe, Madagascar. *Ecological Economics,* 69, 1262–71.

Southgate, D., Haab, T., Lundine, J. & Rodríguez, F. (2007) Responses of poor, rural households in Ecuador and Guatemala to payments for environmental services. Unpublished report. Ohio State University, Columbus.

Sunderlin, W.D., Dewi, S.D. & Puntodewo, A. (2007) *Poverty and Forests: Multi-Country Analysis of Spatial Association and Proposed Policy Solutions.* CIFOR Occasional Paper 47. CIFOR, Bogor, Indonesia.

Turpie, J.K., Marais, C. & Blignaut, J.N. (2008) The working for water programme: evolution of a payments for ecosystem services mechanism that addresses both poverty and ecosystem service delivery in South Africa. *Ecological Economics,* 65, 788–98.

Wunder, S. (2008) Payments for environmental services and the poor: concepts and preliminary evidence. *Environment and Development Economics,* 13, 279–97.

Wunder, S. & Albán, M. (2008) Decentralized payments for environmental services: the cases of Pimampiro and PROFAFOR in Ecuador. *Ecological Economics,* 65, 685–98.

Wunder, S., Bui Dung The, and E. Ibarra. (2005) Payment is good, control is better: why payments for environmental services so far have remained incipient in Vietnam. CIFOR, Bogor, Indonesia.

(15)

Pastoralism and Conservation – Who Benefits?

Katherine Homewood[1], Pippa Chenevix Trench[2] and Dan Brockington[3]

[1]Department of Anthropology, University College, London, UK
[2]Independent Consultant, Washington D.C., USA
[3]School of Environment and Development, University of Manchester, UK

Introduction

Conservation is big business in East Africa. Tourism is regularly among the top three contributors to GDP and to foreign exchange earnings in Kenya, accounting for US$884 million in 2010 (KShs 73.7 billion).[1] Despite the global financial crisis, Tanzania earned US$1.16 billion from tourism in 2009.[2] In both countries, tourists are largely drawn by the appeal of wildlife alongside other attractions. Conservationists see tourists' dollars as one of the principal means to generate meaningful income for the rural poor. Government policies (United Nations Development Programme *et al.*, 2005; United Republic of Tanzania (URT), 2005), conservation NGO projects (African Wildlife Foundation, 2005), entrepreneurial initiatives (Nelson, 2004; Lewa Wildlife Conservancy, 2012) and research publications (Pearce & Moran, 1994; Hutton *et al.*, 2005) all promote wildlife-based tourism. Maasailand, the region of Kenya and Tanzania dominated by Maa-speaking pastoralists, is a hotspot of conservation,

[1] See MTW (2008) and Kenya National Bureau of Statistics (economic survey and leading economic indicators), http://www.knbs.or.ke.
[2] See http://www.tanzania.go.tz/economicsurveyf.html.

Biodiversity Conservation and Poverty Alleviation: Exploring the Evidence for a Link, First Edition.
Edited by Dilys Roe, Joanna Elliott, Chris Sandbrook and Matt Walpole.
© 2013 John Wiley & Sons, Ltd. Published 2013 by John Wiley & Sons, Ltd.

poverty and new initiatives to redistribute tourist income, and a good place to explore the dynamics and distribution of revenues in the pastoral context.

In Kenya, Maasailand and other pastoral areas are among the fastest growing tourism destinations (33% growth in bed-nights 2004–5: Ministry of Tourism and Wildlife (MTW), 2006). In Tanzania in 2009, 16 national parks earned US$43.8 million, Ngorongoro Conservation Area earned US$22.6 million and tourist hunting US$14.9 million.[3] In both Kenya and Tanzania, the highest earning protected areas are situated within, and effectively excised from, Maasailand (Figure 15.1), as is a high proportion of the two countries' conservation estate overall. Parts of Kenyan Maasailand have shown rapid economic growth driven by wildlife conservation, rising domestic and export markets for crops and rising land values (Norton-Griffiths and Said, 2010). However, pastoral areas including many Maasai communities in both Kenya and Tanzania continue to display wide and deep poverty with respect to international and national rural poverty thresholds (Oxfam, 2006; Kenya: Thornton *et al.*, 2006; Boone *et al.*, 2011; Tanzania: Tenga *et al.*, 2008).

Kenyan and Tanzanian governments see pastoralist livestock management (mobile transhumance on unfenced, unmodified rangelands) as unproductive and environ-mentally damaging (e.g. URT, 1997; Ministry of Livestock and Fisheries Development, 2006). Pastoral migration to south Tanzania is perceived, without good data, to be driven by pastoralists' own degradation of their rangelands (Brockington, 2006). Regional and district governments impose draconian confiscations of cattle and fines, constraining pastoral activities while benefiting from their productivity. Wildlife tourism is portrayed as a means for pastoral groups to diversify, generate revenues and improve well-being.

This chapter explores the role that livestock play in rural Maasai household economies, and the contribution of wildlife tourism to poverty reduction and local livelihoods. Taking a comparative approach across Kenyan and Tanzanian study sites, these data allow evaluation of conservation and poverty reduction policies and practices.

Approach and methods

Multisite studies (reported in detail in Homewood *et al.*, 2009) sought a balanced view of the contribution of wildlife conservation to local livelihoods in rural Maasailand by asking:

- What do people do to meet their day-to-day and longer term livelihood needs?
- How well are they doing?

---

[3] Ibid.

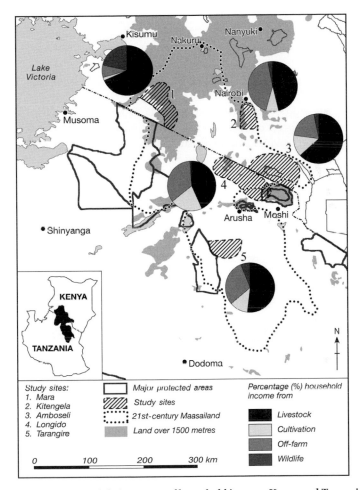

Figure 15.1   **Study sites and their sources of household income, Kenya and Tanzania.**

- What factors influence people's choice of income-earning activities?
- What factors influence how well they do?

We summarise detailed findings from independent researchers[4] working in three Kenyan and two Tanzanian sites (Homewood *et al.*, 2009). The five sites represent

---

[4] Without detailing all those involved in data collection and analysis, lead researchers included (besides the present authors) David Nkedianye and Patti Kristjanson (Kitengela), Michael Thompson (Mara), Shauna BurnSilver (Amboseli) and Hassan Sachedina (Tarangire). See Homewood *et al.* (2009) for full details.

very different circumstances ranging from populations adjacent to high-earning conservation areas (Mara: Thompson *et al.*, 2009), to minimal- (or zero-) earning areas (Longido: Chenevix Trench *et al.*, 2009). They also range from remote rural areas where grazing, farming and wildlife tourism are the main options (Amboseli: BurnSilver, 2009; and Longido), to mining areas (Tarangire: Sachedina & Cheneivx Trench, 2009), to peri-urban populations (Kitengela: Nkedianye *et al.*, 2009) where land leasing, sale of produce to urban markets and off-farm and non-farm employment are all significant sources of income. Standardised income and explanatory variables were collected or derived for each of the study sites. Cluster analysis identified groups of households pursuing similar livelihood and income-earning activities in each area; and regression analysis identified significant factors explaining variation in income levels across households. Family portrait studies captured qualitative, household-level pictures of livelihoods and livelihoods change (Homewood *et al.*, 2009). Each case study sought to represent variation in wealth, poverty and environment (Homewood *et al.*, 2009).

Livestock, farming, off-farm work and livelihoods in Maasailand

Most (91–100%) households have livestock, which account for well over half of their income (Figure 15.1). However, a significant proportion of households have too few livestock to fully support household members. Most livestock are concentrated in the hands of a few, with the wealthiest 10–20% owning from one-half to two-thirds of all livestock[5] across all sites. Reliance on non-livestock income is therefore a necessity for most, especially for the poorest, quite apart from being a potentially positive investment option for the well-off. Nonetheless in each site, across all different wealth categories and across most livelihood strategies, people were actively purchasing livestock. Poor households continue to seek to rebuild their herds, while better off households continue to invest in new animals.

In addition to livestock ownership, cultivation is widely practised, despite the semi-arid nature of Maasailand as a whole, and most households' limited access to agro-ecologically favourable sites. In four out of the five sites, over half of households engage in farming – with as many as 88% of households in Tarangire compared to only 13% in Mara. Despite the apparent attraction of cultivation, yields are poor and contribute little to overall incomes. Indeed, in Mara, Longido and Kitengela, over half of cultivating households harvested nothing. Crops account for just 2% of income in Mara, with the maximum contribution being 21% in Longido. There are,

---

[5] Measured in tropical livestock units (TLUs).

however, additional benefits from cultivation. In particular, it is an effective means of staking claim to a plot prior to land privatisation and subdivision (e.g. Mara), and of forestalling the perceived threat of protected area expansion (e.g. Tarangire).

Both direct observation and remotely sensed land cover analyses show extensive large-scale commercial cereal farming around Mara. This is largely driven by a relatively small elite. Commercial cultivation dropped significantly between 1998 and 2004 with the completion of land titling across most of the area studied. Alongside poor rainfall and declining soil fertility, the transaction costs of dealing with multiple smallholders (as opposed to dealing with the group ranch committee for large areas) made large-scale farming in Kenya Maasailand difficult. Large-scale cereal farming has also spread around Tarangire. Maize cultivation has now become lucrative for households able to invest in mechanized farming (Sachedina, 2008). However, such large-scale farming is not widespread in most Longido villages. Formerly village-owned high-potential lands on the slopes of Mt Kilimanjaro (east of Longido) have long since been leased by the state to outside investors.

Off-farm work generally outranks agriculture as a source of income. Half or more households (50–85%) earn off-farm income from petty trade, business, wages or salaried income and remittances. Returns from casual unskilled work are a fraction of those for regular jobs such as teacher, driver or government official. Potentially large but ephemeral income streams from gemstone mining and brokerage, and land leasing, are seen as secondary in importance to livestock and other economic activities. Off-farm work accounts for between 8% (Mara) and 43% (Kitengela) of average household income – second only to livestock in most sites other than Mara. This bears out analyses emphasising the need for off-land work and the willingness of pastoral peoples to pursue these activities (e.g. Sandford, 2006; Tache, 2008).

Income from wildlife – what role in livelihoods?

In comparison to the universal engagement with livestock, and widespread involvement with farming and off-farm work, only a small proportion of households in most sites have wildlife earnings (3–14%). Averaging across those households which do derive income from wildlife, amounts are small in four of the five sites (Amboseli, Kitengela, Longido and Tarangire) contributing <5% of mean annual income. Some positive impacts on household income are arguably invisible – for example, the use of wildlife-related income at the village level to offset village-level taxes in Tanzania. Where data were available, it was clear that village-level benefits were easily captured by local elites and were not having the broader impacts on livelihoods that could influence household decision making (Sachedina, 2008).[6] In Mara, though, two-thirds (64%) of

---

[6] This is not universally the case across Maasailand and there have been instances of community-based tourism resulting in meaningful revenues that were then well distributed (Nelson & Makko, 2003).

households earn some income from wildlife. Wildlife conservation accounts for 21% mean annual income for Mara households in the sample. While overall few Maasai households earn income from wildlife, and the sums they make do not compare with main income streams from livestock, crops and off-farm sources (Figure 15.1), landowning households close to Maasai Mara National Reserve (MMNR) see real benefits from conservation-based enterprises.

MMNR is Kenya's highest earning wildlife tourist destination, taking US$15–20 million annually (Norton-Griffiths, 2007). Numerous additional wildlife enterprises have grown up around the Mara, with landowners[7] on the now-adjudicated, subdivided and privately owned former group ranch lands able to capture wildlife returns both directly (through participation in campsites or other enterprises which pay rent or dividends) and indirectly (through the 19% of gate takings which the MMNR disburses to neighbouring communities). Thompson *et al.* (2009) chart the history of the various revenue-sharing institutions which have evolved around the Mara since the 1970s, including 19% MMNR gate takings paid to the county council; group ranch wildlife associations; post–group ranch, politically constituted wildlife associations and, most recently, conservancy partnerships between tourism investors and landowners. These offer better security of income to landowners (through rent rather than bed-night payments) and require 5-year covenanting of the designated area, during which land sales, homestead construction, cultivation, fencing and grazing are excluded. Conservancy arrangements may offer a better deal for landowners (Thompson *et al.*, 2009) and better conservation outcomes (Western *et al.*, 2006), but remain to be evaluated.

In the Mara, wildlife revenue makes up 15–30% of mean household income from the poorest quintiles to the best off, and is second in importance only to livestock. However, the top 25% Mara households by wealth consistently capture 60–70% of conservation income. The bottom 25% by contrast capture around 5%, rising to 15% if all forms of associated conservation-related employment are included. The middle 50% get around 25% of conservation-related income across the board. Despite significant changes in the volume of tourism returns between 1998 and 2004, there was minimal change in this pattern of distribution across wealth ranks. The poorest 20% of households are consistently more likely to be engaged in cultivation and/or off-farm work, and are significantly less likely to receive wildlife income than other households.

The total volume of tourism returns in Kenya fell significantly between 1998 and 2004, probably largely as a result of the impact of 9/11 on tourism internationally, and tourism collapsed again in 2007–8 following post-election violence. In 2004, mean conservation incomes to households earning from Mara wildlife associations

---

[7] This chapter does not go into the process of privatization, which dispossessed many vulnerable families (Galaty, 1999) but focuses on the impact of conservation business on current, mostly landowning residents.

and campsites averaged just 25% of their 1998 value. Within that changing flow of revenue, the relative proportions captured by the wealthiest, middle and poorest Mara households stayed remarkably constant. However, between 1998 and 2004, the proportion of households receiving income from wildlife associations fell from 55% to 37%. Overall the proportion of households reporting income from wildlife associations and campsites dropped from 55% to 41%.

Despite inequalities, conservation earnings reach most households in the Mara sample, and returns at household level, while very variable, generally make a significant contribution to total income. However, comparison of Mara with the four other sites suggests that these benefits derive from conditions rarely met elsewhere. The MMNR is high earning compared to other tourism destinations. The households sampled are predominantly landowners situated close to the reserve and hence able to command reasonable payments for game viewing or accommodation on that land.[8] By contrast, households in the other four study sites are either remote from tourist attractions or near lower earning sites, do not own the land, are unable to compete for conservation jobs and/or have little access to wildlife income overall. An example of this is the outer Amboseli households – although those close to Amboseli Park reportedly earn significantly more, possibly at levels comparable to those of Mara (David Western, personal communication, 2008).

Conservation, wealth and poverty in Maasailand

These findings underline the lasting importance of livestock to Maasai households. Livestock remain central to subsistence, to pathways out of poverty and to wealth storage and accumulation strategies, alongside the need to diversify into non-livestock activities. Were we to consider the social importance of livestock in maintaining social relations, and not just their economic value, their significance would be all the greater. Cultivation, while widely practised, gives very limited returns. However, besides adding to food security, it may be a tenure strategy, curbing what is perceived locally as the encroachment of conservation on customary rangelands (Sachedina, 2008). Perhaps surprisingly, off-farm income is a very significant component of present-day Maasai livelihoods, usually more so than agriculture, but ranges from poorly paid, insecure and often dangerous work (e.g. miners, watchmen and sex workers) to secure jobs with wider political or economic prospects (e.g. teachers and MPs). These findings also highlight that communities portrayed by some as the wealthiest land and stock owners in East Africa (Norton-Griffiths & Said, 2010) have average incomes far below

---

[8] Households which were not able to secure claim to a private plot have been excluded not only from the possibility of such wildlife income but from the landscape as a whole (Galaty, 1999). By definition, they cannot appear in our sample.

the dollar-per-day international poverty line, and often below national rural poverty thresholds. Given that these average income values are skewed upwards by a small number of well-off households (Homewood *et al.*, 2009), and median incomes are in most cases around half mean values, poverty remains both wide and deep in Maasai rangelands despite potential land values and tourism earnings (Homewood, 2009).

These findings also emphasise the generally limited contribution of income from wildlife conservation to households (other than those under special circumstances as in the Mara). Wildlife generally performs poorly for livelihoods. With the exception of the Mara, wildlife brings little or nothing to the vast majority of Maasai. It is possible that the poor contribution of wildlife to local livelihoods is a factor in the drastic declines in Kenyan savannah wildlife populations over the last 30 years (Ottichilo *et al.*, 2000; Homewood *et al.*, 2001; Western *et al.*, 2006, Ogutu *et al.*, 2011) and declines beyond national park boundaries in some ecosystems in Tanzania (Stoner *et al.*, 2007). If wildlife does not become locally valuable, it may continue to decline (Norton-Griffiths, 2007; Norton-Griffiths & Said, 2010; Ogutu *et al.*, 2011).

Why does wildlife fail to generate local benefits?

Maasai communities have historically captured little of total tourism earnings, with ~95% accruing to tour operators, service industry workers and the state (Norton-Griffiths, 2007; Norton-Griffiths & Said, 2010). Even when tourism revenues are captured, the small amounts are then poorly distributed within and between communities. Proportions captured by local residents are even smaller in Tanzania than in Kenya (Sachedina, 2008) as they trickle through official channels (from central and district government through to the communal level via Wildlife Management Areas (WMAs)).

In part this reflects chronic problems of governance and accountability at local, district and national levels (e.g. Walpole & Leader-Williams, 2001; Homewood, 2009). Wildlife enterprises earned tens of thousands of dollars annually for one village on the edge of Tarangire and yet these revenues were easily dominated by local elites (Sachedina, 2008). Ololoskwan village east of Serengeti was able, briefly, to earn around $50,000 per year from photographic safari operator use of its lands. However the central government feared it would conflict with a hunting block, whose revenues they control. In November 2007, a surprise ministerial declaration criminalised local-level deals for wildlife-related enterprises, capturing all such returns for the state, with no requirement for a set proportion to be returned to the community (Tanzania Natural Resources Forum, 2007). Such unequal contests between the state and local communities for control of conservation enterprises and their returns have become a common occurrence in Tanzania (Nelson, 2004, 2007; Baldus, 2009). They are made the more unequal by the involvement of global investors (Igoe, 2007). In their comparative study of community-based conservation, Nelson and Agrawal (2008) observe that the hunting industry in Tanzania is eminently corruptible, providing easily diverted revenues within a generally impoverished national economy.

The second reason is the historical experience of many Maa-speaking pastoralists. Conservation for them is associated with large-scale eviction and exclusion (from Serengeti, Ngorongoro, Tarangire, Amboseli and Mkomazi) with fines and harassment, compromises and deals that were not honoured and outreach programmes that provided few tangible benefits. Their experience of new revenue-sharing initiatives is rarely positive. The livelihood choices they face now are shaped by decades of such experiences, perceptions and stories as well as by complex communal politics, making it hard to build trust and co-operation.

A third reason results from the efforts of the conservation lobby itself. The financial success of the African Wildlife Foundation led to its growing out of touch with dilemmas in Maasai villages (Sachedina, 2008). This meant that it was poorly equipped to engage effectively with the fierce local politics that surround conservation initiatives in this region. Engaging with communities with such a record is an extremely difficult task.

The fourth reason may be the gloss accorded these conservation enterprises, portrayed by their advocates as win–wins, good for wildlife, good for people, good for the economy, participatory, empowering and liberating (Igoe, 2010). As they wallow in happy sentiment it is difficult for criticism to find purchase. Yet, when examined in detail these schemes rarely produce the benefits they claim. As this chapter underlines, revenues from wildlife rarely begin to compensate for loss of mobility, as well as loss of access to and control over important natural resources, which 'community-based' and other conservation restrictions entail. Even relatively successful schemes produce thoroughly dissatisfied groups marginalised from the lucrative revenue streams flowing past them (e.g. Il Ngwesi, Kenya: Castillo, 2004). In Tanzania, WMAs in Burunge (west of Tarangire: Igoe & Croucher, 2007) and Longido (Homewood et al., 2005) restricted the use of villages' grazing lands while removing their right to control returns, or else caused local displacement and eviction.

Moving forward

Moving forward, the prospects do not seem good. In Kenya, the draft National Land Policy set out innovative and socially equalising reforms which have come up against vociferous challenge by vested interests (Ministry of Lands, 2007a,b; Homewood, 2009). The Wildlife Bill 2007 (MTW, 2007) proposes command and control of wildlife-related activities on private land. Private conservancies buy out some pastoralist landowners, and establish set-aside agreements with others (Thompson *et al.*, 2009). The extent to which they work for people on the one hand and wildlife on the other remains to be shown.

In Tanzania the situation is more alarming still. A strong anti-pastoral environmentalism pervades the country, driving evictions from Usangu in 2007, in which people died; removals from Loliondo; a resumption of attempts to evict pastoralists from the joint land use area of Ngorongoro Conservation Area and large-scale confiscations and fines in Kilosa and Mbarali. Meanwhile, in 2008 President Jakaya Kikwete was feted for his conservation commitments (before the Serengeti Road issue surfaced: Homewood *et al.*, 2010) by US politicians and their sponsors at the International Conservation Caucus Foundation in Washington, DC. These are not incentives likely to make conservation-based enterprises work for Maasai communities in Tanzania. The short-sighted and self-defeating way in which Tanzania has implemented 'community conservation' has been severely criticised by one of the most experienced conservation practitioners in the field (Baldus, 2009). There is a real role for international conservation agencies to use maximum integrity and skill to encourage the state to consider local needs and voices and to foster policies and practices that can genuinely bring benefits to local poor people as well as to international conservation.

Conservation business is booming in East Africa; wildlife-based tourism remains a big earner for some, and conservation NGOs readily sell the idea to their northern support base. This is an arena where considerable profits can be made precisely because the distribution of revenues is so uneven, and local and national governments are so easily compliant. However, rural Maasai land use decisions do not support national- and international-level assumptions about the benefits of wildlife and tourism, nor those about a relatively lower economic importance of livestock production. If wildlife is to remain viable in East Africa, something significant has to change in order to deliver the mutually beneficial relationships sought by conservationists, governments and development organisations.

Acknowledgements

This chapter draws on material and insights from individual studies by BurnSilver, Thompson, Nkedianye, Sachedina, Kristjanson and others, published in detail in Homewood *et al.* (2009). We are grateful to all the collaborating authors who participated in that work, to the many Maasai households who discussed their livelihoods with the researchers, to the Kenya and Tanzania governments who gave permission for the fieldwork and to the Belgian government DGIC fund who made the original collaborative work possible.

A brief preliminary version of this chapter was presented at the 2009 Zoological Society of London symposium on Biodiversity conservation and Poverty reduction. Subsequently a much extended companion paper, encompassing this and other material, has been published as "Pastoralist livelihoods and wildlife revenues: a case for coexistence?" in the journal *Pastoralism: Research, policy and practice*.

References

African Wildlife Foundation (AWF) (2005) *Kilimanjaro Heartlands: A Summary of AWF's Engagement and Related GIS Data*. Presentation given at ILRI Reto-e-Reto Conference, Nairobi, January.

Baldus, R. (2009) A practical summary of experiences after three decades of community-based wildlife conservation in Africa 'What are the lessons learnt?' FAO and CIC, Budapest.

Boone, R.B., Galvin, K.A., BurnSilver, S.B., Thornton, P.K., Ojima, D.S. & Jawson, J.R. (2011) Using coupled simulation models to link pastoral decision making and ecosystem services. *Ecology and Society*, 16, 6. http://www.ecologyandsociety.org/vol16/iss2/art6/ (accessed 1 May 2012).

Brockington, D. (2006) The politics and ethnography of environmentalisms in Tanzania. *African Affairs*, 105, 97–116.

Burnsilver, S.B. (2009) Pathways of continuity and change: Maasai livelihoods in Amboseli, Kajiado District, Kenya. In *Staying Maasai? Livelihoods, Conservation and Development in East African Rangelands*, ed. K. Homewood, P. Kristjanson & P. Chenevix Trench, pp. 161–207. Springer, New York.

Castillo, A.R. (2004) *Sustainable Inequalities? Community-Based Conservation in an 'Inequitable Society'. The Case of Il Ngwesi Group Ranch, Kenya*. Unpublished master's thesis, University of Oxford, Oxford.

Chenevix Trench, P., Kiruswa, S., Nelson, F. & Homewood, K. (2009) Still people of cattle? Livelihoods, diversification, and community conservation in Longido District. In *Staying Maasai? Livelihoods, Conservation and Development in East African Rangelands*, ed. K. Homewood, P. Kristjanson & P. Chenevix Trench, pp. 217–56. Springer, New York.

Galaty, J.G. (1999) Grounding pastoralists: law, politics, and dispossession in East Africa. *Nomadic Peoples*, 3, 56–71.

Homewood, K. (2009) Policy and practice in Kenya rangelands: impacts on livelihoods and wildlife. In *Staying Maasai? Livelihoods, Conservation and Development in East African Rangelands*, ed. K. Homewood, P. Kristjanson & P. Chenevix Trench, pp. 335–67. Springer, New York.

Homewood, K., Brockington, D. & Sullivan, S. (2010) Alternative view of Serengeti Road. *Nature*, 467, 788–9.

Homewood, K., Kristjanson, P. & Chenevix Trench, P. (eds.) (2009) *Staying Maasai? Livelihoods, Conservation and Development in East African Rangelands*. Springer, New York.

Homewood, K., Lambin, E., Coast, E., Kariuki, A., Kikula, I., Kivelia, J., Said, M., Serneels, S. & Thompson, M. (2001) Long-term changes in Serengeti-Mara wildebeest and land cover: pastoralism, population or policies ? *Proceedings of the National Academy of Sciences of the USA*, 98, 12544–9.

Homewood, K., Thompson, M., Chenevix Trench, P., Kiruswa, S. & Coast, E. (2005) Community- and state-based natural resource management and local livelihoods in Maasailand. *Gestione della risorse naturali su base communitaria e statale. Ambiente e sviluppo sostenibile in Africa australe.* Special issue on community-based natural resource management. *Afriche e Orienti*, 2, 84–101.

Hutton, J., Adams, W. & Murombedzi, J. (2005) Back to the barriers? Changing narratives in biodiversity conservation. *Forum for Development Studies*, 2, 341–70.

Igoe, J. (2007) Human rights, conservation and the privatization of sovereignty in Africa – a discussion of recent changes in Tanzania. *Policy Matters*, 15, 241–54.

Igoe, J. (2010). The spectacle of nature in the global economy of appearances: anthropological engagements with the spectacular mediations of transnational conservation. *Critique of Anthropology*, 30, 375–97.

Igoe, J. & Croucher, B. (2007) Conservation, commerce and communities: the story of community-based wildlife management areas in Tanzania's northern tourist circuit. *Conservation and Society*, 5, 534–61.

Lewa Wildlife Conservancy (2012) Lewa Wildlife Conservancy publicity release: web and DVD. www.lewa.org (accessed 1 May 2012).

Ministry of Lands (MoL) (2007a) *Draft National Land Policy*. MoL National Land Policy Secretariat, Nairobi, Kenya.

Ministry of Lands (MoL) (2007b) *National Land Policy: The Formulation Process*. MoL National Land Policy Secretariat, Nairobi, Kenya.

Ministry of Livestock and Fisheries Development (MoLF) (2006) *Draft National Livestock Policy 2006*. MoLF, Nairobi, Kenya.

Ministry of Tourism and Wildlife (MTW) (2006) *Statistical Analysis of Tourism Trends*. MTW Central Planning Unit, Nairobi, Kenya. http://www.tourism.go.ke/ministry.nsf/doc/Tourism_Trends_OCT2006_Revised.pdf/$file/Tourism_Trends_OCT2006_Revised.pdf (accessed 1 May 2012).

Ministry of Tourism and Wildlife (MTW) (2007) Draft Wildlife Bill 2007. MTW, Nairobi, Kenya.

Ministry of Tourism and Wildlife (MTW) (2008) *Facts and Figures*. MTW, Nairobi, Kenya. http://www.tourism.go.ke/ministry.nsf/pages/facts_figures (accessed 1 May 2012).

Nelson, F. (2004) *The Evolution and Impacts of Community-Based Ecotourism in Northern Tanzania*. Drylands Issue Paper No. 131. International Institute for Environment and Development, London.

Nelson, F. (2007) *Emergent or Illusory? Community Management in Tanzania*. IIED Issues Paper No. 146. International Institute for Environment and Development, London.

Nelson, F. & Agrawal, A. (2008) Patronage or participation? Community-based natural resource management reform in Sub-Saharan Africa. *Development and Change*, 39, 557–8.

Nelson, F. & Makko, S. (2003) *Communities, Conservation and Conflicts in the Tanzanian Serengeti.* Third Annual Community-Based Conservation Network Seminar: Turning Natural Resources into Assets, Savannah, GA, October.

Nkedianye, D., Radeny, M., Kristjanson, P. & Herrero, M. (2009) Assessing returns to land and changing livelihood strategies in Kitengela. In *Staying Maasai? Livelihoods, Conservation and Development in East African Rangelands*, ed. K. Homewood, P. Kristjanson & P. Chenevix Trench, pp. 115–49. Springer, New York.

Norton-Griffiths, M. (2007) How many wildebeest do you need ? *World Economics*, 8, 41–64.

Norton-Griffiths, M. & Said, M. (2010) The future for wildlife on Kenya's rangelands: an economic perspective. In *Can Rangelands Be Wildlands? Wildlife and Livestock in Semi-Arid Ecosystems*, ed. J.G. Du Toit, R. Kock & J. Deutsch, pp. 367–92. Blackwell, Oxford.

Ogutu, J.O., Owen-Smith, N., Piepho, H.-P. & Said, M.Y. (2011) Continuing wildlife population declines and range contraction in the Mara region of Kenya during 1977–2009. *Journal of Zoology*, 285, 99–109.

Ottichilo, W.K., de Leeuw, J., Skidmore, A.K., Prins, H.H.T. & Said, M.Y. (2000) Population trends of large non-migratory wild herbivores and livestock in the Maasai Mara ecosystem, Kenya 1977–1997. *African Journal of Ecology*, 38, 202–16.

Oxfam (2006) *Delivering the Agenda: Addressing Chronic Under-Development in Kenya's Arid Lands*. Oxfam International Briefing Paper. Oxfam, Oxford.

Pearce, D. & Moran, D. (1994) *The Economic Value of Biodiversity*. Earthscan, London.

Sachedina, H. (2008) *Wildlife Is Our Oil: Conservation, Livelihoods and NGOs in the Tarangire Ecosystem*. Unpublished PhD thesis, Oxford University, Oxford.

Sachedina, H. & Chenevix Trench, P. (2009) Cattle and crops, tourism and Tanzanaite: poverty, land use change and conservation in Simanjiro District, Tanzania. In *Staying Maasai? Livelihoods, Conservation and Development in East African Rangelands*, ed. K. Homewood, P, Kristjanson & P. Chenevix Trench, pp. 263–98. Springer, New York.

Sandford, S. (2006) Foreword. In *Pastoral Livestock Marketing in Eastern Africa: Research and Policy Challenges*, ed. J.G. McPeak & P.D. Little. Intermediate Technology Publications, Rugby.

Stoner, C., Caro, T., Mduma, S., Mlingwa, C., Sabuni, G. & Borner, M. (2007) Assessment of effectiveness of protection strategies in Tanzania based on a decade of survey data for large herbivores. *Conservation Biology*, 21, 635–46.

Tache, B. (2008) *Pastoralism under Stress: Resources, Institutions and Poverty among the Borana Oromo in Southern Ethiopia*. Unpublished PhD thesis, Norwegian University of Life Sciences, Aas.

Tanzania Natural Resources Forum (TNRF) (2007) *New Regulations Signed for All Non-Consumptive Wildlife Use in Game Reserves & Village Lands*. TNRF, Arusha. http://www.tnrf.org/node/6529 (accessed 1 May 2012).

Tenga, R., Mattee, A., Mdoe, N., Mnenwa, R., Mvungi, S. & Walsh, M. (2008) *A Study on the Options for Pastoralists to Secure Their Livelihoods In Tanzania*. CORDS, PWC, IIED, MMM Ngaramtoni, TNRF, UCRT. http://www.tnrf.org/files/E-INFO-RLTF_VOL1_MAIN-REPORT_A_Study_on_options_for_pastoralism_to_secure_their_livelihoods_in_Tanzania_2008.pdf (accessed 1 May 2012).

Thompson, M., Serneels, S., Ole Kaelo, D. & Chenevix Trench, P. (2009) Maasai Mara – land privatization and wildlife decline: can conservation pay its way? In *Staying Maasai? Livelihoods, Conservation and Development in East African Rangelands*, ed. K. Homewood, P. Kristjanson & P. Chenevix Trench, pp. 77–114. Springer, New York.

Thornton, P., BurnSilver, S., Boone, R. & Galvin, K. (2006) Modeling the impacts of group ranch subdivision on agro-pastoral households in Kajiado, Kenya. *Agricultural Systems*, 87, 331–56.

United Nations Development Programme (UNDP), United Nations Environment Programme, International Institute for Environment and Development, International Union for Conservation of Nature and World Resources Institute (2005) *Environment for the MDGs: A Message to the World Summit: Sustaining the Environment to Fight Poverty and Achieve the MDGs: The Economic Case and Priorities for Action.* UNDP, New York.

United Republic of Tanzania (URT) (1997) *Livestock and Agriculture Policy. Section 3: Soil Conservation and Land Use Planning*, and *Section 4: Range Management*, pp. 55–68, 127–31. United Republic of Tanzania Policy Statements. URT, Dar es Salaam.

United Republic of Tanzania (URT) (2005) *Mkukuta: Tanzania's National Strategy for Growth and Reduction of Poverty.* United Republic of Tanzania Vice-President's Office. URT, Dar es Salaam.

Walpole, M.J. & Leader-Williams, N. (2001) Masai Mara tourism reveals partnership benefits. *Nature*, 413, 771.

Western, D., Russell, S. & Mutu, K. (2006) *The Status of Wildlife in Kenya's Protected and Non-Protected Areas.* Paper commissioned by Kenya's Wildlife Policy and Legislation Review. African Conservation Centre, Nairobi. www.tzonline.org/pdf/tourismmasterplan.pdf (accessed 1 May 2012).

16

Local Organisations – An Entry Point for Conservation and Poverty Reduction

David H. L. Thomas

Birdlife International, Cambridge, UK

Introduction

Both poverty and biodiversity loss are problems of global proportion, reflected in global statistics on biodiversity loss (Secretariat of the Convention on Biological Diversity, 2010) and poverty (United Nations Development Programme (UNDP), 2010a). Mechanisms created to deal with these challenges reflect the sense of a global crisis that requires global action. One response has been multilateral environmental agreements such as the Convention on Biological Diversity (CBD), the Ramsar Convention on Wetlands, and the United Nations Convention to Combat Desertification (UNCCD). These are essentially intergovernmental forums, where the only legitimate actors are considered to be nation states (Kothari *et al.*, 2008), and where civil society and levels of organisation and governance below the national scale often struggle to be heard. Many international conservation organisations have also set global conservation priorities, and have been criticised for setting an agenda that fails to take account of local conditions and is divorced from local priorities (Jepson, 2001; Rodríguez *et al.*, 2007). Likewise, the Millennium Development Goals (MDGs) are high-level targets that have done much to mobilise the attention and resources of governments and international agencies behind efforts to address poverty. However, despite the conclusion that 'action at the local level – with local organisations as key actors – underpins the success and sustainability of most environment and development initiatives' (Hazlewood, 2010), there is concern that there has been insufficient attention to local solutions,

Biodiversity Conservation and Poverty Alleviation: Exploring the Evidence for a Link, First Edition.
Edited by Dilys Roe, Joanna Elliott, Chris Sandbrook and Matt Walpole.
© 2013 John Wiley & Sons, Ltd. Published 2013 by John Wiley & Sons, Ltd.

or to the local-level processes and institutions needed to make sure that international development targets deliver on their promises (Satterthwaite, 2003).

Both poverty and biodiversity loss are experienced, and many of the solutions are to be found, at the local level. What is more, for many of the world's poor these two issues are intimately connected since environmental resources make an essential contribution to their livelihood, supplying food, fuel, medicines, shelter and cultural values (Chapter 4, this volume). Wild resources have been shown to provide 20–30% of rural people's income in developing countries, and provide up to 20% of all protein (Kaimowitz & Sheil, 2007). This high degree of dependence of poor people on natural resources has focused attention and effort on the contribution that biodiversity conservation can make to poverty reduction—as this book demonstrates—and also on the need to engage local people in biodiversity conservation (e.g. Ancrenaz *et al.*, 2007). There are diverse views on the relationship between biodiversity conservation and poverty reduction (e.g. Adams *et al.*, 2004; Gilbert, 2010) but both conservation and development sectors have advocated the importance of local participation and of partnership between international or national agencies and local people (Vermeulen & Sheil, 2007). There are many examples which demonstrate the central place for local actors in natural resource management, and the effectiveness with which they can carry out this role (e.g. Roe *et al.*, 2009).

This chapter examines the place for local organisations in addressing the linked challenges of poverty and biodiversity loss. It describes the important role played by local organisations, the international policy context that supports that role and the emerging frameworks for environmental management that make support to local organisations essential.

What are local organisations?

Local organisations include any group that is working locally to provide goods and services of value to local people—defined this broadly it includes branches of government providing services to local people, agencies responsible for law and order (e.g. the police) as well as community-based organisations (Satterthwaite & Sauter, 2008). In this chapter we are mainly interested in grassroots institutions whose members belong or have strong links—be they economic, cultural, social or recreational—to a specific geographical place and community. This means that they generally understand local culture as well as local power dynamics, and members are affected personally, and collectively, by their activities and their outcomes. They form part of civil society, and are generally independent of the state. Even so, such local organisations show great variety, differing for example in terms of their origins, membership, governance, scale, staffing and legal status.

The importance of local organisations

Whilst high-profile institutions created to deal with conservation and poverty alleviation (e.g. the CBD and the MDGs) are international, many of the barriers to conservation and development are local—local power structures, land-owning patterns and inequities in the way that costs and benefits are shared within and between communities. Equally many of the key decisions affecting resource management are made locally, and many of the changes needed to achieve effective development and sustainable natural resources management, relating for example to education and awareness, rules governing access to resources as well as land-planning decisions, need to be delivered locally. International organisations working locally may founder because of cultural disconnect and a poor understanding of local cultural perceptions and expectations (West, 2006). Local organisations, managing local resources, with an intimate knowledge of local social, cultural, economic and environmental contexts (e.g. Sagoff, 2011), and with close connections to a range of local actors, have a major role in addressing these issues and identifying culturally relevant conservation and development strategies. Grassroots organisations are able to mobilise ideas and local resources in ways that would not be possible for government, non-governmental organisations (NGOs) or the corporate sector. They also have the potential to translate large-scale national policy into locally relevant action, through experimentation and locally specific adaptation which reflect the diversity of real-life situations. Local organisations therefore have a key role to play in bridging the international-national-local divide.

Local organisations also provide an extremely efficient approach to conservation and development. Much of the input from their members is voluntary, and individuals are often highly motivated and are able to use community connections to leverage social capital supporting sustainable management. Being part of local society, the members of local organisations are also more likely to take the holistic approach that is needed to address environment and development issues together—issues of health, nutrition, belief and biodiversity are often intimately entwined, and bound by complex and unwritten social rules and cultural norms. People's dependence on their environment means that they have a lot at stake when the goods and services it provides are threatened, and are likely to approach problem solving with a high degree of motivation (Netherlands Environmental Assessment Agency (PBL), 2009). Even where local organisations are employed or contracted, their costs are much lower than those for international experts or national government (Box 16.1). The role of local leadership appears to be especially important in the effectiveness of local organisations (e.g. Kenward *et al.*, 2011). For example, a study of 130 co-managed fisheries has demonstrated the critical importance of community leaders and 'robust social capital' in successfully managing aquatic resources and securing community livelihoods (Gutiérrez *et al.*, 2011).

Box 16.1 Community councils more cost-effective managers of forest than government

Forests in the Indian central Himalayas were nationalised early in the 20th century. In 1930, approximately a decade after nationalisation, villages were permitted to carve out council-managed forests both from common lands not nationalised and from nationalised forests. Using government data on the cost per hectare of managing state forests and surveys to find the cost per hectare of council forest management, this study looked at the effects of devolution on the cost of forest management. It found that state forests cost at least seven times as much per hectare to administer as council-managed forests. A comparison of the extent of degradation in state forests with that in council forests showed only a small and statistically insignificant difference, leading to the conclusion that council management is more cost-effective than state management.

Source: Somanathan *et al.* (2009).

Box 16.2 summarises these and other benefits of working with local organisations.

Box 16.2 The benefits of working with local organisations

- **Local knowledge:** People living near or at a site have an intimate knowledge of the local environment and of local cultural perceptions of nature that can be critical in developing strategies for conservation or sustainable use. Local organisations can translate large-scale national policy into locally relevant action, through experimentation and locally specific adaptation and application.
- **Local networks:** Local people are often well connected, and they have an understanding of local politics, culture, relations and networks. They have a capacity to develop successful partnerships with local governments and are able to mobilise ideas and local resources in ways that would not be possible for government, international NGOs and the corporate sector.
- **Holistic approach:** Environment is not separate from the local economy, society and culture. Local actors understand and incorporate those linkages in their decision making.

- **Motivation**: Many local people are directly dependent on the environment for resources or services and have the most to lose if biodiversity is lost or a site is degraded.
- **Sustainability**: Conservation is a long-term endeavour. Supporting the work of local organisations enhances the prospects that conservation activities will be sustainable, and delivered by motivated and capacitated individuals.
- **Efficiency**: Conservation and development involve complex relations between different stakeholders, resource users and decision makers. Local individuals and institutions understand how to work with this complexity.
- **Cost-effective**: Working with local organisations which effectively address local priorities helps to mobilise complementary resources from households and community organisations.
- **Adaptive**: Local people have an awareness of their surroundings which enables them to make minute, immediate and appropriate changes to management and use in response to factors such as economic (e.g. changes in prices and availability of resources) and environmental change (e.g. climate change).
- **Legitimacy and avoiding conflict**: Decisions and actions led by organisations rooted in the community give social legitimacy to an intervention and may help to avoid conflict.
- **Empowerment and rights**: Working with local organisations recognises and respects people's rights (to resources, to participation and to have a voice) and can be very empowering.

In the face of strong local links to the environment, and a high level of local dependency, initiatives integrating conservation and development that do not have a role for local organisations, but are led by national government and international NGOs, may easily lead to local community resentment, alienation and opposition (e.g. Smith *et al.*, 2009). There is a risk, however, that engaging with local organisations (including building their capacity) can undermine the characteristics for which they are valued. Capacity-building processes, often designed according to the agenda and expectations of an intermediary organisation, may distort the local organisation's unique cultural character. Experience of working with urban community-based organisations (CBOs) suggests that self-empowerment, through a process in which local organisations have ownership of the capacity-building and development processes, and can understand and diagnose themselves, is an important characteristic of the development process (Yachkaschi, 2010).

Empowerment is a critical dimension to local organisations in that it provides the basis through which other qualities and benefits are realised, opening up many opportunities not possible by individuals working alone. Official, legal registration that is often part of local organisation formation can be an important part of this, providing a degree of legitimacy and recognition in the eyes of government and donors, as well as access to funding and resources. Organisations, through their elected representatives and spokespersons, may also provide more effective access to decision makers and planning and policy processes than individuals can achieve on their own (Department for International Development *et al.*, 2002). Another dimension to empowerment is the pooling of resources and knowledge, expanding the options for intervention, as well as the greater impact of coordinated action by a critical mass of community stakeholders. In ways such as these, the empowerment benefits from organising at local level are important outcomes in their own right.

Local organisations: a route to participation and good governance

Local organisations are an important part of the process for delivery of participation and good governance. Both development and conservation practitioners are paying increasing attention to governance—the process of decision making and the process by which decisions are implemented. Governance concentrates attention on the actors and their institutions—and therefore on local organisations. The characteristics of good governance include that it is "participatory, consensus oriented, accountable, transparent, responsive, effective and efficient, equitable and inclusive and follows the rule of law", and, moreover, "Participation needs to be informed and *organised*. This means freedom of association and expression on the one hand and an *organised civil society* on the other hand" (United Nations Economic and Social Commission for Asia and the Pacific, n.d.; emphasis added). Many of these characteristics are met by and through local organisations (Bass, 2008).

Participation has been described as the keystone of good governance, and is now a requirement of many regional and international development and conservation conventions, including the CBD. Although local organisations are not a *requirement* for 'participation', which can be achieved through other means, if effectively constituted and representative of their community they can be an efficient and effective route to achieve it. Greater levels of participation also enhance the delivery of rights, and given that the fulfilment and respect of rights are important aspects of combating multidimensional poverty (Organisation for Economic Cooperation and Development, 2001) there is a clear and direct link between genuine participation (through local organisations) and poverty reduction.

Scaling up and enabling factors

Roe and Bond (2007) argue that the impacts of local organisations in delivering on biodiversity conservation and the MDGs will always remain local, small scale and marginal unless conditions are created to allow scaling up. Despite their knowledge and experience, the potential of local people to manage the environment and deliver development has not been effectively harnessed (UNDP, 2010b). The UNDP Local Capacity Strategy aims to address the factors – *rights, access and finances* – that are limiting local organisations from achieving their full potential. Some of the key changes identified as necessary for local organisations to achieve their potential include:

- **Decentralisation of political, fiscal and administrative power**: For example, giving local organisations rights to collect revenues from resource use, and make budgetary decisions.
- **Land tenure and user rights secured for stakeholders**: Increasing the certainty and security of outcomes, which then increases people's commitment to sustainably manage resources.
- **Capacity, skills and awareness of communities strengthened**: Local-level resource management involves a change in ways of organisation and a shift in relationships of power, and it requires new sets of skills–not just in resource management but also in accounting, conflict resolution, monitoring and negotiation.
- **Capacity built in government**: For many staff in departments of forestry or wildlife, handing over control and responsibility for resource management will represent a major shift in approach and attitude from current practice. Staff will need training in participatory approaches and working with local organisations, and the way that departments are organised and decisions made will need to change.
- **Greater recognition of the role of community organisations in conservation and natural resource management (NRM)**: Decision makers and donors need a better and broader understanding, awareness and recognition of the role of community conservation in ecosystem management.
- **Better integration of the efforts of communities and local organisations**: Local organisations need to be supported to work as key and legitimate partners with the natural resource sector.
- **The efforts of local organisation are adequately and fairly rewarded**: Including through the elimination of perverse incentives such as market-distorting subsidies and other trade interventions. Payments for ecosystem services (see the 'Emerging Issues and Opportunities for Local Organisations' section of this chapter) provide a potential mechanism for achieving this.

An effective process is also important for creating enabling conditions. This includes development of a relationship of mutual trust between local community organisations and government (or NGOs), allowing sufficient time for trust to develop and for lessons to be learned and shared, early implementation of concrete actions and a demonstration of commitment (so that a relationship can be built on something other than words and promises), a process which allows for a progressive build-up of responsibility and commitment (so avoiding pressure), recognition of heterogeneity and conflicting interests *within* communities and mutual respect (Isager *et al.*, 2001).

Support to local organisations—policy and programmes

Some international agencies and donor programmes are beginning to recognise the potential for local organisations to deliver sustainable, rights-based conservation and development at a local level, and are working to help create the conditions in which local organisations can succeed.[1] The Equator Initiative,[2] begun in 2002, brings together international, national and local actors from the United Nations, governments, civil society, businesses and grassroots organisations. Established in response to the observed trend of local leadership and local success in addressing linked environment and development challenges, its goal is to support local efforts and build local-level capacity for poverty reduction through the conservation and sustainable use of biodiversity. As well as offering a high-profile prize to successful local projects, the Equator Initiative also helps to provide opportunities for local leaders to influence national and international policy making (Equator Dialogues), and provides a forum through which local organisations can share experience and best practice (Equator Knowledge).

The Global Environment Facility (GEF), the official financing mechanism of the CBD, was established in 1991. The Small Grants Programme (SGP) was launched a year later, as a mechanism to support communities, NGOs and local organisations in their efforts to address poverty, achieve sustainable livelihoods and deliver global environmental benefits. The GEF's SGP is possibly the largest single funding programme for community-based environmental projects, to date having disbursed US$400 million through 12,000 grants in 122 countries.[3]

UNDP's strategic plan for 2008–11 (UNDP, 2008) makes repeated reference to the need to provide support for local capacities for human development and

---

[1] Decentralisation and local organisation have also become centre stage in the United Kingdom's domestic policy, driven by a desire to find more effective ways to deliver public services. Key aims of the Localism Bill include empowering communities and strengthening accountability to local people (http://services.parliament.uk/bills/2010-11/localism.html).

[2] See http://www.equatorinitiative.org/index.php for more information.

[3] See http://sgp.undp.org/index.cfm for more information.

achievement of the MDGs, and their goal in the area of environment and energy is to "build local capacity to better manage the environment and deliver services" (para. 110). In support of this objective, UNDP has produced a guide to working with local organisations that provides "A vision for how the UNDP Environment and Sustainable Development Practice can approach, prioritize, and integrate its work to create the enabling conditions for effective local action and to significantly speed the scale-up of local-level best practices" (UNDP, 2010b).

Some NGOs too have adopted strategic approaches which aim to empower people at the local level, through support to local organisations. For example, BirdLife International, itself a partnership of 117 national organisations, has been working since 1996 to strengthen organisations and individuals working locally (Box 16.3).

Box 16.3 **BirdLife International's local conservation group approach**

BirdLife is a global network of national civil society organisations, present in over 100 countries worldwide, that is working for biodiversity conservation and sustainable development. Through a participatory process led at the national level, the BirdLife Partnership has identified over 10,000 sites of highest priority for bird and biodiversity conservation.

Working with local communities BirdLife's Local Empowerment Programme[4] aims to mobilise a global network, rooted locally, for the conservation and sustainable management of these Important Bird Areas (IBAs). This growing network is currently present at over 2000 IBAs worldwide. BirdLife calls this its 'Local Conservation Group' (LCG) approach. Institutional structures, governance, membership and specific objectives and activities vary depending on the local situation, reflecting the diversity of cultural, social, economic and legal contexts. The characteristics of BirdLife's approach include:

- **Site and biodiversity focused (shared local-global objectives)**: LCGs are focused at the site level, drawing on local people's attachment to a particular place, for its resources, cultural importance or recreational value for example, and identifying the shared interests between a local agenda for action and global biodiversity concerns.
- **Principles**: Through its work with local organisations, BirdLife aims to reflect the BirdLife Partnership's values and principles as a network of open, democratic, membership-based organisations.

---

[4] See http://www.birdlife.org/community/2010/09/birdlifes-local-empowerment-programme/ for more information.

- **Long term**: Relationships are forged with the intention to be long term, not curtailed by a project time frame.
- **Networks**: The experience of BirdLife shows the value of linking people and institutions in order to share resources and experience. Local Conservation Groups form part of this networking approach.
- **Vertical linkages**: The BirdLife Partnership provides a structure which allows effective two-way communication between local, national, regional and international levels. BirdLife aims to bring local voices to the attention of national and international decision makers.

Through LCGs, BirdLife Partners work alongside local people, helping to integrate conservation with social development and livelihood security for the benefit of people and biodiversity (Thomas, 2011). This often involves extending the partnership to include organisations and government agencies with specialist technical development skills (e.g. in agriculture, marketing and tourism) which can complement BirdLife's biodiversity expertise and the community's local knowledge and networks (BirdLife International, 2008).

Issues and challenges

Even with enabling conditions in place, supporting a role for local organisations in biodiversity conservation and poverty reduction that delivers equitable, sustainable outcomes, and addresses the needs and objectives of local, national and international stakeholders, faces significant challenges. Firstly, it can be important not to have unrealistically high expectations of the interests of local organisations in conservation. Preserving biodiversity may not always be in alignment with local development priorities, and trade-offs are perhaps inevitable (Robinson, 2007; PBL, 2009). As the Millennium Ecosystem Assessment (MA) states, "while 'win-win' opportunities for biodiversity conservation and local community benefits do exist, local communities can often achieve greater economic benefits from actions that lead to biodiversity loss" (MA, 2005). Where there are unequal power relations, it can be important to avoid local community interests from simply being overridden when they are not congruent with those of international partners.

Secondly, local organisations may not always be representative of, and accountable to, local communities, and transferring resource rights from government to local organisations does not necessarily lead to outcomes that are either more accountable or more sustainable. Communities are not homogeneous and many local-level organisations are likely to represent only a subset of interests. It may be important to ensure that support to a local organisation, even if its activities help conserve biodiversity, does not actually worsen local inequities and further concentrate power in the hands of a few.

Thirdly, it is often challenging for local organisations to address issues that are rooted in national or international policy processes. Appropriate enabling frameworks (Roe & Bond, 2007) and networks and links from local organisations to institutions at higher scales can begin to address this issue, but in many cases governments may not be interested in sharing power and decision making with local organisations. Where political conditions fail to provide an effective 'space' for civil society, the potential for local organisations may be limited, at least until there has been political and legislative reform. Therefore, empowering local organisations may need to address questions of power and authority (Veit, 1998), something that may meet resistance from those who currently hold the reins.

Related to the previous point, international conservation NGOs have been accused of using local organisations as mere puppets, co-opting them to deliver the objectives of their global conservation strategies without any real transfer of power or influence (Chapin, 2004). Working with local organisations also needs to address such criticism, and avoid creating a set of relationships and dependencies which lead local organisations to lose the qualities for which they are valued–their grassroots, critical voice and independence (Mavhunga, 2007; Brockington & Scholfield, 2010).

Finally, individuals who are members of local organisations may have limited skills and capacity. Issues such as limited literacy, education and technical capacity, as well as dire poverty and the pressures which that puts on people and their institutions, may need to be addressed for local organisations to achieve their potential (BirdLife International, 2007). Support for local organisations is often built on an assumption that they have a base of knowledge concerning resources and their management, passed down through generations of experience, which puts them in an ideal position to manage resources sustainably. However, globalisation has led to the erosion of local custom and pride in place, and historic centralisation of resource control has often contributed to the loss of traditional skills and knowledge; these may need to be rebuilt (Kothari, 2006).

Emerging issues and opportunities for local organisations

New developments in international approaches to conservation and poverty alleviation are discovering the need–and opening up opportunities–for organisation and action at the local level. Efforts to mitigate climate change through forest conservation (thereby preventing the release of sequestered carbon) are especially significant (see Chapter 18, this volume). Such approaches aim to incentivise forest conservation through financial payments or development assistance to compensate for any profits or revenues forgone. As well as the climate change mitigation objectives, this has the potential to deliver biodiversity conservation and development outcomes. This mechanism is being formalised under the United Nations Framework Convention on Climate Change (UNFCCC) as Reducing Deforestation and Forest Degradation (REDD+, the '+' indicating the inclusion of activities such as forest conservation, sustainable management of forests and the enhancement of forest carbon stocks). Trade in carbon, whether through REDD+ or voluntary carbon markets, has the potential to mobilise significant levels of financing as developed and rapidly developing nations compensate less developed countries for sequestering carbon in their standing forests. Although REDD+ is likely to be an intergovernmental mechanism (under the UNFCCC), it is clear that an effective, efficient and equitable system will require local-level governance and benefit sharing. Tenure rights to many of the world's forests lie with indigenous peoples and local communities (in Asia 23.6% and in Latin America 24.6% of forests are owned by communities and indigenous peoples; Rights and Resources Initiative, 2011), and if the forests are being conserved it is they, and not national government, who both deserve and have the right to be rewarded, and (in some cases) need to be incentivised and compensated for not converting forest into alternative land uses. Management of these forests and the administration of funds and other decision making will require organisation at the local level and new skills of the kind discussed in this chapter.

One particular opportunity for local organisations in REDD+ concerns monitoring. To be effective in providing both biodiversity and carbon benefits, monitoring will be required which verifies that the benefits that are being paid for have been delivered. Studies have found that where there is evidence of trusted community organisation and leadership, monitoring by local community organisations is reliable, effective and cost efficient compared to alternatives (Danielsen et al., 2011; Fry, 2011), bringing the potential to contribute to local livelihoods and the conservation of forest biodiversity.

REDD is just one form of payments for ecosystem services (PES)–involving a global trade in the carbon contained in forests and other ecosystems (for its climate change mitigation service). But other, longer established PES systems exist (Chapter 14, this volume). Like REDD they are facilitated through local organisations which can act as efficient mediators, representing community interests regarding resources held 'in common', or providing critical mass and a collective voice where individually held resources and rights are concerned, negotiating payments and enforcing management conditionalities. For example, many watershed protection schemes compensate catchment communities for protecting forests, making payments to compensate for any profits forgone from timber harvesting or conversion to farmland in order to protect forests which variously increase precipitation, moderate river flows and reduce siltation, benefitting irrigators and hydro-electric power schemes downstream. Experience has shown that key requirements for PES include a sufficient level of social capacity, with a well-organised civil society that is able to receive and disburse ecosystem service payments, and ensure that environmental conditionalities (addressing local threats and pressures on the resource) are met (BioClimate Research and Development, 2010). A clear role for local organisations is indicated.

Another recent development that offers opportunities for linking biodiversity conservation and development and is likely to require effective local organisation is the Nagoya Protocol on Access to Genetic Resources and the Fair and Equitable Sharing of Benefits Arising from Their Utilization (commonly shortened to 'Access and Benefit Sharing', or ABS). Under discussion pretty much since the birth of the CBD, the Nagoya Protocol requires Parties to the Convention to take measures to obtain the free, prior and informed consent of indigenous and local communities for accessing genetic resources (Convention on Biological Diversity (CBD), 2010: Article 6:2) and to support the development of community-level protocols regarding access to genetic resources and sharing of benefits (CBD, 2010: Article 12:3 (a)). Given patterns of resource ownership and stewardship in many developing countries, and also considering that much local knowledge is held by communities rather than individually, local organisations are likely to be critical in the future application of this protocol.

Conclusion

The linked objectives of development and conservation require actions to be designed and implemented at the local level, and participation, benefit sharing and the management and control of natural resources require the existence of local organisations. Whilst there is a long history of sustainable resource management by local organisations delivering sustainable supplies of goods and services to the communities they serve, their importance today is nowhere more apparent than in emerging opportunities for linking biodiversity conservation and poverty reduction through REDD, PES and ABS. Although conceived at the international level, and designed to address global, regional or landscape-scale issues, their implementation is ultimately a local challenge and the outcome of these mechanisms will depend on what happens at particular places and on the behaviour of specific communities and individuals. Credible and respected local organisations will be critical in this regard. Faced with these challenges and opportunities, and the central role for local organisations as potential agents of good governance and participation, the question is not so much 'can biodiversity conservation and poverty be addressed through local organisations' as 'how can local organisations be supported to help address the local and global challenges we face today'?

Processes of democratisation and decentralisation in many developing countries, combined with fiscal measures to reduce government budgets, are providing space and opportunities for local organisations to flourish. However, as noted, there remain significant challenges in terms of capacity, awareness, political commitment and vested interests, which need to be overcome. Donors, decision makers and development agencies need to be doing more to recognise the role and value of local organisations and to provide financial support and enabling mechanisms. The Equator Initiative and UNDP Local Capacity strategy are rare examples of efforts targeted at local organisations. Whilst many donors make reference to the importance of local actors, and their ability to manage resources, catalyse local-level action and foster self-reliance, targeted policy or financial support at the required scale does not always follow. Efficiency measures within aid agencies themselves may mean that they lack the staff to manage grants to local organisations, while aid delivery mechanisms such as direct budgetary support may make it difficult to channel funding beyond the national level. Without appropriate safeguards such modalities for providing development assistance may limit opportunities for engaging local organisations, and prioritise government at the expense of civil society (Menocal & Rogerson, 2006).

But whilst international support is important, an inherent characteristic of local organisations is their foundation on principles of volunteerism and activism, and their ability to emerge without external support as local people take the initiative. Capacity, tools, administrative skills and finance are important, but ultimately communities and their organisations need conditions of rights, tenure and long-term security if they are to invest in local stewardship that delivers both conservation and sustainable livelihoods benefits. Mechanisms which respect and support the diversity of local organisation arrangements, and provide the rights which underpin long-term environmental stewardship, may do most to empower citizens to link conservation and development at the local level.

References

Adams, W.M., Aveling, R., Brockington, D., Dickson, B., Elliott, J., Hutton, J., Roe, D., Vira, B. & Wolmer, W. (2004) Biodiversity conservation and the eradication of poverty. *Science*, 306, 1146–9.

Ancrenaz, M., Dabek, L. & O'Neil, S. (2007) The costs of exclusion: recognizing a role for local communities in biodiversity conservation. *PLoS*, 5, 2443–8.

Bass, S. (2008) *Local Organisations – Key but Neglected Agents 'at the End of the MDG Delivery Chain'?* PEP-14 Meeting, Geneva, April. http://www.povertyenvironment.net/files/PEP14-1Apr-Bass2.pdf (accessed 1 May 2012).

BioClimate Research and Development (BR&D) (2010) *Payments for Ecosystem Services Literature Review: A Review of Lessons Learned, and a Framework for Assessing PES Feasibility.* German Federal Ministry for Economic Cooperation and Development (BMZ) and BR&D, Hawick, UK. http://www.planvivo.org/wp-content/uploads/Framework-for-PES-feasibility_WWF_MorrisonAubrey_2010.pdf (accessed 1 May 2012).

Birdlife International (2007) *Conserving Biodiversity in Africa: Guidelines for Applying the Site Support Group Approach.* Birdlife International, Nairobi, Kenya.

BirdLife International (2008) *Building Partnerships: Working Together for Conservation and Development*. BirdLife International, Cambridge.

Brockington, D. & Scholfield, K. (2010) The work of conservation organisations in Sub-Saharan Africa. *Journal of Modern African Studies*, 48, 1–33.

Chapin, M. (2004) A challenge to conservationists. *World Watch Magazine*, 17, 17–31.

Convention on Biological Diversity (CBD) (2010) Conference of the Parties to the Convention on Biological Diversity, 10th meeting, Nagoya, Japan, 29 October.

Danielsen, F., Skutsch, M., Burgess, N.D., Jensen, P.M., Andrianandrasana, H., Karky, B., Lewis, R., Lovett, J.C., Massao, J., Ngaga, Y., Phartiyal, P., Poulsen, M.K., Singh, S.P., Solis, S., Sørensen, M., Tewari, A., Young, R. & Zahabu, E. (2011) At the heart of REDD+: a role for local people in monitoring forests ? *Conservation Letters*, 4, 158–67.

Department for International Development (DFID), European Commission (EC) and United Nations Development Programme (UNDP) (2002) *Linking Poverty Reduction and Environmental Management: Policy Challenges and Opportunities*. IBRD/World Bank, Washington, DC.

Fry, B.P. (2011) Community forest monitoring in REDD+: the 'M' in MRV ? *Environmental Science and Policy*, 14, 181–7.

Gilbert, N. (2010) Can conservation cut poverty ? *Nature*, 467, 264–5.

Gutiérrez, N.L., Hilborn, R. & Defeo, O. (2011) Leadership, social capital and incentives promote successful fisheries. *Nature*, 470, 386–9.

Hazlewood, P. (2010) *Ecosystems, Climate Change and the Millennium Development Goals (MDGs), Scaling Up Local Solutions: A Framework for Action*. Working paper prepared for the UN MDG Summit, September. World Resources Institute, Washington, DC.

Isager, L., Theilade, I. & Thomson, L. (2001) *People's Participation in Forest Conservation: Considerations and Case Studies*. Proceedings of the South East Asian Moving Workshop on Conservation, Management and Utilization of Forest Genetic Resources. http://www.fao.org/docrep/005/AC648E/ac648e0i.htm#TopOfPage (accessed 8 May 2012).

Jepson, P. (2001) Global biodiversity plan needs to convince local policy makers. *Nature*, 409, 12.

Kaimowitz, D. & Sheil, D. (2007) Conserving what and for whom? Why conservation should help meet basic needs in the tropics. *Biotropica*, 39, 567–74.

Kenward, R.E., Whittingham, M.J., Arampatzis, S., Manos, B.D., Hahn, T., Terry, A., Simoncini, R., Alcorn, J., Bastian, O., Donlan, M., Elowe, K., Franzén, F., Karacsonyi, Z., Larsson, M., Manou, D., Navodaru. I., Papadopoulou, O., Papathanasiou, J., von Raggamby, A., Sharp, R.J., Söderqvist, T., Soutukorva, A., Vavrova, L., Aebischer, N.J., Leader-Williams, N. & Rutz, C. (2011) Identifying governance strategies that effectively support ecosystem services, resource sustainability, and biodiversity. http://www.pnas.org/cgi/doi/10.1073/pnas.1007933108 (accessed 8 May 2012).

Kothari, A. (2006) Community conserved areas: towards ecological and livelihood security. *Parks*, 16, 3–13.

Kothari, A., Balasinorwala, T., International Collective in Support of Fishworkers, Jaireth, H. & Rahimzadeh, A. (2008) *Local Voices in Global Discussion: How Far Have International Conservation Policy and Practice Integrated Indigenous Peoples and Local Communities?* Paper prepared for the Symposium on Sustaining Cultural and Biological Diversity in a

Rapidly Changing World: Lessons for Global Policy. American Museum of Natural History, New York, April.

Mavhunga, C. (2007) Even the rider and a horse are a partnership: a response to Vermeulen & Sheil. *Oryx*, 41, 441–2.

Menocal, A.R. & Rogerson, A. (2006) *Which Way the Future of Aid? Southern Civil Society Perspectives on Current Debates on Reform to the International Aid System*. Overseas Development Institute, London.

Millennium Ecosystem Assessment (MA) (2005) *Ecosystems and Human Well-Being: Biodiversity Synthesis*. World Resources Institute, Washington, DC.

Netherlands Environmental Assessment Agency (PBL) (2009) *How Do Biodiversity and Poverty Relate? An Explorative Study*. PBL, Bilthoven, the Netherlands.

Organisation for Economic Cooperation and Development (OECD) (2001) *The DAC Guidelines: Poverty Reduction*. OECD, Paris.

Rights and Resources Initiative (2011) *Pushback: Local Power, Global Realignment*. Rights and Resources Initiative, Washington, DC.

Robinson, J.G. (2007) Recognizing differences and establishing clear-eyed partnerships: a response to Vermeulen & Sheil. *Oryx*, 41, 443–4.

Rodríguez, J.P., Taber, A.P. & Daszak, P. (2007) Globalization of conservation: a view from the south. *Science*, 317, 755–6.

Roe, D. & Bond, I. (2007) *Biodiversity for the Millennium Development Goals: What Local Organisations Can Do*. IIED Briefing Paper. International Institute for Environment and Development (IIED), London.

Roe, D., Nelson, F. & Sandbrook, C. (2009) *Community Management of Natural Resources in Africa: Impacts, Experiences and Future Directions*. International Institute for Environment and Development, London.

Sagoff, M. (2011) The quantification and valuation of ecosystem services. *Ecological Economic*, 70, 497–502.

Satterthwaite, D. (ed.) (2003) *The Millennium Development Goals and Local Processes: Hitting the Target or Missing the Point?* International Institute for Environment and Development, London.

Satterthwaite, D. & Sauter, G. (2008) *Understanding and Supporting the Role of Local Organisations in Sustainable Development*. IIED Gatekeeper Series. International Institute for Environment and Development (IIED), London.

Secretariat of the Convention on Biological Diversity (2010) *Global Biodiversity Outlook 3*. Montréal, Canada.

Smith, R.J., Verissimo, D., Leader-Williams, N., Cowling, R.M. & Knight, A.T. (2009) Let the locals lead. *Nature*, 462, 280–1.

Somanathan, E., Prabhakar, R. & Mehta, B.S. (2009) Decentralization for cost-effective conservation. *Proceedings of the National Academy of Sciences of the United States of America*, 106, 4143–7.

Thomas, D. (2011) *Poverty, Biodiversity and Local Organisations: Lessons from BirdLife International*. IIED Gatekeeper Series 152. International Institute for Environment and Development (IIED), London.

United Nations Development Programme (UNDP) (2008) *UNDP Strategic Plan, 2008–2011: Accelerating Global Progress on Human Development*. Annual Session 2008, DP/2007/43/Rev.1. UNDP, Geneva.

United Nations Development Programme (UNDP) (2010a) *The Millennium Development Goals Report*. UNDP, New York.

United Nations Development Programme (UNDP) (2010b) *The Local Capacity Strategy: Enabling Action for the Environment and Sustainable Development*. UNDP, New York.

United Nations Economic and Social Commission for Asia and the Pacific (N.d.) *What Is Good Governance?* http://www.unescap.org/pdd/prs/ProjectActivities/Ongoing/gg/governance.asp (accessed 1 May 2012).

Veit, P. (1998) *What Does 'Participation' Mean in Development Discourse?* WRI Content Archive. http://archive.wri.org/page.cfm?id=2558&z=? (accessed 1 May 2012).

Vermeulen, S. & Sheil, D. (2007) Partnerships for tropical conservation. *Oryx*, 41, 434–40.

West, P. (2006) *Conservation Is Our Government Now: The Politics of Ecology in Papua New Guinea*. Duke University Press, Durham, NC.

Yachkaschi, S. (2010) Engaging with community-based organisations – lessons from below: capacity development and communities. In *Capacity Development in Practice*, ed. J. Ubels, N. Acquaye-Baddoo & A. Fowler, pp. 37–41. Earthscan, London.

$$\text{(17)}$$

Poverty Reduction Isn't Just about Money: Community Perceptions of Conservation Benefits

Fikret Berkes

Natural Resources Institute, University of Manitoba, Canada

In conservation–development projects, poverty reduction is often defined and measured in terms of income. However, for many communities the incentive to engage with conservation is often not about money but a mix of economic, political, social and cultural objectives – and empowerment is almost always a key objective. As communities are highly heterogeneous, it is impossible to predict what their priorities will be and thus design a blueprint solution for linking conservation and poverty reduction. With indigenous groups in particular, the political objectives of control of traditional territories and resources are of prime importance because such control is seen as the first step to development.

Introduction

Conserving biodiversity depends on understanding the interactions of societies and the environments in which they live, and understanding what motivates people to conserve. Not all biodiversity can be brought under the umbrella of protected areas; most of the world's biodiversity is not in protected areas but on lands and waters used by people for their livelihoods. It follows, therefore, that long-term biodiversity conservation in much of the world can be achieved only in partnership with the people living with it. This chapter is based on the assumption that in most cases local people

Biodiversity Conservation and Poverty Alleviation: Exploring the Evidence for a Link, First Edition.
Edited by Dilys Roe, Joanna Elliott, Chris Sandbrook and Matt Walpole.
© 2013 John Wiley & Sons, Ltd. Published 2013 by John Wiley & Sons, Ltd.

have to be involved in conservation; it is the only viable option for effective human stewardship of the landscape (Murphree, 2009; Robinson, 2011).

Lack of attention to human rights and livelihoods invites encroachment and poaching; this leads to degradation and conflict, in turn reinforcing the assumption that people do not have the will or capacity to conserve. Coupled by other drives that exacerbate encroachment pressures, the stage is set for failure of both conservation and livelihoods (Roe et al., 2000). The solution is to break the vicious cycle by linking conservation to improved livelihoods, and providing incentives to conserve. For more than two decades, researchers and practitioners have been searching for models to implement this approach, often against trends of centralisation. In recent years, neoliberalisation policies have also been of concern, as they rarely favour the community management necessary to provide local incentives to conserve (Lele et al., 2010).

Salafsky and Wollenberg (2000), among others, have suggested that the aim should be to seek a direct linkage between biodiversity conservation and livelihood to strengthen incentives for conservation. Enhanced biodiversity conservation would lead to increased livelihood benefits relative to old livelihood activities; this would give incentives to the local people to protect resources and mitigate threats, including external threats. The logic of this direct linkage or linked incentives model seems to be robust, but 'the devil is in the details'. For example, can livelihood resources be equated with biodiversity? Who reaps the increased livelihood benefits? What kind of capacity building and legal protection would local people need to counter external as well as internal threats? Some of these questions are addressed elsewhere in this volume, and one of the key questions concerns the nature of benefits – the topic of the present chapter.

Benefits and incentives seem to be much too narrowly conceived in the conservation literature (Berkes, 2007; Murphree, 2009), focussing on the monetary benefits as if following the 'income-poverty model' of the 1960s. More recent models of poverty recognise that poverty not only results from low income but also reflects a deprivation of requirements to meet basic human needs. According to OECD (2001: 8), "poverty encompasses different dimensions of deprivation that relate to human capabilities including consumption and food security, health, education, rights, voice, security, dignity and decent work". Such multidimensional definitions of poverty often seem to be lost in conservation–development debates.

Simplistic definitions of poverty, focussing on the monetary benefits of conservation, have hampered community-based conservation by misdirecting conservationists regarding the question of community wants and needs. *Poverty reduction* or *poverty alleviation* in the simpler income-poverty sense of poverty does not capture the complexity of livelihood- and well-being-related objectives from local points of view. The concept of poverty is relative; its measurement and our ability to link it causally to biodiversity conservation are problematic for a variety of reasons (Agrawal & Redford, 2006). Reliance on simplistic 'poverty reduction' is misleading also because whether conservation is possible and feasible is more often a political issue than not, and the poorest people in an area are also the poorest in terms of power (Blaikie, 2006).

The chapter explores some issues around how to do a better job with the linked incentives model. Specifically, the objective is to seek a better understanding of the nature of incentives, and the benefits that the communities *themselves* consider important: what makes the local people want to conserve resources? The focus is on community-based conservation, and community benefits are seen as a range of benefits. There is no assumption here that the two objectives of biodiversity conservation and community benefits are always compatible (Robinson, 2011). In fact, it is likely that conservation–development will involve trade-offs in most cases (Leader-Williams *et al.*, 2010; McShane *et al.*, 2011).

Following the study background, first I use the evidence from 10 United Nations Development Programme (UNDP) Equator Initiative conservation–development cases to elucidate that there exists a range of economic, environmental, political, social and cultural benefits. Secondly, to explore community objectives and priorities in more detail and to show linkages among what might initially appear to be different kinds of objectives, I use a selection of two UNDP Equator Initiative cases and four additional ones.

Study background and the nature of community benefits

Since 2002, the Equator Initiative has held biennial searches to find and reward entrepreneurship cases that seek to reduce poverty and conserve biodiversity at the same time. The programme focuses on the equatorial region because of its richness of biodiversity and abundance of poverty. Between 2003 and 2009, graduate student researchers from the University of Manitoba, working in partnership with the International Development Research Centre of Canada (IDRC), the UNDP Equator Initiative programme and local agencies, carried out field studies with 10 UNDP Equator Initiative cases. The selection of cases was opportunistic, based on case availability and the language skills, previous experience and interests of graduate students. The researchers applied a standard case study methodology and produced technical reports for each case. Some of the comparative material has been synthesised (Berkes, 2007; Seixas & Davy, 2008; Seixas & Berkes, 2010).

The community benefits identified are based on a detailed understanding of the cases and account for the evolution of the case over the years; they are not based on the original project objectives as listed in the UNDP Equator Initiative database. These changes over the original or 'official' objectives can be substantial, especially for the longer standing cases such as Thailand and Mexico (Orozco-Quintero & Berkes, 2010), both with a record of over 20 years at the time of the study. In many cases, there is evidence of community benefits over and above those originally anticipated;

in other cases, the original objectives and eventual benefits may have been implicit rather than explicit.

Community benefits can be classified as economic, environmental, political, social or cultural (Table 17.1). **Economic benefits** include cash compensation, monetary entitlements to households and to community projects, new employment opportunities, market-related benefits such as opening new markets or circumventing middlemen, livelihood diversification and access to funding and credit. Even though much of the conservation literature seems to conceptualise 'benefits' in terms of financial revenue, the list of economic benefits goes far beyond that and includes non-pecuniary ones (Murphree, 2009).

In some cases, the primary objectives may be stated in terms of **environmental benefits**, such as reversing resource declines, restoring biological productivity and protecting species and habitats important for livelihood resources. Some of these environmental benefits may be closely tied to economic benefits, as in certification. Primary objectives may also be stated in terms of **political benefits**, such as empowerment, including participation in decision making, control of local or traditional lands and resources and better relations with government. Social and cultural benefits are not often among the explicit objectives of a project, but they may be nevertheless important, especially in social entrepreneurship cases (Berkes & Davidson-Hunt, 2007). **Social benefits** may include improvements in social, educational and health services, and in social organisation, such as the formation of women's groups. They may be closely tied to economic benefits (as in the case of women's savings groups) and empowerment. **Cultural benefits** include those related to strengthening or revitalising cultural traditions and cultural identity, protecting traditional values, building community cohesiveness and protecting historical and heritage resources.

The classification of community benefits as economic, environmental, political, social or cultural is somewhat arbitrary, and there are alternative approaches that could be used. For example, Murphree (2009) focuses on benefits (pecuniary and non-pecuniary), empowerment and conservation as the 'three pillars' of communal management. Community people themselves rarely see sharp distinctions between different kinds of benefits. What we might call *economic benefits* are often seen as intimately related to the betterment of living conditions – social benefits. Restoring biological productivity may be a key component of improving livelihoods. Development of leadership has both social and political aspects. Empowerment may result in local control over resources, thus leading to new jobs, additional income, livelihood diversification and the strengthening of local cultural traditions.

Exploring benefits in 10 Equator Initiative cases

The cases summarised in Table 17.1 do support the common perception that economic benefits are important in community-based conservation projects. However, none of

Table 17.1 The nature of community objectives and local benefits in 10 UNDP Equator Initiative cases

Case	Economic	Environmental	Political	Social	Cultural	Reference
Rural communes medicinal plant conservation centre, Pune, India	xxx	xxx	xx	xx	xx	Shukla (2004) and Shukla & Gardner (2006)
Arapaima conservation in the North Rupununi, Guyana	xx	xxx	xxx	xx	x	Fernandes (2004)
Honey Care Africa's beekeeping in Kakamega and Kwale, Kenya	xxx	x		x		Maurice (2004)
Cananeia Oyster Producers Cooperative, São Paulo, Brazil	xxx	xx	x	xx		Haque et al. (2009)
TIDE Port Honduras marine reserve, Belize	xx	xxx	xx	x		Fernandes (2005)
Pred Nai community forestry group mangrove restoration, Trat, Thailand	xx	xxx	xxx	xx		Senyk (2006)
Casa Matsiguenka indigenous ecotourism enterprise, Manu National Park, Peru	xxx		xx	xx	x	Herrera (2006)
Nuevo San Juan forestry management enterprise, Michoacan, Mexico	xxx	xx	xxx	xx	x	Orozco (2006) and Orozco & Berkes (2010)
Wildlife management, Torra Conservancy, Namibia	xxx	xx	xx	x		Hoole (2007)
Pastoralist Integrated Support Program (PISP), Kenya	xx	xx	x	xx	xx	Robinson (2008) and Robinson annd Berkes (2011)

xxx - very strong; xx - strong; x - present.

the 10 cases involved monetary disbursements to households or the community, and only two cases (Peru and Belize) involved compensation for livelihood loss, not in cash but in ecotourism development. New employment opportunities and livelihood diversification were seen in all cases except Guyana. Technical training, capacity building and technology transfer were seen in almost all cases.

For example, the Kenya–Honey Care project involved the use of improved bee-keeping technology, the Kenya–Pastoralist Integrated Support Program (PISP) project tested water conservation technologies and Belize fishers received training in tourism development. Horizontal, community-to-community linkages seem most effective in the transfer of learning. Guyanese fishers of the arapaima (*Arapaima gigas*), a giant Amazonian species, learned and adopted non-intrusive population census methodology from fishers from Mamirauá Sustainable Development Reserve in Brazil, which also involved arapaima (pirarucu) conservation and was also an Equator Initiative nominee (Fernandes, 2004; Fernandes & Berkes, 2008).

The Thailand, India and Brazil projects resulted in the creation of new economic institutions (village savings groups and women's self-help groups) that were also important for self-organisation and for building social capital. These were among the spin-off benefits of the projects and not part of the original objectives. Accessing funds was a major skill developed in all of the projects, often with the help of NGOs. Business networking and marketing were important for the success (in most cases) or failure (in the Peru case) to the extent that the projects were successful or not successful in attracting networking partners. Reaching new markets was key for the Kenya–Honey Care case and the Brazil case, circumventing middlemen and thus increasing the community's share of profits.

In many cases, the benefits sought and obtained by the communities can best be characterised as environmental benefits. Loss or severe decline of livelihood resources was the trigger event leading to the organisation of many of the Equator Initiative projects, including the India, Guyana, Brazil, Belize and Thailand cases (Seixas & Davy, 2008). Restoration of the productivity of these resources and reversing trends of decline were necessary before economic objectives could be achieved.

It is important to point out that, in these cases, the full suite of biodiversity did not come under conservation, at least not deliberately. Some cases involved the conservation of key species (arapaima in the Guyana case) or group of species (medicinal plants in the India case). Some resulted in the restoration of the ecosystem (the mangrove forest in the Thailand case) or in the conservation of the forest ecosystem as a whole in order to protect the valuable resources within (the India and Mexico cases). However, in all cases (except Peru), the project resulted in the development of stewardship or a sense of ownership of the local environment. In the Peru case, the alienation of the indigenous people from their forest in Manu National Park, the ban on hunting and the undermining of what would have been a compensation package (apparently to force them out of the Park) resulted in poaching and the emergence of what Scott (1986) might call *peasant resistance*.

As the Peru case shows, political aspects can be the key in some cases; according to some scholars, politics has to be the key in all cases (Blaikie, 2006). The cases indicate a major political dimension (except perhaps Kenya–Honey Care) and three clusters of political objectives: (i) having a voice in conservation and management decisions (India, Mexico, Thailand, Namibia and Brazil), (ii) better control of land and resources (India, Mexico, Kenya–PISP, Namibia and Thailand) and reduction of outsider poaching and encroachment (Thailand and Belize) and (iii) generally better relations with government agencies (India and Mexico). Leadership is an important factor (Seixas & Davy, 2008; Orozco-Quintero & Davidson-Hunt, 2010). In almost all of the cases, there is evidence that community leaders consider that having a say in management is an objective that enables many other positive developments from the community point of view.

Social objectives are rarely primary motivations of conservation–development projects, but all of the 10 cases indicate some kinds of social benefits. Often, they take the form of better social organisation, as in the formation of women's groups in the Brazil and India cases, or in new community organisations, as in the Kenya–Honey Care case. In almost all cases, the conservation project directly results in the provision of new or better social services, as in Mexico and Peru, and specific social benefits, such as scholarships for education in Belize. Detailed fieldwork in case documentation is important: for example, there is no mention of social objectives in the UNDP documentation of the Mexico case. But interviewing people some 20 years into the project, it is clear that the main legacy of the project is improved health, education and clean water, along with employment and forest conservation (Orozco-Quintero, 2006).

Cultural objectives and benefits are also rarely in the forefront. They are probably more difficult to recognise than social objectives. The India case stands out because the medicinal plant conservation initiative led to the revitalisation of Ayurvedic medical traditions in the area, considered very important by the people themselves (Shukla, 2004). Kenya PISP resulted in strengthening the pastoralist way of life, and the three cases involving indigenous groups ended up strengthening various cultural values. For example, in the Guyana case, the arapaima was once considered by the Makushi people as "mother and father of all the fishes", was associated with myths and stories and was under taboo protection. Although the contemporary Makushi assert that they do not believe in such superstitions, it seems that such cultural values were in the background of conservation (Fernandes & Berkes, 2008).

The striking finding in Table 17.1 is the wide range of motivations or objectives for conservation across the cases. A second finding is that each case has a mix of objectives – never just one objective – and the mix is specific to each case. These multiple objectives often combine economic, ecological, political, social and cultural dimensions of livelihoods and well-being and tend to be interrelated. Economic motives are clearly important but they go far beyond pecuniary incentives and involve, for example, the development of community organisation and social capital as a prelude to meeting livelihood needs. These community objectives and priorities can be explored further by focussing on a smaller number of cases in more detail.

Focussing on local objectives and priorities

To explore community objectives and priorities in more detail and to show the linkages among different kinds of objectives, I expand on two of the Equator Initiative cases (India and Peru) and turn to two additional southern cases (Brazil and Namibia) and two northern cases (Canada–Wemindji, Canada–Whitefeather), as summarised in Table 17.2.

The first example has its origins in problems of sustainable use of medicinal plants in India. Solutions were sought through *in situ* conservation with the partnership of rural communities, the Forest Department and non-governmental organisations (NGOs), involving a network of 13 biodiversity-significant medicinal plant conservation areas in Maharashtra State. The main economic impact of the project has been the conservation of plants used in Ayurvedic medicine for local, low-cost health alternatives. Direct benefits go to the women involved in collecting, processing and marketing the plants. Nurseries supplement *in situ* conservation, and provide opportunities for environmental education and revitalisation of local health traditions. The Forest Department carries out joint forest monitoring with communities, relying on community patrols. The project has resulted in local training and capacity building, new women's groups and the development of a sense of ownership of local resources (Shukla, 2004; Shukla & Gardner, 2006).

The second example, a very different kind of story, involves the Matsiguenka people of south-eastern Peru who live in an area that became a large protected area, Manu National Park. Prevented from hunting and carrying out shifting cultivation for manioc (*Manihot esculenta*), their staple food, the Matsiguenka received compensation in the form of an ecotourism lodge which they own and operate. However, tourism and a cash economy are new concepts to the Matsiguenka, and capacity-building needs are immense. The lodge depends on the private sector for marketing in an area that has many big players. The lodge has not been an economic success in the face of competition from others – the Matsiguenka characterise their operation as "a lizard among the crocodiles" (Herrera, 2006). Despite the dark prognosis, the Matsiguenka take pride in owning a lodge, and say that other groups look at them with more respect. Cash income has led to improved transportation and medical services. The lodge has generated processes of social and political organisation, and produced a dialogue with the Park administration (Herrera, 2006).

Table 17.2 Community objectives and priorities in detail in two UNDP Equator Initiative cases (India and Peru) and in two other southern and two northern community-based conservation cases

Case	Local objectives and priorities	Reference
Rural commune's medicinal plant conservation centre, Pune, India	*In situ* conservation and sustainable use of medicinal plants; low-cost local medicinal plant availability in poorer regions; income for women collecting and marketing; sense of ownership of local resources and conservation partnerships; revitalisation of cultural traditions	Shukla (2004) and Shukla and Gardner (2006)
Casa Matsiguenka indigenous ecotourism enterprise, Manu National Park, Peru	Compensation for restrictions on traditional hunting and other livelihood activities inside the Park; ecotourism for alternative livelihood and jobs; way of defending rights and fighting commercialisation of natural resources by outsiders; source of community pride and social benefits	Herrera (2006)
Peixe Lagoon National Park, Rio Grande do Sul, Brazil	Fight eviction from the Park; getting voices heard; empowerment to defend rights, preserve livelihoods and access technical assistance; recognition of fishers' role in maintaining ecology of Park; job opportunities inside the Park	Almudi and Berkes (2010)
Ehi-rovipuka Conservancy, adjacent to Etosha National Park, Namibia	Economic benefits of wildlife use and ecotourism; employment; meat from wildlife; a vehicle for community organisation and empowerment; participation in wildlife management	Hoole (2007) and Hoole and Berkes (2010)
Canada–Wemindji, Paakumshumwaau-Maatuskaau Biodiversity Reserve, Cree Nation of Wemindji, Quebec, Canada	Biodiversity and landscape conservation; security from hydro-electricity development threat; biodiversity and landscape conservation to safeguard traditional lifestyle; reaffirming land and resource rights; community identity, cohesion and cultural values	Government of Quebec (2008)
Canada–Whitefeather, Pikangikum First Nation, northwest Ontario, Canada	Elders' wisdom to take care of the land; economic and employment opportunities through resource-based tribal enterprises; management that harmonises indigenous knowledge and practice with western science; strong culture through healthy economy	PFN & OMRN (2006)

The third story involves communities, resident in Peixe Lagoon National Park in southern Brazil, that face imminent eviction. These fisher communities are poorly organised and consider themselves weak and disempowered. They have successfully fought eviction in the short term but have no long-term resource access rights, leaving them vulnerable. Obstacles to empowerment are in the form of *missing linkages*, or lack of networking support, for capacity building to defend their rights and stay in the conservation process. There is evidence that traditional fishing activities have maintained the wetland habitat of the migratory birds in the Park. Fishers have traditionally controlled lagoon mouth opening, affecting water levels and movement of species in and out of the lagoon. If their ecological role in maintaining the wetland habitat is recognised, it would be easier to accommodate them in the Park as stewards, park guides and maintenance workers (Almudi & Berkes, 2010).

Namibia's conservancies devolve wildlife rights, use and benefits to local communities, and have often been showcased as a promising model for sharing benefits (see Chapter 12, this volume, or a detailed review of the Namibian conservancy programme). Ehi rovipuka Conservancy itself does not seem to be receiving a great deal of economic benefits, but the arrangement has stimulated community organisation and empowerment for joint wildlife management (Hoole & Berkes, 2010). The conservancy borders on a large protected area, Etosha National Park, previously used by the local people. The community has identified a 'wish list' of potential benefits from the Park: emergency grazing rights during drought periods, joint tourism ventures inside the Park, fences to protect the community from Park animals, rights to visit burial areas and the ability to harvest field foods (such as mopani worms) and medicinal plants inside the Park. None of these would likely compromise Park biodiversity but would provide livelihood benefits (Hoole, 2007). Perhaps more risky, an alternative conservation model came out of community discussions: fencing in communities and fields seasonally and making the Park fence porous. Such a move would restore the link between social and ecological systems, allowing animals and people to move freely in either direction, as the Conservancy is on the natural migration route of many of the Park's species (Hoole & Berkes, 2010).

The two northern examples (Canada–Wemindji and Canada–Whitefeather) come from the boreal forest zone in Quebec and Ontario, and involve two indigenous groups, the Cree and the Anishinaabe (Ojibwa), respectively. The Wemindji Cree people of James Bay requested the creation of a biodiversity reserve in their traditional area, with the objective to save a heritage river from possible hydro-electric development in a region where all of the major rivers have been dammed. In the process of developing the nomination document, the Cree found other reasons for protection: re-assertion of traditional hunting rights, cultural identity, community cohesion and values. Landscape conservation was seen as a way to safeguard traditional lifestyle and values through a locally managed protected area "so our grandchildren can hunt and fish". The Government of Quebec considered these objectives to be consistent with biodiversity conservation, and facilitated the nomination process because of the urgency to satisfy the quota for the area under protection in the Province (Government of Quebec, 2008).

In the final case, Canada–Whitefeather, the Anishinaabe people of Pikangikum First Nation (PFN) are seeking to create economic and employment opportunities through resource-based enterprises, and to develop a land use strategy to undertake resource management (PFN & Ontario Ministry of Natural Resources (OMRN), 2006: 5). The land use strategy is a necessary step in obtaining a commercial forestry permit, even though the Whitefeather Forest in question is PFN's traditional territory. The original objective of economic development is closely intertwined with political, social and cultural objectives. The PFN wants to be in "the driver's seat" (PFN's term) in any development on their land, and they insist on the use of indigenous knowledge, teachings and practices learned firsthand on the land, alongside forestry and wildlife science. They see the land (*ahkee*) as a gift from the Creator that provides for a continued traditional way of life and a source of livelihood, subject to their stewardship responsibilities, "keeping the land" (PFN & OMRN, 2006).

In each of the six cases, livelihood and well-being are in the foreground. The original objectives are diverse: sustainability of medicinal species (India), compensation for livelihood loss (Peru), fishing rights to keep access to livelihood resources (Brazil), share of wildlife benefits (Namibia), security from hydro development (Canada–Wemindji) and ability to engage in forestry development (Canada–Whitefeather). But in each case, there is a mix of objectives, some of them developing as the case proceeds. For example, in the Canada–Wemindji case, cultural and educational objectives developed as the protected area nomination progressed. In Namibia, participation in wildlife management led to the discussion of a whole new way of managing national parks (by making fences porous) and re-integrating people and the environment (Hoole & Berkes, 2010).

The objectives of the six cases go far beyond the income-poverty model of poverty; they illustrate the need to use multidimensional definitions (OECD, 2001). Community-based conservation is found in all six cases; in fact, two of them start with conservation objectives (India and Canada–Wemindji). But it is also clear that local and indigenous understandings of what is to be protected and whether local use should be allowed are different from government views. In all six cases, there is evidence of cultural attachment to land. In the indigenous cases, land tenure is a source of contention. In some cases, the needs of future generations are an integral part of the objectives. Although spiritual or sacred values are not always explicitly stated, cultural values seem to underpin the use of land and resources.

Discussion and conclusions

Community-based conservation has its own logic, not necessarily congruent with that of international conservation science (Berkes, 2004; Robinson, 2011). The main conservation incentives for communities are related to livelihood and well-being, which go far beyond monetary incentives and poverty reduction in the older, simpler sense. The notion of *poverty* has evolved so much over recent decades that it is easy to forget that it is a multidimensional concept that has been made consistent with the views of the poor themselves. Contemporary concepts of poverty not only are about money but also reflect a deprivation of requirements to meet basic human needs such as health and education. Conservation practitioners often overlook these complexities; hence, it is useful to examine conservation incentives from the local point of view. Based on the case studies presented in this chapter, a number of conclusions can be offered:

1. Community objectives that create incentives for biodiversity conservation are complex. For analytical purposes, they can be sorted into economic, environmental, political, social and cultural objectives, but they are interrelated.
2. These community objectives cannot be characterised as *poverty reduction* in the old income-poverty sense, even though economic objectives are in the forefront.
3. In many cases, political, social and cultural objectives can be very important – more important than money – and empowerment is almost always a prime objective.
4. There is almost always a mix of community objectives, but the mix is case specific, making it impossible to design 'blueprint' solutions that can be applied universally.
5. In particular, with indigenous groups, the political objectives of control of traditional territories and resources are of prime importance because such control is seen as the first step to social and economic development and being in 'the driver's seat'.

One caveat is that there are likely to be intra-community differences in perspectives and needs, which are very difficult to deal with. Equally problematic is establishing the actual beneficiaries of poverty alleviation at the community level. In some cases (e.g. India), it is possible to say that the poorer areas and poorer groups are the main beneficiaries of the initiative. In many cases, one can conclude that better social organisation and increased social capital help communities to take care of their poor through sharing networks. But inequities in well-being and power still remain (Blaikie, 2006), and neoliberal policies as a higher level driver may act to exacerbate these differences (Lele *et al.*, 2010).

A second caveat is that the evidence for conservation–development projects in conserving biodiversity is mixed. There is no evidence that these projects deliberately aim to conserve biodiversity in some global sense. Conservation–development is often a trade-off. However, many of the projects aim at reversing resource declines and restoring the biological productivity of some species and ecosystems. In many cases, projects involve the conservation and sustainable use of certain key species (e.g. the arapaima in Guyana) or groups of species (e.g. medicinal plants in India). In some cases, projects may lead to conservation of certain ecosystems (e.g., the mangrove forest in Thailand or the boreal forest in the Canada cases) and thus of all the species therein. The crucial point is that, in all cases, projects may lead to fostering stewardship ethics and building a sense of ownership of the local environment.

A third caveat is that benefits are difficult to quantify. Measuring poverty and conservation outcomes are problematic. First, there is the problem of lack of baseline data. Second, both concepts are contested and fraught with complications (Agrawal & Redford, 2006). In our set of cases, community objectives do not contradict conservation objectives, but neither are they entirely consistent. Focussing on the overlap of objectives, it is feasible to look for trade-offs and win–win outcomes, as with the Mexico and India cases. Win–win outcomes would require taking community objectives seriously, and empowering community conservation (Murphree, 2009). That would necessitate developing a better understanding of local aspirations, refraining from manipulating communities and thinking about trade-offs (Leader-Williams *et al.*, 2010; McShane *et al.*, 2011).

Community-based conservation is not a panacea for rural poverty, nor could it deliver all expectations of biodiversity conservation. However, it has the potential to be more effective than top-down, state-based conservation (Berkes, 2007; Murphree, 2009). Where community-based conservation has not worked, the challenge is to make it work, and not to ignore it. However, this may require adopting a conceptualisation of community-based conservation as a complex systems problem requiring attention to multiple objectives at different scales (Berkes, 2007; McShane *et al.*, 2011). As a complex systems problem, conservation is going to look different from the point of view of each level, from the local to the international. The cases considered here show that community objectives and incentives are case specific, and caution against blueprint solutions or panaceas. Understanding the nature of the benefits that the communities themselves consider important will help those involved in conservation projects sort out the differences and points of agreement among these various levels, and design more appropriate conservation partnerships with a more realistic view of community benefits and conservation outcomes that may be expected.

Acknowledgements

This chapter was made possible by the work of Tikaram Adhikari, Damian Fernandes, Jessica Herrera, Arthur Hoole, Stephane Maurice, Dean Medeiros, Alejandra Orozco-Quintero, Lance Robinson, Jason Senyk and Shailesh Shukla, and their project technical reports or papers cited herein. I thank Dilys Roe for editorial suggestions. The International Development Research Centre of Canada (IDRC) and the Canada Research Chairs program supported the project.

References

Agrawal, A. & Redford, K. (2006) *Poverty, Development and Biodiversity Conservation: Shooting in the Dark?* Working Paper No. 26. Wildlife Conservation Society, New York.

Almudi, T. & Berkes, F. (2010) Barriers to empowerment: fighting eviction for conservation in a southern Brazilian protected area. *Local Environment*, 15, 217–32.

Berkes, F. (2004) Rethinking community-based conservation. *Conservation Biology*, 18, 621–30.

Berkes, F. (2007) Community-based conservation in a globalized world. *Proceedings of the National Academy of Sciences of the USA*, 10, 15188–93.

Berkes, F. & Davidson-Hunt, I.J. (2007) Communities and social enterprises in the age of globalization. *Journal of Enterprising Communities*, 1, 209–21.

Blaikie, P. (2006) Is small really beautiful? Community-based natural resource management in Malawi and Botswana. *World Development*, 34, 1942–57.

Fernandes, D. (2004) *Community-Based Arapaima Conservation in the North Rupununi, Guyana.* Equator Initiative Technical Report. http://www.umanitoba.ca/institutes/natural resources/nri_cbrm_projects_eiprojects.html (accessed 3 May 2012).

Fernandes, D. (2005) *TIDE Port Honduras Marine Reserve, Belize.* Equator Initiative Technical Report. http://www.umanitoba.ca/institutes/natural_resources/nri_cbrm_projects_eiprojects.html (accessed 3 May 2012).

Fernandes, D. & Berkes, F. (2008). 'More eyes watching . . . ': community-based management of the Arapaima (*Arapaima gigas*) in central Guyana. In *El Manejo de las Pesquerías en la Amazonia*, ed. D. Pinedo & C. Soria, pp. 285–305 (in Spanish). Ottawa and Lima: IDRC and Instituto del Bien Comun.

Government of Quebec (2008) *Proposed Paakumshumwaau Maatuskaau Biodiversity Reserve Conservation Plan.* Quebec Strategy for Protected Areas. Quebec City, Canada.

Haque, C.E., Deb, A.K. & Medeiros, D. (2009) Integrating conservation with livelihood improvement for sustainable development: the experiment of an oyster producers' cooperative in southeast Brazil. *Society and Natural Resources*, 22, 554–70.

Herrera, J. (2006) *Casa Matsinguenka Indigenous Ecotourism Project, Peru.* Equator Initiative Technical Report. http://www.umanitoba.ca/institutes/natural_resources/nri_cbrm_projects_eiprojects.html (accessed 3 May 2012).

Hoole, A. (2007) *Lessons from the Equator Initiative: Common Property Perspectives for Community-Based Conservation in Southern Africa and Namibia.* Equator Initiative Technical Report. http://www.umanitoba.ca/institutes/natural_resources/nri_cbrm_projects_eiprojects.html (accessed 3 May 2012).

Hoole, A. & Berkes, F. (2010) Breaking down fences: Recoupling social-ecological systems for biodiversity conservation in Namibia. *Geoforum*, 41, 304–17.

Leader-Williams, N., Adams, W.M. & Smith, R.J. (eds.) (2010) *Trade-offs in Conservation: Deciding What to Save.* Wiley-Blackwell, London.

Lele, S., Wilshusen, P., Brockington, D., Seidler, R. & Bawa, K. (2010) Beyond exclusion: alternative approaches to biodiversity conservation in the developing tropics. *Current Opinion in Environmental Sustainability*, 2, 94–100.

Maurice, S. (2004) *Honey Care Africa Ltd., Kenya.* Equator Initiative Technical Report http://www.umanitoba.ca/institutes/natural_resources/nri_cbrm_projects_eiprojects.html (accessed 3 May 2012).

McShane, T.O., Hirsch, P.D., Trung, T. C, Songorwa, A.N., Kinzig, A., Monteferri, B., Mutekanga, D., Thang, H.V., Dammert, J.L., Pulgar-Vidal, M., Welch-Devine, M., Brosius, J.P., Coppolillo, P. & O'Connor. S. (2011) Hard choices: making trade-offs between biodiversity conservation and human well-being. *Biological Conservation*, 144, 966–72.

Murphree, M.W. (2009) The strategic pillars of communal natural resource management: benefit, empowerment and conservation. *Biodiversity Conservation*, 18, 2551–62.

Organisation for Economic Co-operation and Development (OECD) (2001) *Development Action Committee's Guidelines on Poverty Reduction.* OECD, Paris. http://www.oecd.org (accessed 3 May 2012).

Orozco-Quintero, A. (2006) *Lessons from the Equator Initiative: The Community-Based Enterprise of Nuevo San Juan, Mexico.* Equator Initiative Technical Report. http://www.umanitoba. ca/institutes/natural_resources/nri_cbrm_projects_eiprojects.html (accessed 3 May 2012).

Orozco-Quintero, A. & Berkes, F. (2010) Role of linkages and diversity of partnerships in a Mexican community-based forest enterprise. *Journal of Enterprising Communities*, 4, 148–61.

Orozco-Quintero, A. & Davidson-Hunt, I.J. (2010). Community-based enterprises and the commons: the case of San Juan Nuevo Parangaricutiro, Mexico. *International Journal of the Commons*, 4, 8–35.

Pikangikum First Nation and Ontario Ministry of Natural Resources (PFN & OMRN) (2006) *Keeping the Land: A Land Use Strategy for the Whitefeather Forest and Adjacent Areas.* Pikangikum and Red Lake, Pikangikum First Nation in cooperation with the Ontario Ministry of Natural Resources, Toronto. http://www.whitefeatherforest.com (accessed 3 May 2012).

Robinson, J.G. (2011) Ethical pluralism, pragmatism, and sustainability in conservation practice. *Biological Conservation*, 144, 958–65.

Robinson, L.W. (2008) *Lessons from the Equator Initiative: Institutional Linkages, Approaches to Public Participation, and Social-Ecological Resilience for Pastoralists in Northern Kenya.* Equator Initiative Technical Report. http://www.umanitoba.ca/institutes/natural_resources/ nri_cbrm_projects_eiprojects.html (accessed 3 May 2012).

Robinson, L.W. & Berkes, F. (2011) Multi-level participation for building adaptive capacity: formal agency–community interactions in northern Kenya. *Global Environmental Change*, 21, 1185–94.

Roe, D., Mayers, D., Grieg-Gran, M., Kothari, A., Fabricius, C. & Hughes, R. (2000) *Evaluating Eden: Exploring the Myths and Realities of Community-Based Wildlife Management.* Evaluating Eden Series No. 8. International Institute for Environment and Development, London.

Salafsky, N. & Wollenberg, E. (2000) Linking livelihoods and conservation: a conceptual framework and scale for assessing the integration of human needs and biodiversity. *World Development*, 28, 1421–38.

Scott, J.C. (1986) *Weapons of the Weak: Everyday Forms of Peasant Resistance.* Yale University Press, New Haven.

Seixas, C.S. and Berkes, F. (2010) Community-based enterprises: the significance of partnerships and institutional linkages. *International Journal of the Commons*, 4, 183–212.

Seixas, C.S. & Davy, B. (2008) Self-organization in integrated conservation and development initiatives. *International Journal of the Commons*, 2, 99–125.

Senyk, J. (2006) *Pred Nai Community Forestry Group and Mangrove Rehabilitation, Thailand*. Equator Initiative Technical Report. http://www.umanitoba.ca/institutes/natural_resources/nri_cbrm_projects_eiprojects.html (accessed 3 May 2012).

Shukla, S. (2004) *Medicinal Plants Conservation Centre, Pune, India*. Equator Initiative Technical Report. http://www.umanitoba.ca/institutes/natural_resources/nri_cbrm_projects_eiprojects.html (accessed 3 May 2012).

Shukla, S.R. and Gardner, J.S. (2006) Local knowledge in community-based approaches to medicinal plant conservation: lessons from India. *Journal of Ethnobiology and Ethnomedicine*, 2, 20. http://www.ethnobiomed.com/content/2/1/20 (accessed 3 May 2012).

Part V
Biodiversity and Poverty Relationships in the Context of Global Challenges

(18)

Biodiversity, Poverty and Climate Change: New Challenges and Opportunities

Kathy MacKinnon

International Union for Conservation of Nature (IUCN)/World
Commission on Protected Areas

Introduction

The Millennium Ecosystem Assessment (2005) showed that over the past 50 years, human activities have changed ecosystems more rapidly and extensively than at any comparable period in our history. These changes have contributed to many net economic gains but have also resulted in growing environmental costs including biodiversity loss, land degradation and reduced access to natural resources for many of the world's poorest people. Habitat loss and fragmentation, overexploitation, pollution and the impact of invasive alien species all threaten global biodiversity. Climate change is now also recognised as a major threat to biodiversity that is likely to build on, and exacerbate, other sources of environmental degradation, with devastating consequences for some of the poorest and most vulnerable nations and communities.

Climate change is a serious environmental challenge that could undermine the drive for sustainable development. There is no consensus on how much global temperatures will eventually rise, but even with the expected minimum increase of $2°C$ impacts on food and water security are likely to be substantial, with falling crop yields in many developing countries and increasing water shortages in many regions including southern Africa, South Asia and the Mediterranean (Stern, 2007). Small mountain glaciers will disappear, threatening water supplies in South America and the Indus Plain, while expected sea level rises will threaten major cities (Intergovernmental Panel on Climate Change, 2007). These changes will have the greatest impact on the poor and will

Biodiversity Conservation and Poverty Alleviation: Exploring the Evidence for a Link, First Edition.
Edited by Dilys Roe, Joanna Elliott, Chris Sandbrook and Matt Walpole.
© 2013 John Wiley & Sons, Ltd. Published 2013 by John Wiley & Sons, Ltd.

be further compounded by the increasing frequency and severity of extreme weather events with rising intensity of storms, droughts, floods, forest fires and heat waves.

The impacts of climate change and continuing biodiversity loss will be major constraints to achieving the poverty reduction and development goals enshrined in the Millennium Development Goals (MDGs). Poverty has many dimensions – as has been highlighted in Chapter 1 and elsewhere in this volume. The most common aspects of poverty relate to *income* and *employment opportunities*, but there are other equally important dimensions: *access to natural resources*; *access to healthcare* and *clean water* and *sanitation*; *education* and *access to information*; and *reducing vulnerability to drought*, *famine* and *other natural hazards*. In addition to these aspects of poverty we should add *elements of well-being* such as *opportunity, security*, and *empowerment* (Bucknall *et al.*, 2001). It is often argued that poverty and environmental degradation are linked and this may be true at a local level. Yet some of the most biodiversity-rich areas on the planet are protected by some of the world's poorest nations. What is clear is that the poor suffer disproportionately from environmental degradation, often because of lack of other options.

Other chapters in this book have reviewed different/dimensions of the relationship between biodiversity and poverty. This chapter reviews that relationship in the context of climate change. Although climate change increases the threat to biodiversity, it also provides a unique opportunity to re-emphasize the multiple values of natural ecosystems and to highlight how biodiversity conservation can contribute to reducing the impacts of climate change, especially for the most vulnerable communities. Natural ecosystems, and the goods and services that they provide, can help people to cope by buffering local climate and reducing risks and impacts from extreme events such as storms, floods, droughts and sea-level rise; and by safeguarding ecosystem services essential to human well-being such as water supplies, fish stocks, other wild foods and agricultural productivity (Dudley *et al.*, 2010; World Bank, 2010). In this chapter I review the major climate change impacts and the role that biodiversity conservation (particularly through protected areas) can play in mitigating their effects – particularly for poor people.

Maintaining water supplies

Higher temperatures and more erratic rainfall patterns will result in severe decreases in water availability in many regions. In Africa, for example, severe water stress is expected to affect an estimated 250–700 million people by 2050 and crop production from rain-fed agriculture could decrease by 50% in some countries (World Bank, 2010). Growing concern over water scarcity provides a powerful argument for protecting natural habitats and creating protected areas to maintain water supplies for both agriculture and domestic use (Dudley *et al.*, 2010; Stolton & Dudley, 2010; World

Bank, 2010). Municipal water accounts for less than one-tenth of human water use, but clean drinking water is a critical human need. Today, half of the world's population lives in towns and cities, and one-third of this urban population lacks clean drinking water. These billion have-nots are unevenly distributed: 700 million city dwellers in Asia, 150 million in Africa and 120 million in Latin America and the Caribbean (World Bank, 2010). With expanding urban needs, cities face immediate problems related to access to clean water and mounting problems related to supply.

Maintaining natural vegetation is often the cheapest and most effective way of securing water supplies and water quality through filtration, groundwater renewal and maintenance of natural flows. Water from protected areas is important for domestic use and subsistence agriculture as well as for large-scale irrigation, industrial use and hydro-electric power and as a source of municipal drinking water. Around one-third (33 out of 105) of the world's largest cities including Mumbai, New York, Sofia, Bogotá, Dar es Salaam, Quito, Melbourne, Tokyo and Sydney receive a significant proportion of their drinking water supplies directly from forest protected areas (Dudley & Stolton, 2003). Half of Puerto Rico's drinking water comes from the last sizeable area of tropical forest on the island, in the Puerto Rico National Park, while Mount Kenya, the second highest mountain in Africa, is one of Kenya's five main 'water towers' and provides water to more than two million people (Dudley & Stolton, 2003). Panda reserves in the Qinling Mountains, China, protect the drinking water supplies for Xi'an while the Gunung Gede-Pangrango National Park in Indonesia safeguards the drinking water supplies of Jakarta, Bogor and Sukabumi and generates water with an estimated value of US$1.5 billion annually for agriculture and domestic use. Similarly in Ecuador, two Andean protected areas provide drinking water supplies for 80% of Quito's population (World Bank, 2008a). In South Africa, the recognised value of the mountains of the Cape Peninsula and Drakensberg in providing water supplies for Cape Town, Johannesburg and Durban has led to considerable national investments in the Working for Water programme to remove invasive alien species (Box 18.1).

Box 18.1 **Removing invasive species in South Africa for cost-effective water supplies**

Invasive alien plants are estimated to affect 10 million hectares (8.28%) of South Africa with significant ecological and economic costs. With high evapotranspiration rates, invasive trees are an immense burden to already water-scarce regions and reduce the amount of water available to reservoirs. In 2002 the South African government approved R 1.4 billion (US$173.5 million) for the proposed Skuifraam Dam on the Berg River near Franschhoek to help address the looming water crisis in the Western Cape and Cape Town. A feasibility study demonstrated that water delivery would cost 3 cents less per kilolitre if invasive species were managed in the catchment area. It was estimated that clearing invasive plants from the Theewaterskloof catchment would deliver additional water at only 10.5% of the cost of delivery from the new Skuifraam scheme if no clearance was carried out. Accordingly, large-scale programmes to clear invasive trees are being undertaken as part of management for the new Berg Dam. Such clearance programmes provide important biodiversity and social benefits by removing invasives in rare fynbos habitats and providing large-scale employment programmes to poor and disenfranchised communities. The Working for Water Programme and its sister programmes which focus on wetlands and reducing fire risks are now major employers with annual government budgets exceeding US$100 million per year – a win-win for poverty alleviation and biodiversity.

Source: Pierce *et al.* (2002).

Food security

The decreases in water supplies discussed here are expected to have significant impacts on agricultural productivity, especially in arid and semi-arid regions (Stern, 2007). Agriculture is the primary user of freshwater sources: up to 50% in most countries and as much as 90–95% in many developing countries. Water shortages will affect crop production, food security and human health, further depressing the livelihoods and welfare of the poor as well as national economic growth in the least developed countries. Wild natural resources can become an increasingly important source of food for poor people especially if crops fail (Chapter 4, this volume) but can often provide only short-term relief – filling gaps or acting as insurance in an emergency. In many regions with chronic hunger, reducing poverty will require better land and water management to maintain ecosystem services and to rehabilitate degraded lands and natural resources critical for expanding agricultural productivity and achieving food security.

Increasing temperatures, drought, the spread of invasive species and land taken out of food production to produce biofuels will push poor farmers into ever more marginal lands, leading to further land degradation and biodiversity loss (World Bank, 2008b). Expanding the area of land under agricultural production is not necessarily the solution to ensuring food security – especially in a changing climate. Intact natural ecosystems perform a vital role which outweighs the short-term benefits of land conversion, safeguarding essential ecosystem services and the adaptive capabilities of the diverse genetic resource base that underlies agricultural biodiversity.

Irrigated agriculture produces over one-third of the global food harvest but is dependent on sustainable water supplies. Natural systems safeguard downstream water supplies for agricultural fields and act as natural reservoirs for medicinal plants, wild crop relatives, pollinators and pest control (Stolton & Dudley, 2010). A study of Madagascar, for example, showed that economic benefits from forest protection were much greater than those from converting areas to agriculture with at least 50% of benefits accruing from watershed protection, primarily from averting the impacts of soil erosion on downstream smallholder irrigated rice production (World Bank, 2006, 2010). Protected areas help to protect crop wild relatives, vital for providing genetic material for crop breeding, while local communities and traditional agricultural practices maintain endangered landraces and old crop varieties, many of them better adapted to drought conditions (Box 18.2; Chapter 4, this volume).

Box 18.2 **Adapting to climate change: exploiting agro-biodiversity in Yemen**

The Republic of Yemen is an important primary and secondary centre of diversity for cereals. This local agro-biodiversity is threatened by global, national and local challenges, including land degradation, climate change, globalization, anthropogenic local factors and loss of traditional knowledge. Communities in the highlands of the Republic of Yemen retain old crop varieties and traditional knowledge related to the use of these agro-biodiversity resources. Knowledge and practice have evolved over more than 2000 years to increase agricultural productivity in areas of limited rainfall. The construction and management of terraces, for instance, help to improve the efficient use of water and to minimize land degradation. Most of the landraces and local crop varieties have been selected to meet local needs and have adaptive attributes for coping with adverse environmental and climatic conditions.

Source: World Bank (2010).

Climate change threatens fishery stocks as well as agricultural productivity. It is already modifying the distribution of both marine and freshwater species and altering marine and freshwater food webs, with unpredictable consequences for fish production (Food and Agriculture Organization, 2009). These climate-induced threats exacerbate pressures from habitat loss and overexploitation to reduce reef productivity and fisheries stocks, affecting some 250 million people globally who depend on small-scale fisheries for their protein. A growing body of empirical evidence suggests that protected areas can rejuvenate depleted fish stocks in a matter of years when they are managed collaboratively with the resource users (World Bank, 2010) – and as a result have a positive impact on poverty alleviation (Chapter 9, this volume). A review of 112 studies in 80 marine protected areas (MPAs) found strikingly higher fish populations and larger fish inside the reserves compared with surrounding areas (or the same area before an MPA was established) and these additional fish provide valuable 'spill-over' into, and replenish, adjacent fished areas (Halpern, 2003).

Globally many coral reefs already show extensive damage from rising ocean temperatures, ocean acidification and overexploitation (Stern, 2007). In Indonesia many of the archipelago's coral reefs and the small-scale fisheries they support have reached a level and mode of exploitation where the only way to increase future production and local incomes is to protect critical habitats and reduce fishing effort. The Coral Reef Management Programme has contributed to reef protection and management in 12 coastal districts off Sulawesi, Aru and Indonesian Papua, including 1500 coastal villages with more than 500,000 residents. The centrepiece of these efforts is collaboratively managed marine reserves, many within existing marine parks, contributing towards a government target of protecting 30% of the total area of coral reefs in each participating district (World Bank, 2006).

Health

Water shortages associated with climate change will affect not just agricultural productivity but also potable water supplies, sanitation and human health, further depressing the livelihoods and welfare of the poor (World Bank, 2010). Sickness and mortality associated with diarrheal disease, primarily associated with floods and droughts, are expected to rise. According to the World Health Organisation, 23–25% of the global disease burden could be avoided by improved management of environmental conditions (Stolton & Dudley, 2010). One in five people in the developing world live without a reliable water supply, and two billion city dwellers do not have adequate sanitation. Lack of clean water increases infant mortality and water-borne diseases; it reduces productivity, strains health services and causes millions of deaths every year, making dirty water one of the world's largest killers, particularly of children. Several MDGs relate to health, but the role that natural ecosystems can play in delivering healthcare is often overlooked.

Natural ecosystems help to maintain community health through provision of clean water and as the source of local and global medicines, especially as an accessible source of healthcare for the poor. Locally collected herbs and other native species are major resources for meeting primary healthcare needs in Asia, Latin America and Africa;

in Africa up to 80% of the population use traditional medicines for their primary healthcare (Stolton & Dudley, 2010). More species of medicinal plants are harvested than any other natural products; over one-quarter of all known plants have been used medicinally at some time (Stolton & Dudley, 2010). Loss and destruction of natural habitats, species and traditional knowledge are threatening both rural health and future health solutions. As natural ecosystems shrink in the wider landscape, many local communities are looking to arrangements with protected area agencies to access medicines on a sustainable basis. In Colombia, for example, the Alto Orito Indie-Ande Medicinal Plants Sanctuary was established by Parques Nacionales in collaboration with the indigenous Kofán community to protect biodiversity, medicinal plants and the traditional ecological knowledge of local healers. In Ethiopia, more than 340 species of medicinal plants are recognised in the Bale Mountains National Park, a spectacular area of Afro-montane habitat, where 95% of neighbouring households use medicinal plants to treat common ailments and for pre- and post-natal care (United Nations Development Programme & World Bank, 2007). Yet these resources are threatened by increased livestock grazing and firewood collection within the park, pressures that are likely to increase as climate change intensifies degradation of surrounding land and resources.

Reducing vulnerability to natural disasters

One anticipated impact of climate change is an increasing incidence and severity of natural disasters (Stern, 2007). Worldwide the impact and human cost of natural disasters are rising steadily. Climate instability and more extreme weather events are combining with rising human populations and land tenure issues to put more people at risk, especially in marginal and vulnerable habitats (Dudley *et al.*, in press.). Ecosystems such as coral reefs, mangrove swamps and coastal wetlands play a key role in protecting human settlements against typhoons, hurricanes and tidal surges from storms and tsunamis. Well-managed ecosystems including forests and wetlands can buffer against many flood and tidal events, landslides and storms (Sudmeier-Rieux *et al.*, 2006; Stolton *et al.*, 2008). Evidence of this protective effect can be clearly found in the case of Hurricane Katrina, which hit New Orleans in 2005. The loss of coastal wetlands was identified as a major contributory factor in the devastating scale of impacts, and as a result a major restoration programme is now underway in the Louisiana wetlands (Stolton & Dudley, 2010). Ecosystem-based approaches not only help to mitigate the impacts of extreme weather events such as hurricanes, storms and flood but also can be highly cost-effective (Box 18.3). As a result, there is increasing international interest in so-called *ecosystem-based approaches to adaptation* (EBAs), as discussed in the 'Ecosystem-Based Adaptation' section of this chapter.

Box 18.3 **Natural habitats: protecting communities from storms and floods**

Since 1994, local communities have been planting and protecting mangrove forests in Vietnam as a way to buffer against storms. An initial investment of US$1.1 million saved an estimated US$7.3 million a year in sea dike maintenance and significantly reduced the loss of life and property from Typhoon Wukong in 2000 in comparison with other areas (International Federation of Red Cross and Red Crescent Societies, 2002). Conversely, loss of mangrove area has been estimated to increase the costs of storm damage on Thailand's coast by US$585,000 or US$187,898 per square kilometre (in 1996 dollars) (Stolton *et al.*, 2008).

Integrating protection of natural riverine forests and wetlands into flood management strategies has proven to be a cost-effective way to complement early-warning systems and hard infrastructure along the Parana River in Argentina, protecting more than 50% of Argentina's birds, mammals, amphibians and reptiles in the floodplains as well as human settlements (Quintero, 2007). Similarly the Whangamarino wetland protected area and Ramsar site in New Zealand was originally protected for its biodiversity but has also been valued at US$601,037 per year (2003 prices) on account of its flood control utility (Dudley *et al.*, in press.).

Many mountain habitats and protected areas can be justified due to their role in protecting downstream and vulnerable communities from natural hazards such as floods and unstable hillsides. Forested watersheds can help mitigate any but the largest floods. After the catastrophic floods of 1998, the Chinese government introduced a moratorium on logging and re-assessed its forestry programmes. Under the Natural Forest Protection Programme approximately 50 million hectares, more than half the country's natural forests, were re-designated as nature reserves, forest parks, watershed forests or areas for selective logging according to their biological and protection values, to ensure the long-term protection of forests in watershed catchments and to reduce the vulnerability of downstream villages and towns to flooding (World Bank, 2006). Forested slopes are better able to withstand avalanches and landslips, including the after-effects of earthquakes. Research in Pakistan found that villages below vegetated mountain reserves fared better after earthquakes than those under bare slopes (Stolton *et al.*, 2008).

Drought, desertification and land degradation can diminish biological diversity and many of the ecosystem goods and services on which human societies depend. Drought is uncontrollable, but desertification and land degradation can be reduced by improving land management practices; protecting watersheds and wetlands; maintaining natural habitats and stabilising dunes to stop advance of deserts; restoring habitats, including reforestation and natural regeneration of watersheds and degraded pastures; introducing more sustainable livestock and pasture management; promoting agroforestry systems; improving forest management and managing invasive alien species. Mortimore describes (in Chapter 7 of this volume) the complexity of dryland management and the sophisticated systems that pastoralists in the Sahel have developed over thousands of years for maximising the benefits from the dynamic and heterogeneous environment in which they live. Similarly herders in the Lake Hovsgol area of Mongolia have been encouraged to introduce rotational grazing to reduce pressure on natural vegetation and slow the rate of permafrost melt caused by climate change (World Bank, 2010). In the Middle East, desert countries like Kuwait and Jordan are planning protected area systems to maintain desert vegetation and stop desertification and the dune formation and dust storms that are yearly growing more frequent (Al-Awadi *et al.*, 2005).

Looking forward – opportunities for enhancing the synergies between biodiversity, climate change and poverty alleviation

Climate change increases threats to biodiversity – and hence the natural safety net on which so many poor people depend – but it also provides an opportunity to re-evaluate natural solutions as well as technological fixes for mitigation and adaptation. In this section I review the opportunities presented by increasing international interest in, and attention to, (i) ecosystem-based approaches to adaptation and (ii) reducing emissions from deforestation and forest degradation.

Ecosystem-based adaptation

There is growing consensus between the parties to the international conventions on climate change (United Nations Framework Convention on Climate (UNFCC)) and biological diversity (Convention on Biological Diversity (CBD)) on the need to strengthen management of natural ecosystems as part of climate change response strategies. Natural ecosystems help to mitigate climate change by storing and sequestering carbon (Dudley *et al.*, 2010). Forests are the world's largest terrestrial carbon stock, covering some 30% of the world's land area but storing about 50% of terrestrial carbon, including soil carbon, and continuing to sequester carbon in old-growth phases. Inland wetlands, particularly peat, also contain huge carbon stores as do grass-lands which are estimated to hold 10–30% of global soil carbon (Parish *et al.*, 2008). With poor management or conversion, however, these habitats can easily degrade and switch to become net sources of carbon. About 20% of total greenhouse gas (GHG) emissions are caused by deforestation and land use changes; in tropical regions, emissions attributable to land clearance are much higher, up to 40% of national totals in Brazil and Indonesia (World Bank, 2010). Coastal and marine habitats, especially salt marshes, mangroves, kelp and sea grass beds, are all important carbon sinks, sequestering carbon more efficiently than terrestrial ecosystems of equivalent area (Laffoley & Grimsditch, 2009). Improved protection and management of these coastal systems can contribute significantly to reducing global carbon emissions as well as protect nurseries and fishing grounds important to artisanal fisheries.

Adaptation is an increasingly important part of the development agenda, especially in developing countries most at risk from climate change. Biodiversity and natural ecosystems provide multiple goods and services which reduce vulnerability to dis-aster yet have been largely unrecognised in national accounting. Ecosystem-based approaches, including protected areas and more sustainable management of natural resources, contribute to adaptation strategies by protecting and enhancing vital ecosys-tem services such as water flows and water quality; conserving habitats that maintain nursery, feeding and breeding areas for fisheries, wildlife and other species on which human societies depend; reducing land degradation and protecting water sources by preventing, and controlling, invasive alien species; maintaining coastal barriers and natural mechanisms of flood control and pollution reduction and protecting reservoirs of wild crop relatives to enhance agricultural productivity and crop resilience (World Bank, 2010).

Reducing emissions from deforestation and forest degradation

Under the climate change agenda, Reducing Emissions from Deforestation and Forest Degradation (REDD) is the mitigation option with the greatest potential for maintaining carbon stocks in standing forests. Strategies to reduce deforestation could also contribute to biodiversity and poverty reduction by protecting forest resources on which many local communities depend. Nevertheless a key challenge remains: how to reward countries and communities that conserve natural habitats. Only a few countries, such as Costa Rica, have developed effective programmes to make payments to landholders for ecosystem services (see Chapter 14, this volume).

At the Bali conference of UNFCCC in 2007, Parties adopted the principle of providing financial incentives to developing countries for REDD, but the original REDD concept excluded payments to indigenous reserves, community forests and conservation areas on the basis that they were 'protected' already. Subsequently

REDD has been expanded to REDD+ to include conservation in protected areas and indigenous reserves, sustainable forest management (e.g. through forest certification) and enhancement of forest stocks through regeneration and reforestation, including plantations. Substantial international finance for REDD+ could afford an exciting opportunity to protect forests for multiple benefits including biodiversity conservation and greater community benefits from native forest management and watershed protection. Nevertheless some key challenges remain, especially when competing land uses may be highly profitable (Box 18.4). High-priority sites for tackling deforestation to reduce emissions may not always reflect other forest values such as biodiversity conservation, livelihood benefits or water delivery (Miles & Kapos, 2008). One obvious risk associated with REDD+ is the displacement of pressures resulting from continuing demand for land for agriculture, timber and even biofuels, to ecosystems with low carbon values, either less carbon-rich forests or non-forest ecosystems such as savannahs or wetlands.

Box 18.4 **Can carbon markets save Sumatra's tigers?**

Riau Province in central Sumatra harbours populations of endangered Sumatran tigers and Asian elephants. Riau has already lost 65% of its original forest cover and has one of the highest rates of deforestation in the world, due to conversion of forest for agriculture, for pulpwood plantations and for industrial oil palm plantations to serve the surging biofuels market. If the current rate of deforestation continues, estimates suggest that Riau's natural forests will decline from 27% today to only 6% by 2015. All of this comes at a global cost. The average annual carbon dioxide (CO_2) emissions from deforestation in Riau exceed the emissions of the Netherlands by 122% and are equivalent to 58% of Australia's annual emissions.

Can carbon trading provide a new economic incentive to protect Riau's forests, especially the carbon-rich peat swamp forests? Although new programmes are under consideration to reduce deforestation, the prevailing price of carbon may be too low to shift incentives from forest clearance for oil palm or pulpwood to conservation. Moreover, high-biodiversity forests may not be given priority over other forests with higher carbon sequestration potential because the systems pay only for carbon and give little attention to biodiversity. Even if carbon markets do help forest conservation, there are still major questions about how, and where, those benefits will accrue to local communities rather than big business.

Nevertheless carbon markets may have potential to promote conservation in less productive lands and bring real benefits to the poor. In parts of South Asia, the returns (present value) of arable land are often as low as US$100–150 per hectare. Clearing a hectare of tropical forest could release 500 tons of CO_2. At an extraordinarily low carbon price of even US$10 per ton of CO_2, an asset worth US$5000 per hectare is being destroyed for a less valuable use. A modest payment through avoided deforestation schemes could be sufficient to shift incentives in some of the unproductive arable land in South Asia.

Source: Damania et al. (2008).

Another key concern is the question of who owns carbon and who should benefit from any carbon credits: national governments or the local communities and indigenous groups who manage and protect those forests and are dependent on them for their livelihoods. Traditional indigenous territories, for example, encompass up to 22% of the world's land surface and coincide with areas that harbour some 80% of known biodiversity (World Resources Institute, 2005). Tropical forest protected areas, and especially those managed by indigenous peoples, lose less forest cover than surrounding lands (Campbell et al., 2008; Nelson & Chomitz, 2011). Assuring the equitable distribution of revenues gained from carbon credits to communities affected by improved forest protection may prove to be a key challenge of REDD implementation (Peskett et al., 2008; Sandbrook et al., 2010).

Several initiatives, often NGO led, are piloting schemes to provide financial benefits to communities for protecting high-biodiversity forests through funding from voluntary carbon markets. In north-eastern Madagascar, where the Makira Protected Area includes more than 370,000 ha of largely intact rainforest and 22 of the world's 71 lemur species, the Wildlife Conservation Society has marketed 140,000 tons of pre-verified emission reduction credits which will bring carbon revenue back to local communities. Other projects are working with communities in Cambodia, Aceh and indigenous territories in the greater Madidi-Tambopata landscape in Bolivia to reduce deforestation rates and establish carbon revenue distribution mechanisms that benefit local communities and biodiversity (Wildlife Conservation Society, 2009).

While carbon funds can provide extra incentives for protecting forests, they will not be a 'silver bullet' for conserving biodiversity or alleviating poverty. Experiences with integrated conservation and development projects (ICDPs) provide useful and cautionary lessons in managing expectations about the benefits of carbon funds and REDD+ resources (Wells et al., 1999; MacKinnon, 2001; McShane & Wells, 2004).

Nevertheless there is growing understanding of the need to strengthen the conservation and management of natural ecosystems as part of climate change response strategies at both national and local levels. It is time to fully recognise the social and economic benefits that enhanced protection of habitats and biodiversity can bring to benefit local communities as cost-effective, proven and sustainable solutions to reduce vulnerability to climate change.

Conclusion

Climate change has become the key environmental concern of the decade. It will exacerbate other global challenges such as biodiversity loss, food security and water shortages, further threatening the livelihoods of the poor and undermining efforts to achieve poverty alleviation and sustainable development. Healthy intact ecosystems can reduce vulnerability to climate shocks, protect the goods and services on which human livelihoods and welfare depend and increase local and national resilience to climate change. There is strong evidence that protected area systems are effective tools for optimizing the contribution of natural ecosystems in climate change response strategies (Dudley et al., 2010).

Some 15% of global terrestrial carbon, for instance, is stored in protected areas (Kapos et al., 2008). Already more than 12.7% of the earth's terrestrial surface is included in legally designated protected areas, under a number of management and governance regimes, ranging from strict no-access areas to protected landscapes and indigenous reserves that include human settlements and cultural management. In October 2010, state signatories to the CBD agreed to expand the number and area of protected areas to at least 17% of terrestrial and inland water habitats and 10% of marine areas (up from just over 1% currently). Protected areas are the cornerstones of biodiversity conservation, but generating support for more conservation areas will require stronger social and economic arguments to engender the political will to move fine words to effective action – especially given the real and perceived negative social impacts associated with some protected areas (Chapter 10, this volume). Meeting the Aichi targets will also require more recognition of alternative governance models including indigenous and community conservation areas (ICCAs) and private reserves to complement and connect state-managed parks and protected area systems.

Biodiversity conservation and protection of natural ecosystems can help people to cope with, and adapt to, the impacts of climate change. Protecting natural ecosystems, and the goods and services that they provide, is a smart investment option which could strengthen national and local responses to climate change, benefit poor communities and help to address some of the greatest challenges facing us in the 21st century.

References

Al-Awadi, J.M., Omar, S.A.S. & Misak, R.F. (2005) Land degradation indicators in Kuwait. *Land Degradation and Development*, 16, 163–76.

Bucknall, J., Kraus, C. & Pillai, P. (2001) *Poverty and the Environment*. Environment Strategy Background Paper. World Bank, Washington, DC.

Campbell, A., Kapos, V., Lysenko, I., Scharlemann, J.P.W., Dickson, B., Gibbs, H.K., Hansen, M. & Miles, L. (2008) *Carbon Emissions from Forest Loss in Protected Areas*. United Nations Environment Programme and World Conservation Monitoring Centre, Cambridge.

Damania, R., Seidensticker, J., Whitten, A., Sethi, G., MacKinnon, K., Kiss, A. & Kushlin, A. (2008) *A Future for Wild Tigers*. World Bank, Washington, DC.

Dudley, N., MacKinnon, K. & Stolton, S. (In press) Reducing vulnerability: the role of protected areas in mitigating natural disasters.

Dudley, N. & Stolton, S. (2003) *Running Pure: The Importance of Forest Protected Areas to Drinking Water*. World Bank and WWF, Washington, DC and Gland, Switzerland.

Dudley, N., Stolton, S., Belokurov, A., Krueger, L., Lopoukhine, N., MacKinnon, K., Sandwith, T. & Sekhran, N. (eds.) (2010) *Natural Solutions: Protected Areas Helping People to Cope with Climate Change*. IUCN-WCPA, TNC, UNDP, WCS, World Bank and WWF, Gland, Switzerland.

Food and Agriculture Organization (FAO) (2009) *The State of World Fisheries and Aquaculture*. FAO, Rome.

Halpern, B.S. (2003) The impact of marine reserves: do reserves work and does reserve size matter? *Ecological Applications*, 13, 117–37.

Intergovernmental Panel on Climate Change (IPCC) (2007) *Climate Change 2007: Synthesis Report*. IPCC, Geneva, Switzerland.

International Federation of Red Cross and Red Crescent Societies (IFRC) (2002) *Mangrove Planting Saves Lives and Money in Vietnam*. World Disasters Report Focus on Reducing Risk. IFRC, Geneva, Switzerland.

Kapos, V., Ravilious, C., Campbell, A., Dickson, B., Gibbs, H., Hansen, M., Lysenko, I., Miles, L., Price. J., Scharlemann, J.P.W. & Trumper, K. (eds.) (2008) *Carbon and Biodiversity: A Demonstration Atlas*. United Nations Environment Programme and World Conservation Monitoring Centre, Cambridge.

MacKinnon, K. (2001) Integrated conservation and development projects – can they work? *Parks*, 11, 1–5.

McShane, T.O. & Wells. M. (eds.) (2004) *Getting Biodiversity Projects to Work: Towards More Effective Conservation and Development*. Columbia University, New York.

Miles, L. & Kapos, V. (2008) Reducing greenhouse gas emissions from deforestation and forest degradation: Global land use implications. *Science*, 320, 1454–55.

Millennium Ecosystem Assessment (2005) *Ecosystems and Human Well-Being: Biodiversity Synthesis*. World Resources Institute, Washington, DC.

Nelson, A. & Chomitz, K.M. (2011) Effectiveness of strict vs. multiple use protected areas in reducing tropical forest fires: a global analysis using matching methods. *PLoS One*, 6, e22722.

Parish, F., Sirin, A., Charman, D., Joosten, H., Minayeva, T., Silvius, M. & Stringer, L. (eds.) (2008) *Assessment on Peatlands, Biodiversity, and Climate Change: Main Report*. Wetlands International, Kuala Lumpur, Malaysia.

Peskett, L., Huberman, D., Bowen-Jones, E., Edwards, G. & Brown, J. (2008) *Making REDD Work for the Poor*. A Poverty Environment Partnership (PEP) Report. Overseas Development Institute, London.

Pierce, S.M., Cowling, R.., Sandwith, T. & MacKinnon, K. (eds.) (2002) *Mainstreaming Biodiversity in Development: Case Studies from South Africa*. World Bank, Washington, DC.

Quintero, J.D. (2007) *Mainstreaming Conservation in Infrastructure Projects: Case Studies from Latin America*. World Bank, Washington, DC.

Sandbrook, C., Nelson, F., Adams, W. & Agrawal, A. (2010) Carbon, forests and the REDD paradox. *Oryx*, 44, 330–3.

Stern, N. (2007) *The Stern Review on the Economics of Climate Change*. Cambridge University Press, Cambridge.

Stolton, S. & Dudley, N. (eds.) (2010) *Arguments for Protected Areas*. Earthscan, London.

Stolton, S., Dudley, N. & Randall, J. (2008) *Natural Security: Protected Areas and Hazard Mitigation*. WWF International, Gland, Switzerland.

Sudmeier-Rieux, K., Masundire, H., Rizvi, A. & Rietbergen, S. (2006) *Ecosystems, Livelihoods and Disasters: An Integrated Approach to Disaster Risk Management*. International Union for Conservation of Nature, Gland, Switzerland.

United Nations Development Programme (UNDP) and World Bank (2007). *Reducing Threats to Protected Areas: Lessons from the Field: A Joint UNDP and World Bank GEF Lessons Learned Study*. World Bank, Washington, DC.

Wells, M.S., Guggenheim, A., Khan, Wardojo, W. & Jepson, P. (1999) *Investing in Biodiversity: A Review of Indonesia's Integrated Conservation and Development Projects*. World Bank, Washington, DC.

Wildlife Conservation Society (WCS) (2009) *Carbon for Conservation: Climate Change, Biodiversity Conservation, and Local Livelihoods*. Wildlife Conservation Society, New York.

World Bank (2006) *Mountains to Coral Reefs: The World Bank and Biodiversity 1988–2005*. World Bank, Washington, DC.

World Bank (2008a) *Biodiversity, Climate Change, and Adaptation*. World Bank, Washington, DC.

World Bank (2008b) *World Development Report: Agriculture for Development*. World Bank, Washington, DC.

World Bank (2010) *Convenient Solutions to an Inconvenient Truth: Ecosystem-Based Approaches to Climate Change*. World Bank, Washington, DC.

World Resources Institute (WRI) (2005) *World Resources Report: The Wealth of the Poor – Managing Ecosystems to Fight Poverty*. WRI, Washington, DC.

Conservation in the Anthropocene: Biodiversity, Poverty and Sustainability

William M. Adams

Department of Geography, University of Cambridge, Cambridge, UK

Conservation, community and poverty

As many chapters in this book show, projects that seem to combine conservation and poverty alleviation can be successful, but are complex to plan and slow to have the effects desired of them. The integration of conservation and poverty reduction has proved more difficult and more expensive than many had hoped. The idea of integrating conservation and development has been criticised in terms of the cost and effectiveness of conservation outcomes (Oates, 1999; Hutton *et al.*, 2005), and conservation projects have been critiqued for their negative social impacts (Brockington *et al.*, 2008). Barrett and Arcese (1995: 1081) dismiss integrated conservation and development projects as "no more than short-term palliatives". Therefore, while the rhetoric of win–win has been widely deployed by conservationists in speaking both to those outside their organisations and in conversations with each other (McShane *et al.*, 2011), win–win outcomes have proved elusive, particularly in conservation–poverty programmes (Barrett *et al*, . 2011): linking conservation and development has proved "extremely difficult, even at a conceptual level" (Brandon & Wells, 1992: 567).

This chapter reflects on this challenge of attempting to link conservation and development, in what Sanderson and Redford (2004: 146) describe as "the growth-development-poverty alleviation-conservation nexus". It argues that thinking about conservation and poverty has been limited by being side-tracked into just one corner of the wider debate about human poverty and biodiversity loss, focusing exclusively on rural poverty and the protection of biodiverse pristine habitat.

Biodiversity Conservation and Poverty Alleviation: Exploring the Evidence for a Link, First Edition.
Edited by Dilys Roe, Joanna Elliott, Chris Sandbrook and Matt Walpole.
© 2013 John Wiley & Sons, Ltd. Published 2013 by John Wiley & Sons, Ltd.

Wider questions about the nature of development itself need to be addressed in discussions of conservation and the reduction of poverty, for several distinct reasons. Firstly, pursuit of development aimed (in part at least) at the reduction of poverty also causes biodiversity loss (oil palm versus old growth forests, dams and downstream impacts, industrialisation and pollution – the list is potentially endless). Moreover, many development projects have mixed impacts on the poor. Lastly, and most importantly, the conventional model of the developed economy (in industrialised countries) everywhere involves unsustainable patterns of consumption and resource use, and waste production. Globally, this consumption is the chief driver of biodiversity loss, and ignoring it makes no sense for conservation.

Understanding the links between conservation and anti-poverty action needs to extend across levels, up to the world economy. Rurally constrained debate about conservation and development is not enough.

Away from the Last Frontier

Conservationists may take a range of positions to the question of poverty (Adams *et al.*, 2004), yet in practice, arguments about conservation and poverty are highly selective in their spatial frame. Debate has focussed overwhelmingly on the small 'local' scale, and on 'specific sites and communities' (Sanderson & Redford, 2004: 256). The concern of those social critics who explore conservation's negative socio-economic impacts is almost always a specific project, such as a protected area or a resource exploitation zone (e.g. a forest, grazing area or inshore fishery) where conservation regulations are enforced.

Locally, conservation action can reduce poverty (e.g. through cleverly designed ecotourism projects, or schemes to pay directly for biodiversity or for ecosystem service), and it can also increase it, for example by evicting or otherwise displacing people from parks (Brockington *et al.*, 2008; Dowie, 2009).

But what is the broader relationship between conservation and poverty? It is recognised that there are geographical overlaps between areas of high poverty and high biodiversity (Barrett *et al.*, 2011). The links between biodiversity loss and poverty are not limited to rural locations where ecosystems are little changed by human action. Mittermeier *et al.* (2003) define these areas, highly problematically, as *wilderness* (cf. Cronon, 1995). Caro *et al.* (2011: 1) prefer the term *intact ecosystems*, where "the majority of native species are still present in abundances at which they play the same functional roles as they did before extensive human settlement or use, where pollution has not affected nutrient flows to any great degree, and where human density is low". They argue that such areas are, and should remain, the main focus of biodiversity conservation activity.

The important point for the conservation and poverty debate is that the vast majority of the world's poor live elsewhere, in and around urban areas, and in landscapes that have been extensively transformed by people, for example through agriculture (Redford *et al.*, 2008). Of the 28% of the global population that may be considered the 'most poor', 70% live in such transformed landscapes (Redford *et al.*, 2008). Only 16.3 million people (0.95% of the 'most poor', and 0.25% of the global population) live in areas with anything like 'intact ecosystems', which biodiversity conservationists see as their main target for action. From this, Redford *et al.* (2008) conclude that biodiversity conservation organisations (who work mostly in such areas) are likely to be able to contribute relatively little to overall global poverty alleviation efforts, although they can (and should) address poverty in the remote areas where they work.

Whether or not Redford *et al.* are correct in their argument, there is no doubt that the tight framing of the conservation and poverty debate within rural contexts is in danger of localising and marginalising it – literally to the sites identified by conservation planners, on the geographical margins of the mainstream world economy. In such locations, conservation is often the dominant actor, whose decisions have huge implications for welfare, economy and cultural survival. There are important debates and challenges for conservation and poverty policy makers in these zones, at the capitalist resource frontier and around protected areas, as this book attests.

It might be argued (as Caro *et al.*, 2011, do) that conservation should focus on remaining intact ecosystems. However, pursuing the shrinking frontier of relatively intact ecosystems limits the potential for conservation to address biodiversity loss. At best, this strategy will lead to the creation of protected areas that are small biodiverse islands on a fraction of the earth's surface (perhaps 17%), with the rest of the earth (to which humanity is restricted) radically transformed, and perhaps permanently impoverished in diversity. As Kaimowitz and Sheil (2007: 569) note, "the future of many, if not most species, depends on what happens outside strictly protected areas". What, they ask, about the biodiversity and productivity of the transformed lands where far larger numbers of people live? Even those areas that are protected will change, faster or slower, in response to climate change and other human-induced pressures. That hardly promises anything like conservation success.

Such 'frontier' conservation ignores the impacts of industrialisation, consumption and waste production on global biodiversity, and constraining thinking about 'conservation and poverty' to what happens around biodiversity hotspots artificially isolates it from wider debates about poverty and development. This is a mistake. Ignoring the wider economic and social dimensions of global change and development consigns conservation to a marginal position on the side-lines of debate about the world economy and the sustainability, or unsustainability, of global development. If conservation is confined to protecting intact ecosystems, it will always be pitched against human material interests, be they those of the powerless poor or those of the acquisitive rich. The 2004 World Parks Congress put forwards a very different vision for protected areas, as "the building blocks of biodiversity in an ocean of sustainable human development, with their benefits extending far beyond their physical boundaries" (Steiner, 2003: 21). This suggests that conservation must engage with the biodiversity that remains in and around where people live, as part of an engagement with their welfare and aspirations. If conservation turns its back on the vast tracts of the earth that are transformed by human action, it will surely fail to meet the needs and aspirations of coming generations.

The 'frontier' framing of debate about conservation and poverty ignores the biggest driver of biodiversity loss, which is economic activity, or rather the growth in natural resource and energy consumption that accompanies it. And it ignores the fact that it is precisely this that has been, since the end of the Second World War, the means by which people escape poverty. The economic development model to which industrialised and developing country governments have long been committed has sought (with some success) to reduce poverty and increase wealth through economic growth. However, as people escape poverty, others become rich, and global consumption of raw material and energy (and production of wastes) has risen inexorably. Poor countries pursue the model of the rich, and poor people dream of emulating the wealthy. Meanwhile, biodiversity shrinks before the combined onslaught of people and wealth.

It should be a matter of great concern to biodiversity conservationists that so little of their effort is focussed on the processes that drive biodiversity loss. The economic machine that consumes biodiverse habitat has its foundation not in the communities residing in and around biodiversity hotspots, but in the wider world economy and the production and consumption that are part of 'developed' lifestyles. Development is the agreed strategy to reduce global poverty, but it is also the process that has everywhere reduced the diversity of life. This conundrum should be at the centre of debates about poverty and conservation. Yet it is not, for they are too tightly framed within the familiar analytical terrain of intact ecosystems and rural communities.

Development, iPads and ivory

The inexorable rise of human demand on the biosphere (Vitousek *et al.*, 1997) is an inherent feature of capitalism. Human enterprise, harnessed by businesses and markets, transforms and refashions nature, just as it does economies and societies. It is the world economy that drives human demand beyond planetary boundaries (Rockström *et al.*, 2009). Seventy-five per cent of the habitable earth is disturbed by human activity (Hannah *et al.*, 1994), and humans appropriate 32% of global net primary production (more in places; Rojstaczer *et al.*, 2001; Imhoff *et al.*, 2004). The fuel for the engine of the world economy is nature and human work. Nature cannot be conserved except by coming to terms with the way that engine works.

Global consumption patterns cause biodiversity loss in many different ways: demand for cheap tropical timber, chipboard and pulp; for fish and cheap beef; for soya and palm oil; for long-haul flights and automobiles; for big houses, central heating and air conditioning; for computers and endless web connectivity and for limitless electric power at the touch of a fingertip. These products and services are almost universally regarded as a good thing by consumers, politicians and pundits in industrialised economies, as adding to human freedoms (Sen, 1999) and human well-being. For the poor, they are the promised fruits of the process of 'development', which has dominated global thinking about the future since the middle of the 20th century.

The conventional model of 'development' is designed to deliver escape from poverty, but only as a waypoint on a longer journey to prosperity. The idea of *development* in the 20th century cannot be separated from the Euro-American world where it was coined, or the idea of modernity that characterised that world (Escobar, 2004). After the Second World War, the meaning of development narrowed to economic growth, on the Western capitalist model. The purpose of 'development' was widely understood as the replication of the experiences of 'developed' countries across the world, classically conceived in terms of successive *stages of economic growth* from traditional society, pre-conditions for take-off, take-off, maturity and the age of high mass consumption (Rostow, 1960). Development planning is the organised and coherent attempt to overcome constraints on economic growth, through processes of industrialisation, urbanisation, democracy and capitalism. Attempts to measure 'development' have long focussed on quantitative economic indices, such as Gross Domestic Product (GDP), a measure critiqued in Chapter 3 of this book.

Hubris about the quick achievement of this outcome of 'development' in the sense of rapid economic growth had faded by the end of the 20th century, as the less ambitious poverty alleviation objectives of the Millennium Development Goals (MDGs) seemed aspirational enough to wealthy governments. However, focussed on basic needs as they are, the MDGs offer no new development model and no new destination, just a more clearly articulated concentration on the lowest rungs of the ladder, and a renewed commitment to helping more people scale them. In the 21st century, development is still conceived of as involving industrial production and increased consumption of manufactured goods, accompanied by increased energy and raw material consumption. The speed with which governments retreated to a simplistic debate about how to restore economic growth after the banking crisis of 2008 shows how powerful the old economistic ideology of development is.

Where poverty has retreated fastest, in China and India, material and energy consumption has risen fastest. Like the United States, Europe and Japan, economic growth in the newly industrialising countries (especially India and China) is being built on surging demand for consumer goods: white goods, cars, electronics, new kinds of food (fish, meat or the soya to supply intensive feedlots) and luxury products (such as elephant ivory; Wittemyer, 2011). The same tide of cheap products feeds both the extraordinarily rapid growth in demand of an aspirant middle class in newly industrialising countries and the rising consumption in developed economies of cheaply produced manufactured products from iPads to disposable fashion.

Sweatshops and low wages ensure that, numerous though their recruits are, the middle classes in these countries remain a minority, but the engine of the consumer economy ensures that significant numbers become less poor. The reduction of poverty is accompanied by a growth in energy and material consumption. This is not an accidental link or a by-product; it is how capitalism works.

Conservation's development paradox

Conservation therefore confronts a basic contradiction in the way it engages with debates about development. Most attempts to combine conservation and poverty reduction at the local level involve development in some form (e.g. ecotourism and payments for ecosystem services). This links local poor rural clients into wider processes of development by connecting them to the money economy, with both its undoubted benefits (e.g. acquisition of consumer goods such as torches, bicycles and mobile phones) and its risks (e.g. dependence on fickle international tourism numbers). Such local actions do not challenge the material basis of that wider economy or address its impacts on global biodiversity. The extra consumption enabled by the trickle of money to the rural poor is not matched by a reduction in consumption anywhere else. This is development: aggregate consumption rises, as demand stimulates industry elsewhere; and the rising tide of economic growth floats all boats, poverty shrinks and lives get better. The problem is that the 20th-century dream of an endlessly growing world economy is unsustainable. So in helping to reduce poverty in one place, conservation makes the wider problem just that little bit worse. There is resounding silence about the impacts of the larger processes of economic and ecological change unleashed by the conventional development model.

Biodiversity conservation offers no systematic critique of the developed world's material and energy consumption (that is left to broader environmental activists), or its poor record in delivering significant reductions in poverty at a national or international scale (that is left to development activists). Although individuals protest or abstain, most biodiversity conservation organisations operate quietly within 'an empire built on an industrial economy based on consumerism' (McPherson, 2011). The world of conservation is diverse (Redford, 2011), but it offers little challenge to the shape of the world crafted by neoliberal capitalism. Indeed, it is to a large extent complicit with it. Critics point out that conservation has become an integral and essential part of neoliberal capitalism, developing products and markets that facilitate capture of nature, and securing their value through certification. In the process, conservation organisations become financially dependent on and complicit with resource-exploiting corporations (Brockington *et al.*, 2008; Brockington & Duffy, 2010).

The most ubiquitous conservation 'solution' to poverty at the local level, the development of local 'wildlife tourism' or 'eco-tourism' activities as a source of income for poor rural people around biodiverse sites, is entirely dependent on consumption by members of a global elite. Wildlife tourism is a classic high-order good, consumed by people (backpackers and middle-class tourists as well as millionaires) whose lifestyles involve global travel, specialised consumer goods and luxury foods. Their holidays, from which some trickle of revenue or other benefit may well reach the poor, contribute significantly to their already heavy use of fossil fuels, as do all the possessions they carry. Furthermore, tourist visits are not without their cultural impact, for they create justifiable hunger for the wealth and material goods that tourists display. Tourists are powerful living advertisements for modernity and the high levels of consumption achieved in rich countries in the 20th century.

Ecotourism is not alone in its mixed impacts. Payments for forest-based carbon credits allow corporations, governments and rich concerned individuals to continue to burn fossil hydrocarbons without changing their lifestyles or the material basis of their enterprises (in terms of material throughput or energy use), meanwhile paying others to forgo the immediate economic benefits of forest conversion. Revenue that provides a conservation-linked solution to a local poverty problem is obtained only by plugging into a model of development that is unsustainable.

This is the paradox that conservation faces. A possible solution at one level (the rural community's need for income linked to a biodiverse habitat) ignores and condones wider problems of the world economy's growing hunger for energy and natural resources. To local people, even if they start to escape poverty, development holds out a model where the wealth of the tourist is available to all. Poverty alleviation is never an acceptable policy end point, but a step to realising the promise of development: wealth for all.

The challenge of sustainability

This is the terrain missing from much debate about conservation and poverty. Much is now understood about how strategies for the rural poor and frontier biodiversity can (and cannot) be meshed (Redford & Sanderson, 2008). The choices for conservationists about how to think about the links between conservation and poverty reduction are widely discussed (e.g. Adams *et al.*, 2004). These debates are important: the people in and around the last areas of biodiverse habitat beyond the grip of organised capital are among the poorest in the world in material terms, and development's engagement in these zones (e.g. biofuels, oil palm, logging and mining) is often drastically negative in its impacts on the environment, and often local people. Here conservation organisations have an important role to play.

Yet the more powerful dynamics of biodiversity loss are driven by the engine of a world economy said by national governments to be designed and run in the name of human welfare and the eradication of global poverty.[1] And, on this, biodiversity conservationists tend to be silent. They see their job as saving nature in its last fastnesses, and not as considering the wider picture of the world economy.

There are important questions for conservation regarding not just how it can engage with poverty at the frontier, but also how it can engage with the development processes set up and run to eradicate poverty as a whole. Conservation is not looking systematically at this problem. This detachment is surprising.

Historically, conservation in the 20th century was effectively a subset of environmentalism. Many contemporary concerns (e.g. rare species and extinction) long pre-dated the 'new environmentalism' of the 1960s, but from the WWF Panda to *Silent Spring*, from the Sierra Club to the *World Conservation Strategy*, conservation's DNA lies deep within 20th-century environmentalism (Adams, 2004, 2009).

However, conservation's engagement with environmentalism has always been partial. Since the 1960s, a central issue for environmentalism has been the need of humankind for the living biosphere. In modern terms, this essentially anthropocentric concern might be expressed as the need for enough biodiversity to maintain a functioning biosphere, and the right kinds of biodiversity to deliver the services and material flows that are optimal for aggregate human welfare. This concern underlay the eco-catastrophism of 1970s 'limits to growth', just as it does contemporary fears about anthropogenic climate change. It dominated debate about sustainable development since that term was coined in the early 1970s, at the time of the UN Conference on the Human Environment in Stockholm in 1972. Decades later, after the Brundtland Report (1987), the Rio Conference (1992) and the Johannesburg World Summit on Sustainable Development (2002), mainstream thinking about sustainable development is led by an essentially neoliberal agenda of ecological modernisation and 'green growth' (Adams, 2009).

However, an important strand of environmentalism was more radical, and specifically concerned with the poor (Guha & Martínez-Alier, 1997) – with a concern for intra-generational as well as inter-generational equity. Alongside its neo-Malthusianism, significant strands of 1970s environmentalism radically critiqued the standard development model, from Schumacher's *Small Is Beautiful* (1973) through green anarchism (Bookchin, 1982) to socialism (Bahro, 1982), as well as calls for a zero-growth economy (Daly, 1973). Ecosocialism and 'ecological Marxism' have persisted in critiques of capitalism (Eckersley, 1992; Igoe et al,. 2010).

In the 2000s, the centrality of growth has been challenged once more. Tim Jackson (2009) writes of *Prosperity without Growth*, neatly separating the end from the means.

---

[1] Of course the crash of 2009, the public bailouts of the global private banks and the debt crisis in the European Union and United States in 2011 call into question the extent to which the global economy is really managed, let alone managed for these admirable public goals.

There is a growing interest in the idea of 'degrowth' or *décroissance* (Latouche, 2004). 'Socially sustainable economic degrowth' (Martínez-Alier *et al.*, 2010: 1741) stands against the dominant economic paradigm that rewards "more instead of better consumption and private versus public investment in man-made rather than natural capital". It is not, however, a formal theory, but what Latouche sees as "a political slogan with theoretical implications" (Martínez-Alier *et al.*, 2010: 1742).

Conservation's challenge

In contrast to such dreamy radicalism, conservation's place within mainstream sustainable development has avoided environmentalism's harder questions. Biodiversity conservation has become comfortably conformist, at ease with capitalism. In the 1980s, conservation plugged into the sustainable development debate through its new big idea of 'community conservation' based largely on wishful thinking, and the superficial appeal of rhetorical claims for the efficacy of grassroots citizen action, neo-populism and political neutrality (influenced by a parallel discourse of participation and decentralisation in development). The 'new conservation' sought to protect biodiversity not only through the community but also through the market (Hulme & Murphree, 1999). This approach evades the larger questions.

How should humanity respond to the challenge of the *Anthropocene*, the era of humankind? The term refers to the current geological epoch of extensive human modification of ecological and geological processes, which started in about 1800, and whose transformations have accelerated since 1950 (Crutzen, 2002; Steffen *et al.*, 2007). The fundamental debate about conservation and poverty does not concern the frontier, important though that is, but the world's urban heartlands; not the beleaguered forest dweller but the home-owning, holiday-booking, SUV-driving hyper-consumer of internet advertisement and television soaps, and the vast numbers of the urban and rural poor who dream, justifiably, of achieving the same lifestyle. On the current model, the only escape from poverty leads to an environmentally destructive lifestyle.

One question suggests itself: why does conservation not address the standard 20th-century model of development? The answer is multi-stranded. The characteristic path dependence of thinking in conservation constrains innovative thinking (Adams, 2010). Most conservationists are trained to know about biology, not capitalism. Their instinct and their 'mission-driven' discipline of conservation biology both lead them to focus on immediate drivers of biodiversity loss, and so they are preoccupied with matters of greater urgency. Moreover, they operate within the capitalism system, are the beneficiaries of the conventional development model and in many cases are dependent on the taxes or donations of the wealthy for their income and jobs. Why question a system you do not understand, which supports the world as you know it and from which you derive all prospect of future benefit? The sustainability of the world economy is surely someone else's problem.

There has been some progress in addressing that problem in recent decades, challenging the standard development model of unchecked economic growth and associated energy use, resource consumption and waste production. The issue of sustainability has considerable buy-in from governments and businesses, at least at the rhetorical level. The many achievements of 'green' production and consumption are impressive given what went before. However, they are still trivial compared to the scale of the problem. To even survive the Anthropocene, humanity needs to dramatically reduce carbon use and increase technical efficiency in all industrial processes of production and consumption, delink energy generation from carbon production and delink energy consumption from economic growth (Adams & Jeanreneaud, 2008). We need to do all these things while enabling poor countries (and poor people) to produce and consume more. This demands an agenda of contraction and convergence, of redistributing wealth and resource use, the 'goods' and 'bads' of development.

The political challenges of conservation and poverty are deep. We do not know how to deal with the outright selfishness of capitalism or the ignorance and short-sightedness of planners and resource users. We do not know how to take hard decisions, and avoid the endless tragic farce of non-decision as a result of the interaction of consumer, capitalist, voter and politician. We need to re-integrate conservation with a broader environmentalism, and yet we do not know enough to do this with any confidence – about ecosystem function or how brittle ecosystems are (e.g. how many species we need in different ecosystems before they start to unravel). We have too few ways of explaining the value of nature beyond the 20th century's appeal to the powerful but meaningless ideas of wilderness or 'the pristine', or recourse to the language of monetary value.

Conservation needs to take the implications of the Anthropocene seriously, and respond to them with greater ambition. As David Orr wrote in 2007 "we are not told that the consumer way of life will have to be rethought and redesigned to exist within the limits of natural systems and better fitted to our human limitations" (Orr, 2007: 1393). That rethinking and redesign comprise the greatest challenge for both biodiversity conservation and poverty alleviation in the 21st century.

References

Adams, W.M. (2009) *Green Development: Environment and Sustainability in a Developing World*. Routledge, London.

Adams, W.M. (2010) Path dependence in conservation. In *Trade-offs in Conservation: Deciding What to Save*, ed. N. Leader-Williams, W.M. Adams & R.J. Smith, pp. 292–310. Wiley-Blackwell, Oxford.

Adams, W.M., Aveling, R., Brockington, D., Dickson, B., Elliott, J., Hutton, J., Roe, D., Vira, B. & Wolmer, W. (2004) Biodiversity conservation and the eradication of poverty. *Science*, 306, 1146–9.

Adams, W.M. & Jeanrenaud, S.J. (2008). *Transition to Sustainability: Towards a Humane and Diverse World*. International Union for Conservation of Nature, Gland, Switzerland.

Bahro, R. (1982) *Socialism and Survival*. Heretic, London.

Barrett, C.B. & Arcese, P. (1995) Are integrated conservation-development projects (ICDPs) sustainable? On the conservation of large mammals in sub-Saharan Africa. *World Development*, 23, 1073–84.

Barrett, C.B., Travis, A.J. & Dasgupta, P. (2011) On biodiversity conservation and poverty traps. *Proceedings of the National Academy of Sciences of the United States of America*, 108, 13907–12.

Bookchin, M. (1982) *The Ecology of Freedom: The Emergence and Dissolution of Hierarchy*. Cheshire, Palo Alto, CA.

Brandon, K.E. & Wells, M. (1992) Planning for people and parks: design dilemmas. *World Development*, 20, 557–70.

Brockington, D. & Duffy, R. (2010) Capitalism and conservation: the production and reproduction of biodiversity conservation. *Antipode: A Radical Journal of Geography*, 42, 469–84.

Brockington, D., Duffy, R. & Igoe, J. (2008) *Nature Unbound: The Past, Present and Future of Protected Areas*. Earthscan, London.

Caro, T., Darwin, J., Forrester, T., Ledoux-Bloom, C. & Wells, C. (2011) Conservation in the anthropocene. *Conservation Biology*. doi:10.1111/j.1523-1739.2011.01752.x, 26, 185–188.

Cronon, W. (1995) The trouble with wilderness, or, getting back to the wrong nature. In *Uncommon Ground: Toward Reinventing Nature*, ed. W. Cronon, pp. 69–90. W.W. Norton, New York.

Crutzen, P.J. (2002) Geology of mankind: the Anthropocene. *Nature*, 415, 23.

Daly, H.E. (ed.) (1973) *Towards a Steady-State Economy*. W.H. Freeman, New York.

Dowie, M. (2009) *Conservation Refugees: The Hundred-Year Conflict between Global Conservation and Native Peoples*. MIT Press, Cambridge, MA.

Eckersley, R. (1992) *Environmentalism and Political Theory: Toward an Ecocentric Approach*. UCL Books, London.

Guha, R. & Martínez-Alier, J. (1997) *Varieties of Environmentalism: Essays North and South*. Earthscan, London.

Hannah, L., Lohse, D., Hutchinson, C., Carr, J.L. & Lankerani, A. (1994) A preliminary inventory of human disturbance of world ecosystems. *Ambio*, 23, 246–50.

Hulme, D. & Murphree, M. (1999) Communities, wildlife and the 'new conservation' in Africa. *Journal of International Development*, 11, 277–86.

Hutton, J., Adams, W.M. & Murombedzi, J.C. (2005) Back to the barriers? Changing narratives in biodiversity conservation. *Forum for Development Studies*, 32, 341–70.

Igoe, J., Neves, K. & Brockington, D. (2010) A spectacular eco-tour around the historic bloc: theorising the convergence of biodiversity conservation and capitalist expansion. *Antipode*, 42, 486–512.

Imhoff, M.L., Bounoua, L., Ricketts, T., Loucks, C., Harriss, R. & Lawrence, W.T. (2004) Global patterns in human consumption of net primary production. *Nature*, 429, 870–3.

Jackson, T. (2009) *Prosperity without Growth: Economics for a Finite Planet*. Earthscan, London.

Kaimowitz, D. & Sheil, D. (2007) Conserving what and for whom? Why conservation should help meet basic human needs in the tropics. *Biotropica*, 39, 567–74.

Latouche, S. (2004) Why less should be so much more: degrowth economics. *Le Monde Diplomatique*. http://mondediplo.com/2004/11/14latouche (accessed 4 May 2012).

Martínez-Alier, J., Pascual, U., Vivien, F.-D. & Zaccaï, E. (2010) Sustainable de-growth: mapping the context, criticisms and future prospects of an emergent paradigm. *Ecological Economics*, 69, 1741–7.

McPherson, G. (2011) Going back to the land in the age of entitlement. *Conservation Biology*, 25, 855–7.

McShane, T.O., Hirsch, P.D., Trung, T.C., Songorwa, A.N., Kinzig, A., Monteferri, B., Mutekanga, D., Thang, H.V., Dammert, J.L., Pulgar-Vidal, M., Welch-Devine, M., Brosius, J.P., Coppolillo, P. & O'Connor, S. (2011) Hard choices: making trade-offs between biodiversity conservation and human well-being. *Biological Conservation*, 144, 966–72.

Mittermeier, R.A., Mittermeier, C.G., Brooks, T.M., Pilgrim, J.D., Konstant, W.R., da Fonseca, G.B. & Kormos, C. (2003) Wilderness and biodiversity conservation. *Proceedings of the National Academy of Sciences*, 100, 10309–13.

Oates, J. (1999) *Myth and Reality in the Rainforest*. California University Press, Berkeley.

Orr, D.W. (2007) Optimism and hope in a hotter time. *Conservation Biology*, 21, 1392–5.

Redford, K.H. (2011) Misreading the conservation landscape, *Oryx*, 45, 324–30.

Redford, K., Levy, M., Sanderson, E. & de Sherbinin, A. (2008) What is the role for conservation organizations in poverty alleviation in the world's wild places ? *Oryx*, 42, 516–28.

Rockström, J., Steffen, W., Noone, K., Persson, A., Chapin, F.S. III, Lambin, E.F., Lenton, T.M., Scheffer, M., Folke, C., Schellnhuber, H.J., Nykvist, B., de Wit, C.A., Hughes, T., van der Leeuw, S., Rodhe, H., Sörlin, S., Snyder, P.K., Costanza, R., Svedin, U., Falkenmark, M., Karlberg, L., Corell, R.W., Fabry, V.J., Hansen, J., Walker, B., Liverman, D., Richardson, K., Crutzen, P. & Foley, J.A. (2009) A safe operating space for humanity. *Nature*, 461, 472–5.

Rojstaczer, S., Sterling, S.M. & Moore, N.J. (2001) Human appropriation of the products of photosynthesis. *Bioscience*, 294, 2549–52.

Sanderson, S. & Redford, K. (2004) The defence of conservation is not an attack on the poor. *Oryx*, 38, 146–7.

Schumacher, E.F. (1973) *Small Is Beautiful: Economics as if People Mattered*. Blond and Briggs, London.

Sen, A. (1999) *Development as Freedom*. Oxford University Press, Oxford.

Steffen, W., Crutzen, P.J. & McNeill, J.R. (2007) The Anthropocene: are humans now overwhelming the great forces of nature ? *Ambio*, 36, 614–21.

Steiner, A. (2003) Trouble in paradise. *New Scientist*, 180, 21.

Vitousek, P.M., Mooney, H.A., Lubchenco, J. & Melillo, J.M. (1997) Human domination of Earth's ecosystems, *Science*, 277, 494–9.

Wittemyer, G. (2011) Effects of economic downturns on mortality of wild African elephants. *Conservation Biology*, 25, 1002–9.

(20)

Tackling Global Poverty: What Contribution Can Biodiversity and Its Conservation Really Make?

Dilys Roe[1], Joanna Elliott[2], Chris Sandbrook[3] and Matt Walpole[3]

[1]International Institute for Environment and Development, London, UK
[2]African Wildlife Foundation, Oxford, UK
[3]United Nations Environment Programme World Conservation Monitoring Centre, Cambridge, UK

Introduction

There is an explicit assumption that conserving biodiversity (or reducing the rate of biodiversity loss) can help in efforts to tackle global poverty. Evidence of this assumption lies in the target that the Parties to the Convention on Biological Diversity (CBD) set in 2002: "to achieve by 2010 a significant reduction of the current rate of biodiversity loss at the global, regional and national level *as a contribution to poverty alleviation* and to the benefit of all life on earth" (CBD, 2002; emphasis added). The development community also bought into this assumption, adopting the CBD '2010 Target' as part of the Millennium Development Goals (MDGs) (United Nations, 2006). The reduction in the rate of biodiversity loss anticipated in the 2010 target was not achieved (Butchart *et al.*, 2010; Mace *et al.*, 2010). This continued loss of biodiversity is lamented not just for its own sake but also for its implications for continued human well-being and poverty alleviation. The 2010 progress report on the MDGs, for example, noted that continued biodiversity loss would hamper efforts to meet other MDGs, "especially those related to poverty, hunger and health, by increasing the vulnerability of the poor and reducing their options for development" (United Nations, 2010a: 55). A high-level meeting at the September 2010 UN General Assembly further stressed the linkage, claiming that "preserving biodiversity is inseparable from the fight against poverty" (United Nations, 2010b). The CBD's new Strategic Plan (2011–2020), agreed at the 10th Conference of Parties in Nagoya, Japan in October 2010, continues to emphasise the link between achieving conservation goals and reducing poverty: its mission is to "take effective and urgent action to halt the loss of biodiversity in order to ensure that by 2020 ecosystems are resilient and continue to provide essential services, thereby securing the planet's variety of life, and contributing to human well-being, and poverty eradication" (CBD, 2010).

Progress towards meeting the MDGs has also been slow – particularly for some goals and in some regions of the world (United Nations, 2010a). But would this situation be different if biodiversity had been conserved better? The majority of the world's poor live in rural areas (International Fund for Agricultural Development, 2010) where their day-to-day lives are intertwined with biodiversity (Chapter 4). The well-being of these people – who depend on local ecosystem goods and services – may be closely coupled to the fate of biodiversity. Likewise, efforts to conserve biodiversity in areas of poverty may influence, and be influenced by, the needs and aspirations of the poor.

However, there is growing concern that much of what is known about these relationships is based on myth and assumption with little empirical exploration or evidence (Barrett *et al.*, 2011). In this book we have sought to better understand linkages between biodiversity conservation and poverty alleviation by reviewing what the current state of knowledge really is, what evidence claims and counterclaims are based on and what is assumption rather than fact. The chapters of this book have examined the empirical evidence for the dependency of the poor on biodiversity and for the effectiveness of various conservation mechanisms in alleviating poverty. Different ecological settings and different types of interventions provide very different contexts for exploring biodiversity–poverty relationships and reveal very different outcomes. Thus, while some international policies and proclamations assert that conserving biodiversity and eliminating poverty are two sides of the same coin (e.g. United Nations, 2010b), the chapters in this book suggest the need for a much more nuanced approach.

Do biodiversity and its conservation contribute to poverty alleviation?

Despite a limited evidence base, the chapters in this book confirm that (i) the poor often depend disproportionately on biodiversity for their subsistence needs in terms of both income and insurance against risk (e.g. Chapters 4, 5 and 8) and (ii) biodiversity conservation can be a route out of poverty under some circumstances, but a route into poverty in others (e.g. Chapters 10, 11 and 15).

The contribution that biodiversity and its conservation make to alleviating poverty at the individual or household level varies hugely from context to context. In some cases it is biodiversity itself that makes a direct contribution (Box 20.1). In other cases, organizations and institutions that act to conserve biodiversity can contribute – through interventions that are intended to generate income and other benefits, improve governance or increase accountability. Conservation of biodiversity – and hence the maintenance of critical ecosystem services – can contribute to poverty alleviation for *some* people in *some* places. It is not, however, a given that conserving biodiversity will inevitably be beneficial for poor people: all sorts of politics, power relationships, global environmental and economic pressures and governance issues mediate the relationship.

Box 20.1 **Different contributions of biodiversity to poverty alleviation**

Biodiversity "has key roles at all levels of the ecosystem services hierarchy" (Mace *et al.*, 2012), and therefore makes a range of contributions to human well-being. It regulates underlying processes such as nutrient cycling, can itself be a final ecosystem service (e.g. a crop species) and can be a good for consumption, such as a charismatic species visited by tourists (Mace *et al.*, 2012). Many components of biodiversity – such as fish, wildlife and crops – are freely available and can be harvested and used with little processing and with low-cost technologies. Biodiversity thus acts as a form of natural capital – which is particularly important for individuals and households with little financial or physical capital. The World Bank estimates that forest products provide roughly 20% of poor rural families' 'income' – of which half is cash and half is in the form of subsistence goods (Vedeld *et al.*, 2004).

Biodiversity provides the poor with a form of cost-effective and readily accessible insurance against risk, particularly food security risks, health risks and risks from environmental hazards (Chapter 4). The evidence suggests that, as the poor have few alternative sources for protecting themselves, they have a higher dependency on biodiversity for dealing with risk. As such, harvesting wild biodiversity provides a safety net whereby the benefits provided by forest resources stop rural dwellers from becoming poorer and provide cash income at critical times of the year, particularly during times of low agricultural production (Angelsen & Wunder, 2003; Ros-Tonen & Wiersum, 2005).

Biodiversity underpins the delivery of a range of ecosystem services – clean water, soil fertility and stabilisation, pollination services and pollution management – on which we are all dependent, but the poor more so because of their inability to purchase technological substitutes (water filters, chemical fertilisers etc.) (Millennium Ecosystem Assessment (MA), 2005). For example, as natural habitat for bees, bats and other critical taxonomic groups, forests provide pollination services to adjacent agricultural areas.

Under the right conditions, conserving areas of high priority for ('wild') biodiversity can deliver well-being benefits for significant numbers of poor people (Chapters 2, 3 and 14). At the same time, (managed) biodiversity in more densely settled and transformed areas can in different ways be just as important for sustaining and supporting the livelihoods of the rural poor (Chapters 7 and 8). However, there are some important caveats to this conclusion, as highlighted in the chapters in this book:

1. It is often the goods and services from biodiversity of low commercial value that are most significant to the poorest members of the community. Resources of higher commercial value attract the attention of more affluent groups – for example, those with access to assets such as land or cattle – often crowding out the poor in the process unless conscious efforts are made to target the poor and marginalised (as in some conservation enterprises). This suggests that where biodiversity resources have poverty alleviation potential (especially when measured in monetary terms), it is not likely to be the poorer members of the community that directly benefit from this. Thus biodiversity more often acts as a *safety net* to prevent people from falling deeper into poverty, and occasionally it can become a poverty trap (Chapters 4, 5 and 7).

2. In the short term, it is often not so much the *diversity*, or variety, of biological resources that makes an important contribution to poor peoples' livelihoods – particularly in terms of fulfilling immediate needs (e.g. for food and fuel) and for generating cash – but rather their *abundance* or mass (Chapter 9). In Chapter 4, Vira and Kontoleon note: "the term 'nature's resources' better captures the generic categories of resources that have been studied in this literature. These include forests, both in terms of wood-based and non-timber forest products (NTFPs); mangroves; fish; wild animals (bushmeat) and wild plants (including herbs); and common pool resources (CPRs) more generally". Diversity does, however, underpin the ecological systems of which these species are part (Box 20.1) and provides poor people with a strategy for risk management – particularly in terms of the ability to switch to alternative resources in the face of changing conditions – whether climate change, harvest failure or the like (Chapters 6, 7, 8 and 18). This is particularly true for agricultural biodiversity. Furthermore,

biodiversity can underpin biomass production in some ecological systems – for example fisheries (Chapter 6).

3. Even when biodiversity conservation can be shown to make a contribution to poverty alleviation, the scale of impact may be limited (Chapter 11). Few conservation interventions specifically target the poorest – indeed, households with higher assets and higher levels of social capital are more likely to participate in conservation initiatives (Chapters 9, 13 and 15). Despite some good intentions, many conservation interventions just do not lend themselves well to poverty alleviation.

4. A focus on cash benefits alone obscures the real poverty alleviation potential of biodiversity conservation (Chapters 12, 13 and 17). Benefits and incentives are generally too narrowly conceived in the conservation literature, focusing on monetary benefits akin to the income-poverty model of the 1960s. Yet it is widely documented that communities have a diversity of objectives for engaging in conservation – economic, environmental, political, social and cultural – and this is consistent with the idea that poverty not only results from low income but also reflects a deprivation of the diverse requirements for meeting basic human needs. Expanding the scope of analyses to incorporate social well-being may enhance an understanding of poverty–conservation dynamics (Coulthard *et al.*, 2011).

5. Conservation interventions can *exacerbate* poverty. Because the rural poor depend on biodiversity for their day-to-day livelihoods, it is logical to conclude that if biodiversity is conserved (i.e. maintained, enhanced and/or restored) it can continue to provide livelihood support functions. However, the *intervention* that is employed to reach that conservation outcome may itself have a negative poverty impact. For example, strict enforcement of protected areas and other exclusionary land uses may actually increase local incidence of poverty through the loss of resource access. The way conservation interventions are designed and implemented is a key determinant of their poverty impacts – whether they are protected areas (Chapter 10), payments for ecosystem services (PES) schemes (Chapter 14) or any other mechanisms (Chapter 9).

6. Scale matters: the relationship between biodiversity and poverty changes with the spatial scale of analysis (e.g. zooming in from Chapter 2 through to Chapter 4). Appropriate interventions can change too. Grassroots engagement with local organizations (Chapter 16) and 'enterprise' approaches (Chapter 13) may work in the local context, but at a broader scale they can do little to challenge or tackle the global economic system that helps to drive both poverty and biodiversity loss (Chapter 19).

So should development pay more attention to biodiversity?

Although there may be far more (and far more effective) pathways out of poverty, biodiversity does seem to have a particular role to play in *poverty avoidance* and in *risk reduction* for the rural poor – which should be key components of anti-poverty strategies. In many cases it may be the only option for poverty avoidance, for example for people in remote rural areas living beyond the reach of national social protection programmes. At the very least, the co-location of chronic rural poverty with areas of biodiversity interest provides a strategic opportunity to explore linked solutions (Chapter 2). Given that biodiversity represents a natural asset to the rural poor – and one of the few they have – its potential role in poverty alleviation and/or prevention should be given serious consideration in development planning. Equally, biodiversity strategies and plans should take into account the needs – and rights – of the rural poor when considering the types of conservation interventions that are employed, the components of biodiversity that are targeted and the types of benefits that are delivered. A critical first step in this regard is more and better communication and collaboration between those who 'do' poverty alleviation and those who 'do' biodiversity conservation in order to (i) learn from each other regarding what works and what does not, (ii) identify where seemingly different interventions and objectives can be mutually supportive and where they can't and (iii) increase the effectiveness of individual efforts to tackle the governance issues that affect both poverty and biodiversity outcomes.

At the national level, biodiversity is typically addressed by environment ministries. Maximising the contribution of biodiversity to poverty alleviation requires acknowledgement that it also requires serious engagement by finance and planning ministries. More widespread application of concepts such as 'GDP of the poor', championed by TEEB (The Economics of Ecosystems and Biodiversity; Chapter 3), would provide much greater recognition of the importance of ecosystem services and the links between biodiversity and poverty alleviation. This would also allow for a more balanced assessment of the social and environmental impacts of rapidly growing investment in mainstream development interventions, such as infrastructure projects, in biodiversity-rich areas. Private sector institutions also have a crucial role to play in the development process, and must recognise the role of biodiversity in their businesses and the lives of their customers and employees (TEEB, 2010a). The language of sustainability and ecosystem services is becoming more common amongst large-scale companies, some of which have made strong public declarations regarding their commitment to biodiversity conservation (e.g. Shell, detailed in Bishop *et al.*, 2008), although Robinson (2011) queries the extent to which they are likely to be able to play anything more than a 'supporting role' to an enabling environment led by government and civil society.

It is critical to realise, however, that both biodiversity conservation and poverty alleviation are inherently political processes. Decisions about how resources are used and by whom are often made independently from any concern with either poverty alleviation or biodiversity conservation but rather reflect the political or commercial interests of the decision maker. Promoting pro-poor biodiversity conservation thus implies addressing perverse incentives, structures and processes: promoting good governance (or 'good enough' governance) at all levels, from international to local.

The need for improved evidence of causal linkages and the results of actions

Undermining efforts to analyse links between biodiversity and poverty is the state of the evidence base. Even if we can agree on common understandings of key words and concepts, what evidence is there that reducing the rate of biodiversity loss can make a contribution to poverty alleviation? Despite the wealth of case studies on biodiversity–poverty linkages, the existing body of work suffers from an overload of conjectural and anecdotal assertion rather than evidence (the links have not been considered important in policy, and so incentives and procedures to monitor and assess them have understandably been weak). In many cases the evidence is constrained because case study outcomes are measured differently and are therefore difficult to compare or aggregate (Agrawal & Redford, 2006). In other cases the empirical evidence is weak in terms of both quantity and quality. For example, gaps remain in our understanding of ecosystem dynamics (particularly in non-equilibrium systems) and how they interact with social systems (e.g. Chapter 7). Equally, we still know relatively little about how different kinds of biodiversity underpin ecosystem resilience and regulating services such as natural hazard protection (Chapter 4). Full understanding of the links between biodiversity conservation and poverty alleviation in any given context requires systematic data collected using robust methods, including appropriate historical baselines and counterfactual 'control' sites (Ferraro, 2009; Barrett et al., 2011) – and, as the chapters by Vira and Kontoleon (Chapter 4) and Leisher et al. (Chapter 9) point out, very few studies are able to do this.

Although this book has synthesized a wide-ranging body of evidence, some clear research gaps remain to be addressed with targeted policy analysis:

1. *Dependency links*: 'mapping' local links between biodiversity, ecosystem services and the well-being of the poor;
2. *Impact links*: assessing the aggregate poverty (and not only livelihood) impacts of conservation programmes and
3. *Risk assessment*: due diligence on the implications of failure of development programmes to address declines in biodiversity and ecosystem services.

We note, however, that 'scientific' evidence is not the only source of useful information on the relationship between biodiversity and poverty. Other forms of evidence and knowledge can also have validity, and how to incorporate traditional knowledge and anecdotal evidence into 'scientific' assessments and analyses remains a major challenge (e.g. see Fazey *et al.*, 2006). This problem is particularly acute when thinking about social issues such as the cultural dimensions of human well-being, which are unlikely to be amenable to the quantitative scientific method. As Holmes and Brockington point out in Chapter 10, the sheer complexity of the relationship between poverty and biodiversity in particular contexts may also limit the explanatory and predictive power of even the most sophisticated analysis.

Implications: improving policy and practice

Chapter 19 provides a sobering reminder of the scale of the problem we face if we are truly to achieve sustainable development – of which biodiversity-based poverty alleviation is only a small part. The focus of attention on the links between conservation and poverty has been largely at the local level in rural areas with relatively well-preserved ecosystems. Meanwhile at the broader scale, biodiversity loss and acute poverty are both caused by the same driving forces of overconsumption, unfair distribution systems and the development paths chosen by already wealthy countries. A growing body of literature points to the risk of exceeding safe ecological and social limits, or 'planetary boundaries' if current unsustainable development paths are not transformed (Rockström *et al.*, 2009; Brundtland *et al.*, 2012; Raworth, 2012).

Challenging the dominant paradigms of consumerism and economic growth means going beyond localised biodiversity–poverty linkages and attempting to influence mainstream economic policy processes. One example – by no means a panacea – is the various efforts to take better account of the value of biodiversity and ecosystem services in decision making. Initiatives such as the MA (2005), the study on The Economics of Ecosystems and Biodiversity (TEEB, 2010b) and the Wealth Accounting and Valuation of Ecosystem Services (WAVES) Partnership launched at CBD CoP 10 recognize the contribution that biodiversity and ecosystem services make to poor people's livelihoods but also highlight that because they are public goods they are often undervalued – if valued at all – in national economies. The MA and TEEB warn that until the values of biodiversity and ecosystem services are properly taken into account and alternative metrics for 'growth' developed, they will continue to be depleted and their potential to support poor people jeopardised.

As a result of these and other efforts, there is increasing international interest in green economy principles and in means to incorporate the values of natural capital in national accounting (e.g. Department for Environment, Food and Rural Affairs (Defra), 2011) and private sector decision making (Hanson *et al.*, 2012). Ecosystem assessments incorporating economic and social values offer the potential to underpin national biodiversity strategies and action plans and to enhance their relevance to central government planners and budget holders, as well as to a range of other public and private sector decision makers whose policies and processes influence biodiversity and ecosystem services, often unintentionally. To ensure the interests of the poor are not marginalized, however, such assessments must incorporate distributional effects, identifying who wins and who loses as ecosystems change. They must also consider the social dimensions of linked social-ecological systems, to ensure that institutional changes, in governance and land tenure for example, that affect benefit distribution are included.

We recognise that biodiversity loss and persistent poverty are symptoms of much greater development challenges. Nevertheless this does not mean that we should ignore these symptoms in order to concentrate our efforts on higher order problems. We conclude by highlighting a number of policy implications – relevant at all levels and to a variety of different actors – in Box 20.2. These are but a small contribution to the challenge of sustainable development. However, without these kinds of approaches, biodiversity and ecosystem services will continue to be depleted and their potential to act as a safety net for the poor – let alone to contribute to poverty alleviation – will remain in jeopardy.

Box 20.2 **Linking biodiversity conservation and poverty alleviation: pointers for policy and practice**

1. Ensure absolute clarity about how different definitions of poverty, biodiversity and conservation are being used and interpreted in different contexts to ensure that complex issues are not confused and misrepresented.
2. Give more policy attention to how biodiversity can help *prevent* poverty. In many cases its contribution to poverty *reduction* has tended to be overstated but its major contribution to poverty *prevention* has been somewhat overlooked.
3. Include safeguards in the design of conservation policy and projects to ensure that poor people are not made worse off, or their rights infringed.
4. Ensure that pro-poor conservation efforts give as much attention to less visible biodiversity (e.g. 'production' biodiversity – such as microbes and invertebrates – and also plants) as to endangered species – which are most often prioritised by conservation interventions.
5. Conserve ecosystem services *in the places where poor people live*. This implies much greater attention to dryland ecosystems (arid, semi-arid, and dry

sub-humid zones) and their biodiversity, which are particularly important to the one billion poor or severely poor people living in these zones.

6. Promote good governance (or 'good enough' governance) at all levels – international, national and local – in order to address perverse political structures and processes. In particular, (i) clarify and strengthen local land and resource rights;[1] (ii) improve local participation in, and transparency over, decision making and (iii) strengthen national and local resource management institutions.

7. Improving national and international policy frameworks, combined with conducive governance and institutional arrangements, can help scale up the benefits that poor people derive from biodiversity. At the national level, the integration, or 'mainstreaming', of biodiversity-friendly objectives into different sectors constitutes a key opportunity for integrating the maintenance of biodiversity with local economic development and poverty reduction.

8. Be clear on the balance of global and local pressures driving biodiversity loss in any given context in order to design effective conservation interventions. It may be easier to use policy and legislative mechanisms to target those drivers fed by local poverty, but the impact of global consumption patterns cannot be overlooked, and should also be addressed (Chapter 19).

9. Learn from good practice in preventing increased levels of development and consumption from affecting biodiversity. Draw on the existing good examples of countries increasingly using their biodiversity as a development advantage.

10. Be realistic about the fact that it is not possible to achieve 'win–win' from all situations. A more realistic aim is to 'win more' and 'lose less'. Be prepared to manage the trade-offs inherent in many conservation–poverty interventions – for example through locally appropriate compensation schemes that help mitigate the negative impacts of protected areas or human–wildlife conflict.

Source: Adapted from Roe *et al.* (2011).

---

[1] It must be recognized that clarification of land tenure can represent a threat as well as an opportunity to the poor, if more powerful actors are able to obtain legal tenure for land (and hence resources) previously accessible to and used by the poor. Clarification and strengthening of land and resource rights must therefore be done in a way that does not put the poor at risk.

References

Angelsen, A. & Wunder, S. (2003) *Exploring the Forest–Poverty Link: Key Concepts, Issues and Research Implications.* CIFOR Occasional Paper No. 40. Centre for International Forestry Research, Bogor, Indonesia.

Barrett, C.B., Travis, A.J. & Dasgupta, P. (2011) On biodiversity conservation and poverty traps. *Proceedings of the National Academy of Sciences of the United States of America,* 108, 13907–12.

Bishop, J., Kapila, S., Hicks, F., Mitchell, P. & Vorhies, F. (2008) *Building Biodiversity Business.* Shell International Ltd and IUCN, London and Gland, Switzerland.

Brundtland, G.H., Ehrlich, P., Goldemberg, J., Hansen, J., Lovins, A., Likens, G., Lovelock, J., Manabe, S., May, B., Mooney, H., Robert, K.-H., Salim, E., Sato, G., Solomon, S., Stern, N., Swaminathan, M.S. & Watson, B. (2012) *Environment and Development Challenges: The Imperative to Act.* Blue Planet Laureates Paper presented at the UNEP Governing Council, February. http://www.thenaturalstep.org/sites/all/files/Blue_Planet_Laureates_Environment_and_Development_Challenges_The_Imperative_to_Act.pdf (accessed 5 May 2012).

Butchart, S.H.M., Walpole, M., Collen, B., van Strien, A., Scharlemann, J.P.W., Almond, R,E.A., Baillie, J.E.M., Bomhard, B., Brown, C., Bruno, J., Carpenter, K.E. Carr, G.M., Chanson, J., Chenery, A.M., Csirke, J., Davidson, N.C., Dentener, F., Foster, M. Galli, A., Galloway, J.N., Genovesi, P., Gregory, R.D., Hockings, M., Kapos, V., Lamarque, J.-F., Leverington, F., Loh, J., McGeoch, M.A., McRae, L., Minasyan, A., Hernández Morcillo, M., Oldfield, T.E.E., Pauly, D., Quader, S., Revenga, C., Sauer, J.R., Skolnik, B., Spear, D., Stanwell-Smith, D., Stuart, S.N., Symes, A., Tierney, M., Tyrrell, T.D., Vié, J.-C. & Watson, R. (2010) Global biodiversity: indicators of recent declines. *Science,* 328, 1164–8.

Convention on Biological Diversity (CBD) (2002) *Decision VI/26: Strategic Plan for the Convention on Biological Diversity.* Secretariat of the Convention on Biological Diversity, Montreal.

Convention on Biological Diversity (CBD) (2010) *Decision X/2: Strategic Plan for Biodiversity 2011–2020.* Secretariat of the Convention on Biological Diversity, Montreal.

Coulthard, S., Johnson, D. & McGregor, J.A. (2011) Poverty, sustainability and human wellbeing: a social wellbeing approach to the global fisheries crisis. *Global Environmental Change,* 21, 453–63.

Department for Environment, Food and Rural Affairs (Defra) (2011) *The Natural Choice: Securing the Value of Nature.* HM Government, London.

Fazey, I.R.A., Fazey, J.A., Salisbury, J.G., Lindenmayer, D.B. & Dovers, S. (2006) The nature and role of experiential knowledge for environmental conservation. *Environmental Conservation,* 33, 1–10.

Ferraro, P.J. (2009) Counterfactual thinking and impact evaluation in environmental policy. *New Directions for Evaluation,* 2009, 75–84.

Hanson, C., Ranganathan, J., Iceland, C. & Finisdore, J. (2012) *The Corporate Ecosystem Services Review: Guidelines for Identifying Business Risks and Opportunities Arising from Ecosystem Change.* World Resources Institute, Washington, DC.

International Fund for Agricultural Development (IFAD) (2010) *The Rural Poverty Report 2011.* IFAD, Rome.

Mace, G.M., Cramer, W., Diaz, S., Faith, D.P., Larigauderie, A., Le Prestre, P., Palmer, M., Perrings, C., Scholes, R.J., Walpole, M., Walther, B.A., Watson, J.E.M. & Mooney, H.A. (2010) Biodiversity targets after 2010. *Current Opinion in Environmental Sustainability (COSUST),* 2, 3–8.

Mace, G.M., Norris, K. & Fitter, A.H. (2012) Biodiversity and ecosystem services: a multilayered relationship. *Trends in Ecology and Evolution,* 27, 19–26.

Millennium Ecosystem Assessment (MA) (2005) *Ecosystems and Human Well-Being: Biodiversity Synthesis.* World Resources Institute, Washington, DC.

Raworth, K. (2012) *A Safe and Just Space for Humanity – Can We Live within the Doughnut?* Oxfam Discussion Paper. Oxfam, Oxford.

Robinson, J. (2011) Corporate greening: is it significant for biodiversity conservation ? *Oryx,* 45, 309–10.

Rockström, J., Steffen, W., Noone, K., Persson, A., Chapin, F.S., Lambin, E., Lenton, T.M., Scheffer, M., Folke, C., Schellnhuber, H.J., Nykvist, B., de Wit, G.A., Hughes, T., van der Leeuw, S., Rodhe, H., Sörlin, S., Snyder, P.K., Costanza, R., Svedin, U., Falkenmark, M., Karlberg, L., Corell, R.W., Fabry, V.J., Hansen, J., Walker, B., Liverman, D., Richardson, K., Crutzen, P. & Foley, J. (2009) Planetary boundaries: exploring the safe operating space for humanity. *Ecology and Society,* 14, 32. http://www.ecologyandsociety.org/vol14/iss2/art32 (accessed 10 May 2012).

Roe, D., Thomas, D., Smith, J., Walpole, M. & Elliott, J. (2011) *Biodiversity and Poverty: Ten Frequently Asked Questions – Ten Policy Implications.* Gatekeeper 150. IIED, London.

Ros-Tonen, M.A.F. & Wiersum, K.F. (2005) The scope of improving rural livelihoods through non-timber forest products: an evolving research agenda. *Forests, Trees and Livelihoods,* 15, 129–48.

The Economics of Ecosystems and Biodiversity (TEEB) (2010a) *The Economics of Ecosystems and Biodiversity Report for Business – Executive Summary 2010.* TEEB, Geneva.

The Economics of Ecosystems and Biodiversity (TEEB) (2010b) *The Economics of Ecosystems and Biodiversity: Mainstreaming the Economics of Nature: A Synthesis of the Approach, Conclusions and Recommendations of TEEB.* TEEB, Geneva.

United Nations (2006) *Report of the Secretary General on the Work of the Organization.* United Nations, New York.

United Nations (2010a) *The Millennium Development Goals Report.* United Nations, New York.

United Nations (2010b) *Secretary-General, at High-Level Meeting, Stresses Urgent Need to Reverse Alarming Rate of Biodiversity Loss, Rescue 'Natural Economy'.* Press release, 22 September. http://www.un.org/News/Press/docs/2010/ ga10992.doc.htm (accessed 10 May 2012).

Vedeld, P., Angelsen, A., Sjaastad, E. & Kobugabe Berg, G. (2004) *Counting on the Environment: Forest Incomes and the Rural Poor.* Environmental Economics Series No. 98. World Bank, Washington, DC.

Index

Note: Page references in *italics* refer to Figures; those in **bold** refer to Tables and Boxes